Practical Control of Mosquito Disease Vectors

Practical Control of Mosquito Disease Vectors

Edited by

Jacques Derek Charlwood

CABI

Frontispiece

'We search for simplicity and hope for truth' Goodman

'The folly of mistaking a metaphor for a truth, a torrent of verbiage for a spring of capital truths, and oneself as an oracle, is inborn in us' according to Valery, quoted by the book burning, but very well read, Captain Beatty in *'Fahrenheit 451'*

'False facts are highly injurious to the progress of science for they often long endure: but false views, if supported by some evidence, do little harm, for everyone takes a salutary pleasure in proving their falseness; and when this is done, one path towards error is closed and the road to truth is often at the same time opened.' Darwin, The Descent of Man.

My contributors, and my editors, must have viewed me like Mr Dick in *David Copperfield,* 'Every day of his life he had a long sitting at the Memorial, which never made the least progress no matter how hard he laboured. For if anything were certain under the sun, it was certain that the Memorial would never be finished'. Three major wars have been started since these were submitted. My apologies to all concerned for the delay. I would like to thank them all, but especially Carla Sousa of the IHMT in Lisbon and Lucy Pritchard and Helen Elliott of CABI, for their patience and trust that this project might eventually find its way into print.

CABI is a trading name of CAB International

CABI
Nosworthy Way
Wallingford
Oxfordshire OX10 8DE
UK

CABI
200 Portland Street
Boston
MA 02114
USA

Tel: +44 (0)1491 832111
E-mail: info@cabi.org
Website: www.cabi.org

T: +1 (617)682-9015
E-mail: cabi-nao@cabi.org

A catalogue record for this book is available from the British Library, London, UK.

ISBN-13: 9781789248821 (hardback)
9781789248838 (ePDF)
9781789248845 (ePub)

DOI: 10.1079/9781789248845.0000

Commissioning Editor: Jamie Lee
Editorial Assistant: Helen Elliott
Production Editor: James Bishop

Typeset by Exeter Premedia Services Pvt Ltd, Chennai, India
Printed and bound in the UK by CPI Group (UK) Ltd, Croydon, CR0 4YY

Contents

Contributors

Gabriel Carrasco-Escobar: InnovaLab, Institute of Tropical Medicine 'Alexander von Humboldt', Universidad Peruana Cayetano Heredia, Lima, Peru and School of Public Health, University of California San Diego, USA. Email: gabriel.carrasco@upch.pe

Jacques Derek Charlwood: Global Health and Tropical Medicine, GHTM, Instituto de Higiene e Medicina Tropical, IHMT, Universidade NOVA de Lisboa, UNL, Rua da Junqueira 100, 1349-008 Lisboa, Portugal. Email: jdcharlwood@gmail.com

Jan E. Conn: The Wadsworth Center, New York State Department of Health, Albany, New York, USA and Department of Biomedical Sciences, School of Public Health, State University of New York at Albany, NY, USA.

Kimberly Fornace: Faculty of Infectious and Tropical Diseases and Centre for Climate Change and Planetary Health, London School Hygiene and Tropical Medicine, London, UK.

Amir Galili: Westham Co., Ha-Banai St 5, Hod Hasharon, Israel. Email: amir@westhamco.com

Manuela Herrera-Varela: Grupo de Investigación en Entomología, Facultad de Medicina, Universidad Nacional de Colombia, Bogotá D.C., Colombia.

Robert T. Jones: Department of Disease Control, London School of Hygiene & Tropical Medicine, Keppel Street, London, WC1E 7HT, UK. Email: Robert.Jones@lshtm.ac.uk

Emmanuel Kaindoa: Environmental Health & Ecological Sciences, Ifakara Health Institute PO Box 53, Ifakara, Tanzania.

Kevin Kobylinski: Mahidol Oxford Research Unit, Bangkok, Thailand. Email: Kobylinskikevin@gmail.com

Marta Maia: Kenya Medical Research Institute (KEMRI) PO Box 54840 00200 Off Raila Odinga Way. Nairobi, Kenya. Email: MMaia@kemri-wellcome.org

Edgar Manrique: InnovaLab, Institute of Tropical Medicine 'Alexander von Humboldt', Universidad Peruana Cayetano Heredia, Lima, Peru.

Louisa A. Messenger: Parasitology and Vector Biology (PARAVEC) Laboratory, School of Public Health, University of Nevada, Las Vegas, Nevada; Department of Environmental and Occupational Health, School of Public Health, University of Nevada, Las Vegas, Nevada. Email: louisa.messenger@unlv.edu

Meir Morag: Westham Co., Ha-Banai St 5, Hod Hasharon, Israel.

Marta Moreno: Faculty of Infectious and Tropical Diseases and Centre for Climate Change and Planetary Health, London School Hygiene and Tropical Medicine, London, UK.

Fredros Okumu: Environmental Health & Ecological Sciences, Ifakara Health Institute PO Box 53, Ifakara, Tanzania and School of Biodiversity, One Health and Veterinary Medicine, University of Glasgow, Scotland. Email: Fredros@ihi.or.tz.

João Pinto: Global Health and Tropical Medicine, GHTM, Associate Laboratory in Translation and Innovation Towards Global Health, LA-REAL, Instituto de Higiene e Medicina Tropical, IHMT, Universidade NOVA de Lisboa, UNL. Rua da Junqueira 100, 1349-008 Lisboa, Portugal. Email: JPinto@ihmt.unl.pt

Perran A. Ross: Pest and Environmental Adaptation Research Group, Bio21 Institute and the University of Melbourne, Parkville, Victoria, Australia. Email: perran.ross@unimelb.edu.au

Carla A Sousa: Global Health and Tropical Medicine, GHTM, Associate Laboratory in Translation and Innovation Towards Global Health, LA-REAL, Instituto de Higiene e Medicina Tropical, IHMT, Universidade NOVA de Lisboa, UNL. Rua da Junqueira 100, 1349-008 Lisboa, Portugal. Email: casousa@ihmt.unl.pt

1 Introduction

J. D. Charlwood*
Global Health and Tropical Medicine, Instituto de Hygiene e Medicina Tropical, Universidade NOVA de Lisboa, Lisbon, Portugal

Abstract
Over 80% of the global population live in regions of the world at risk from one vector-borne disease, and more than half the world's population live in areas where at least two different vector-borne diseases pose a threat to health. In the present book, a number of topics relevant for the control of mosquito vectors of disease are covered. In particular, novel techniques that have shown 'proof of principle' but which have yet to be used on a wide scale are written by experts in the field. A number of chapters explore the possibilities of improving health in general by reducing inequalities between people. This is, ultimately, the only way of achieving sustainable improvements to people's lives. How to achieve this in the most 'biodiversity-friendly' way should be the goal of all scientists, including those involved in vector control. It is a thread that runs through much of this book.

> Life is an adventure in a world where nothing is static; where unpredictable and ill-understood events constitute dangers that must be overcome, often blindly and at great cost; where man himself, like the sorcerer's apprentice, has set in motion forces that are potentially destructive and may someday escape his control.
>
> (Dubos, 1987)

*Corresponding author: jdcharlwood@gmail.com

DOI: 10.1079/9781789248845.0001

Introduction

We live in troubling times. Global warming, or rather global heating or, even, global boiling, is already under way with a vengeance, and the loss of biodiversity is accelerating. Biological systems are becoming unable to absorb the excess carbon emitted through the burning of fossil fuels, with the result that in the past 3 years global temperatures have increased at a faster rate than recently. If nothing is done, even subjects like the ones covered in this book (except perhaps self-help topics) will become irrelevant. In addition to global heating and the loss of biodiversity, global pandemics also pose existential threats to humanity, and potentially all life on earth. These crises cannot be compartmentalized and must be tackled both immediately and simultaneously. We need to heal the planet. The priority of all humans and all systems, as pointed out by Attenborough (2022) in his 'Witness statement', should be the rewilding of the world. In every walk of life, and in every system, approaches which are most in line with this goal should be preferred over others. This includes tropical disease and vector control. The orator Cato was famous because he always concluded every speech he made with the words, 'Carthage must be destroyed' (it was eventually), so, similarly, Attenborough's message needs to be given prominence, but at the start, in every book like this and at every scientific meeting. Floods and droughts and the scale of hurricanes tell us that nature is no longer an inert background from which resources are extracted for human activities; rather, it has reclaimed its role as an active agent in the fate of the planet. As the philosopher Latour (1998) says, the choice is 'to modernise or to ecologize'.

One way of ecologizing is to educate people and give them more freedom to make informed choices. One way of doing that is to improve their health. How to achieve this in the most 'biodiversity-friendly' way should be the goal of all scientists, including those involved in vector control. It is, we hope, a thread that runs through much of this book. Although it will be hard, such an approach will, ultimately, improve people's lives. Grants should be written with this in mind. In his book, *Simplicity* (1998), Edward de Bono shows that in order to explain something simply, one needs a good understanding of the subject, and by explaining something in simple terms (think $E = mc^2$), the need for an interpreter is removed. As he points out, 'there is also the underlying fear that the role of interpreter will no longer be necessary if the matter is simple enough for ordinary people to understand and use'. We do not belong to those academics who 'feel a great insecurity about this'.

In the field of vector-borne diseases, the signals concerning the future are mixed. While there is still no vaccine for dengue and the geographical range of the vectors – *Aedes aegypti* and *Aedes albopictus* – is increasing every year, malaria cases and deaths due to malaria have decreased dramatically since the start of the millennium (Bhatt *et al.*, 2015), and a vaccine that really works has recently been approved for use in high-transmission regions by the WHO. Despite an increasing frequency of insecticide resistance among the vectors, between 2000 and 2017, the worldwide annual incidence of malaria declined by 36%, and the annual death rate declined by 60%. Thus, the push to eradicate malaria continues. Malaria eradication is defined by WHO as 'the permanent reduction to zero of the worldwide annual incidence of malaria infection caused by all species of human malaria parasites: *Plasmodium vivax*, *Plasmodium falciparum*, *Plasmodium malariae* and *Plasmodium ovale*' (although *Plasmodium knowlsei* should now perhaps be added to that list).

Nevertheless, over 80% of the global population live in regions of the world at risk from one vector-borne disease, and more than half the world's population live in areas where at least two different vector-borne diseases pose a threat to health. The two most important of these diseases are malaria and dengue, both transmitted by mosquitoes. The fragility of the situation is illustrated by the increase in the number of cases of malaria in 26 of the 29 high-burden countries between 2015 and 2017 (but still a long way from pre-intervention levels). The potential spread of the urban malaria vector, *Anopheles stephensi*, recently introduced into Africa, is also a major threat (Charlwood, 2023).

Dengue is the fastest increasing infectious vector-borne disease in the world, and the number of estimated global cases has risen sharply from 8.3 million reported cases in 1990 to 58.4 million in 2013 (Stanaway *et al.*, 2016). In 2024, Latin America, especially Brazil, experienced a major epidemic, associated with increasing temperatures (PAHO, 2024). Rising temperatures, extreme weather events, the El Niño phenomenon, rapid population growth and unplanned urbanization play a crucial role. Poor housing conditions and inadequate water and sanitation services create mosquito breeding sites through discarded objects that can collect water. The lack of a vaccine or treatment against dengue and other vector-borne viruses means that their control is limited to the control of the vectors. This is not easy, although major advances in genetic control as well as the use of *Wolbachia* have recently been achieved.

Similarly, the number of cases of West Nile virus is increasing in Europe as the number of potential vectors increases. For example, in 2022, where the majority of the more than 770 cases have occurred, the number of potential vectors in the region increased by 27% (Weaving, 2022).

For all vector-borne diseases, as pointed out by Feachem *et al.* (2019), in a review entitled 'Malaria eradication within a generation: ambitious, achievable, and necessary', it is operational obstacles rather than technical ones that largely limit the success of disease control programmes in many countries. These obstacles include ineffective management, poor use of data, a staff that is apathetic and local communities that are not involved either in the decisions that might affect them or in the outcome of any control efforts. The solution to many of these problems does not involve the development of new technologies but rather requires strong and determined leadership. Clause IV of the Alma-Ata declaration stated, 'People have a right and duty to participate individually and collectively in the planning and implementation of health care' (WHO, 1978). The improved use of data for decision making and the targeting of interventions should increase their effectiveness. At the same time, there are biodiversity-friendly, locally possible, ways to improve people's lives that also reduce exposure to disease.

An old Chinese proverb states: 'I hear, and I forget; I see, and I remember; I do, and I understand'. A pragmatic learning-while-doing approach that tackles local problems should be encouraged. Some techniques do not really require much testing in order for them to be recommended. People will always want to improve their houses. Many of these improvements almost certainly reduce exposure to vectors. The job of the scientist/professional is to enable the kind of improvements that people prefer to be achieved in the most biodiversity-friendly way, such as being able to build with pressed rather than burnt bricks. This, in itself, would significantly reduce deforestation and would help in the reduction of global heating. How this might be achieved is discussed in the chapter on housing in the present volume.

Implementation research, to find practical solutions to local operational problems, and not just large-scale randomized or other formalized trials, to answer operational questions, is also important.

In addition to highlighting the need for improved management, Feachem *et al.* (2019, pp. 1056–1057) described the need for 'new technologies that provide long durations of efficacy, do not require difficult or protracted compliance from individuals and households, and drive down disease in high-transmission or otherwise problematic settings'. Disease-specific tools would obviously help in driving down transmission of any vector-borne disease and might enable the goal of elimination to be achieved. According to the World Health Organization, 'mosquito control is the only intervention that can reduce malaria transmission from very high levels to close to zero' (WHO, 2014).

With respect to malaria, we are reaching a stage where a number of new techniques that supplement so-called conventional techniques (especially impregnated mosquito nets) are being developed and tested. Many of these may act synergistically and so have a greater effect when used in combination. In the present volume, we hope to produce a state-of-the-art statement regarding several such techniques.

A book is always going to be out of date as soon as it is published. Nevertheless, a good book becomes a reference source, a keystone in different people's utilization and investigation. Such books need not be written by authorities on the subject, but it helps when they are. We are lucky that we have been able to obtain contributions from a number of the leading experts on these new techniques. We hope that it will be useful as a starting off point to explore the methods in greater detail and will enable the reader to follow developments as they occur. We hope not just to merely instruct but, as Goethe requested, to 'invigorate' people's activities.

So Why This Book? Why Now?

The situation in the world is rapidly changing, and it is uncertain what will happen now that the Covid pandemic has become more manageable, and, indeed, once the wars in Ukraine, the Middle East and Sudan are over (they must surely end sometime – unless this is written shortly before the Third World War). More powerful and dangerous effects of global heating, climate change and, in particular, an increased loss of biodiversity, especially the extinction of endemics, will almost certainly make themselves felt with increasing force. An Intergovernmental Science-Policy Platform on Biodiversity and Ecosystem Services report in 2019 (IPBES, 2019) found that nearly 75% of all freshwater resources on earth are used by crop and livestock production; human activities have significantly altered 75% of all ice-free terrestrial environments and 66% of marine environments. According to the Copernicus Climate Change service, the average global sea surface temperature in March 2024 was 21.07°C, which means that the ocean has been breaking temperature records every day for more than a year. The global average temperature in the 12 months between April 2023 and March 2024 is the highest on record at 0.70°C above the 1991–2020 average.

Major tipping points in the climate, such as the slowing of the Gulf stream (Caesar *et al.*, 2021) or the collapse of the Atlantic Meridional Overturning Circulation (AMOC) – a large system of ocean currents that carry warm water from the tropics into the North Atlantic – any time from 2025 onwards (Ditlevsen and Ditlevsen, 2023), means that we are entering into unknown territory – postmodernism, a period without history.

How the world reacts to these changes will affect the likelihood of maintaining the status quo of interventions and research or will lead to a change in the way that they are conducted.

A survey conducted in 2023 of 9220 researchers around the world, from a range of scientific and academic disciplines, found that more than 90% agree that 'fundamental changes to social, political, and economic systems' are needed (Grossman, 2024), and in a speech outlining the prospects for the year given to the United Nations in early February 2023, Antonio Guterres, the secretary general, said, 'We have started 2023 staring down the barrel of a confluence of challenges unlike any in our lifetimes'. He called for a 'radical transformation' of global finance, saying that 'Something is fundamentally wrong with our economic and financial system'. One year later, the situation has only become worse. We remain like the man who fell from the top storey of the Empire State Building who, as he passed the fifth floor, was heard to remark 'So far, so good'. The urgency of these issues cannot be overstated.

Our economic and financial system – capitalism – shapes a range of behaviour patterns incompatible with sustainability. This directly affects the future of humanity on our planet. Inherent to it is the need for constant, infinite, exponential growth, which is unsustainable in a finite world. In addition to creating the conditions for boom-and-bust cycles, the system leads to short-term thinking, concentrates wealth and erodes equality, which promotes selfish and non-collaborative behaviours. As Scott-Fitzgerald said, 'The rich get rich and the poor get children'. 'Staggering degrees of inequality and wealth, both intra- and inter-nationally, generate ethically and ecologically untenable distributions of consumption and waste' (Buller, 2022). Gutteres has sounded the warning bell so many times before that he must feel like Cassandra, who could accurately predict the future but whom nobody believed or listened to. As the philosopher Alain Badiou wrote in *Infinite Thought*, 'For the process of a truth to begin, something must happen. What there already is - the situation of knowledge as such - generates nothing other than repetition'.

The Situation of Knowledge

A quick review of the 'situation of knowledge as such', even though it is repetition, may be worthwhile: The top 1% own half of the world's wealth, the top 10% own 85%, while the remaining 90% hold the remaining 15%. Over the past 2 years, the world's super-rich 1% have gained nearly twice as much wealth as the remaining 99% combined, and in 2019, they emitted as much carbon dioxide as the poorest two-thirds of humanity (Riddell *et al.*, 2024). In that time, more than five million people were added to those living below the poverty line. 'The world can no longer afford two things: first, the costs of economic inequality; and second, the rich' (Wilkinson and Pickett, 2024). This inequality is greatest in the USA. The top 1% holds 42.5% of national wealth (in no other OECD country does the richest 1% hold more than 28% of their country's wealth). It is estimated that in a few years, the world will see the first trillionaire (Riddell *et al.*, 2024). The highest wealth concentrations are found in the Middle East, Africa and North America. The top 1% of households in these countries account for 70–90% of all their respective countries' private financial and real-estate ownership. Currently, 26 billionaires hold as much wealth as the bottom 50% of the global population combined. Bigger gaps between rich and poor are accompanied by higher rates of homicide and imprisonment. They also correspond to more infant mortality, obesity, drug abuse and Covid-19 deaths, as well as higher rates of teenage pregnancy and lower levels of child well-being, social mobility and public trust.

'Equality is essential for sustainability. The science is clear – people in more-equal societies are more trusting and more likely to protect the environment than are those in unequal, consumer-driven ones' (Wilkinson and Pickett, 2010, 2024). Research contends that inequality reduces social cohesion, a dynamic that leads to more stress, fear and insecurity for everyone. Countries with lower income gaps do better along almost all relevant social and ecological dimensions than countries with large income gaps. As pointed out by Bill Gates in his blog of 14 August 2023: 'As temperatures go up and extreme weather events become more common, it will get harder to do things like provide bed nets, get rid of malaria-carrying mosquitoes, and offer basic health care in the world's most vulnerable communities'. 'Confronting profound inequalities in both wealth and power within the global economy is neither optional nor a distraction from the challenge at hand' (Buller, 2022). As pointed out by the director of Oxfam, Behar (2024), 'To decolonize global development, status quo or slow change is not an option. Transformation is the only viable alternative'.

It is ironic that people working on climate change are responsible for the most 'air-miles'. Scientists working on large tropical medicine projects may not be far behind. International travel will need to be curtailed. Institutions in the west have to realize that they and their staff serve as a model that students in developing countries want to emulate (i.e. become an 'expert' who flies around the world visiting different projects and who stays in four- or five-star hotels for a couple of weeks once or twice a year). The earlier entomologists lived on-site. Perhaps that is what makes their work so good. The editor of this book believes that the days when projects are run (i.e. funds given to and, so, controlled) by people outside of the area where the work will actually be undertaken are numbered. The easiest way for scientists in these countries to obtain independence, and get off this treadmill, is by not doing large-scale trials but much smaller, therefore affordable, pieces of work. The coordination and dissemination of such work (with artificial intelligence helping to determine what the best 'mix' of techniques might be) may be the responsibility of anyone with access to the internet. 'Virtual' meetings, which became popular during the Covid pandemic, are obviously the way of the future.

The real challenge on the ground is education and community involvement. There is a huge amount of research that could be done with limited resources if such involvement were present. Mobile technology (smartphones) can make a difference. For example, the photos taken on an old model phone in France were of sufficient quality to enable identification of the species doing the biting (Fig. 1.1) (www.ihmt.unl.pt >mosquitoweb).

Used in this way, for both public health and agricultural pests, the establishment of databases that include geographical information on the insects can potentially be a game changer in terms of our understanding of mosquito ecology. The rise in the number of smartphones available is, however,

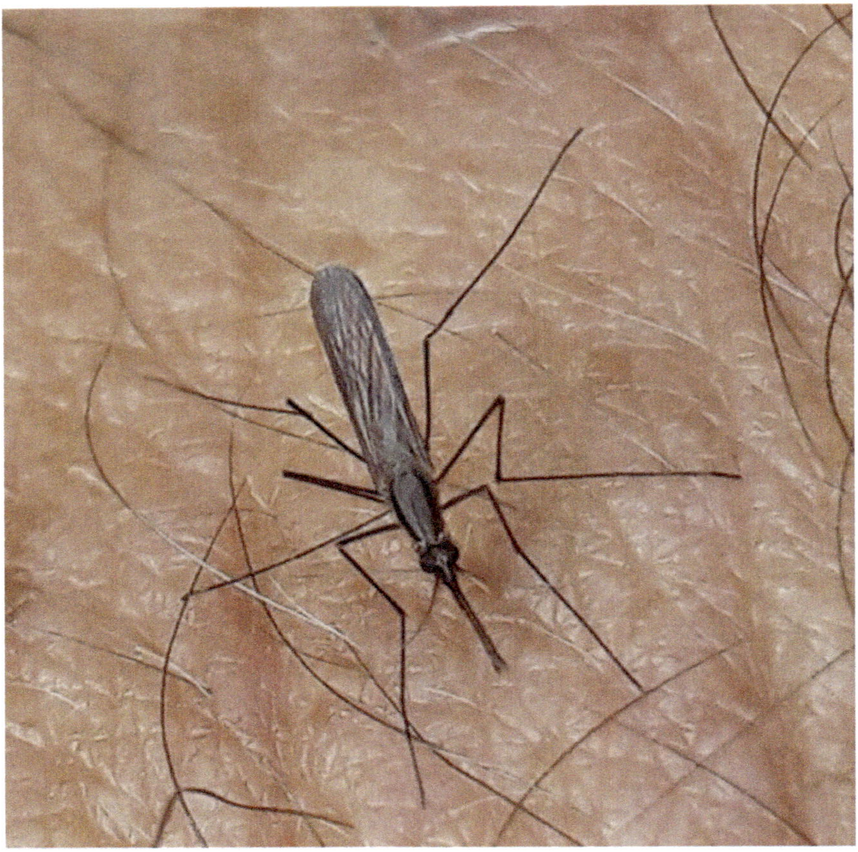

Fig. 1.1. *Anopheles plumbeus* biting the author's arm in the Pyrenees taken with a phone and identified from the photograph (is she not beautiful?). (Photo: J.D. Charlwood.).

a two-edged sword. On the one hand, they enable anyone to contribute to knowledge, but the other edge of the sword is dangerous. 'Falsehood flies, and the truth comes limping after it', wrote Jonathan Swift in 1710. He also said, 'If a lie be believed, if only for an hour, it has done its work'.

People need to be educated so that they can distinguish between fact and fiction. As pointed out in *The Guardian* in an editorial on 17 February 2023, 'The healthy functioning of democracies depends on the quality of the information that frames debate within them. But digitalisation, the rise of social media and increasingly sophisticated forms of artificial intelligence are delivering new opportunities to poison the well of public discourse'. 'Critical acumen is clearly essential to a democracy' (Solomon, 2003).

Conspiracy theories are not new. Following the Fire of London in 1663, a rumour started that it had been a French plot, and several houses owned by Frenchmen were destroyed by a mob. There is always likely to be a proportion of any population (these days especially among those exposed to social media) who believe any conspiracy theory, no matter how outlandish. 'No smoke without fire' would be their response, even though the smoke is an illusion. Such conspiracy theories already occur in health issues: vaccination refusal, for example, originally based on the misinformation that vaccination caused autism. Fewer people believe in science as 'truth'. Indeed, this view was reinforced because during the Covid-19 pandemic, the prestigious scientific journal *The Lancet* had to retract at least one 'major' article about the virus.

Entomology and vector control are not going to be immune from such challenges. Potential new techniques for control are especially likely to be open to misinterpretation if balanced information is not available. In the present volume, we hope to help provide such information on a number of such 'new' techniques.

Just as there used to be a 'divide' between Francophone publications (and research) and English ones, a divide exists between medical entomologists working on malaria vectors and those working on dengue vectors (not forgetting those who might be working with all the other vectors of neglected diseases). The integration of entomological activities among people working on different mosquito groups, however, makes sense, especially if malaria elimination (and the virtual elimination of the vectors) succeeds. With the approach of elimination, the people involved become the equivalent of 'firemen' reacting to fires that can quickly get out of control (as has been shown in Australia, the USA, Greece, the Amazon, Siberia and Portugal) and just like a wildfire, the devastation that such fires (epidemics) can have on communities goes far beyond simple effects of the disease. Control (as opposed to elimination) is much more like refuse (garbage) collection. You enact control, and do not stop doing so, for the same reason that you do not give up on garbage collection. You do not want the situation to get any worse, which it will do if you are not always doing something about it. At the same time, an omission in collection can be corrected reasonably easily, whereas a fire can quickly get out of control. If the money spent on arms was instead spent on fire-control techniques and equipment, perhaps some of the fires presently engulfing the world could be extinguished for good.

The editor believes that only mosquitoes that act as vectors of disease, or are nuisance biters, need to be controlled. Many disease vectors belong to species complexes or groups of closely related species. They are often the only member of the complex, or group, that is a significant vector. In the absence of control, it is often the dominant member (it feeds on a host without fur, feathers or scales and has a delightful place to rest after eating). Following successful control operations, directed at the adult vectors, the larval niche may be left vacant. Emptying the larval niche of the vector means that other species with similar requirements (but in general less-efficient competitors) can occupy it. Thus, another member of the complex may take the vector's place and occupy the larval habitat. In areas where a different member of a complex or group has supplanted the vector species, control of that mosquito by larviciding (the technique often quoted to help achieve Integrated Vector Management – IVM) may be counterproductive, even if they are incidental vectors themselves. Following indoor residual spraying (IRS) campaigns, the replacement of *Anopheles funestus*, a major vector, with *Anopheles rivulorum*, at most an incidental vector (Wilkes *et al.*, 1996), is one such example. If they are killed by a new intervention, then the larval niche will be left unoccupied. Should the vector species develop resistance, as *An. funestus* does, they would be able to recolonize the habitat without opposition. Were the secondary species there in large numbers, then the niche could not so easily be occupied again. If the original vector changes its biting behaviour, to feed on alternative hosts outdoors, then that is also useful. In a study of secondary vectors of malaria in Tanzania that show this kind of behaviour, Gillies (1964) concluded that none of them was likely to be of any importance once transmission by the main vectors had been interrupted. This is unlikely to have changed. Their poor vectorial capacity appeared to be linked with a naturally occurring high mortality rate, which may itself be linked to the habit of resting outdoors. In Mozambique, the only mosquitoes with sporozoites among the large numbers of anopheline species tested were the primary vectors (Charlwood *et al.*, 2013).

As pointed out by Jeremy Farrar (quoted in Maxmen, 2021), 'Everything starts with smarter surveillance. If you don't look, you don't see. If you don't see, you will always respond too late'. It is also said that if we do not learn from the mistakes of history, we are bound to repeat them. This may be particularly true with respect to the development of insecticide resistance and its effect on disease transmission. Across all WHO regions with ongoing malaria transmission, malaria vectors have developed resistance to the four main classes of insecticides commonly used in adult malaria vector control (WHO, 2018). A case in point is the use of pyrethroids for insect control. Originally obtained from the plant *Pyrethrum (Chrysanthemum)* (family Compositae), natural pyrethroids have a long history of use against insects. However, they break down rapidly in ultraviolet light and so need to be frequently reapplied, which made them expensive. They were largely replaced by cheaper, synthetic insecticides (such as DDT) in the 1940s. An important consequence of this was that relatively few insects became resistant to them. The development of several photostable synthetic pyrethroids led to a resurgence in their importance, and they now account for about one-third of all insecticides in use today. They are the only insecticides suitable for treating fabric that comes into close contact with humans (in other words, mosquito bed nets). At the

Delight
Despair
Think of something else

Fig. 1.2. The three phases of malaria control, according to Charlwood (1984).

start of their wide-scale use, the market for bed net treatment was small and was not considered to be important by the chemical companies producing the insecticide. They instead either aggressively marketed the insecticides to be used in agriculture or as an insecticide for IRS. Not surprisingly, this led to the development of resistance: from 10 species in 1970 to over 80 in 2003. This has reduced the efficacy of nets treated with pyrethroids in many parts of the world and may be one reason why progress in malaria reduction has stalled since 2017. On the other hand, the presence of resistant populations did not appear to affect the incidence of malaria in a multi-country study involving four million health centre visits (Kleinschmidt *et al.*, 2018). Recent articles have, indeed, focused on the utility of the net, even without it being treated with an insecticide and on retreatment tablets (Okumu, 2020; Lindsay *et al.*, 2021; Charlwood, 2024).

The main aim of this book, therefore, is to provide a balanced position on a number as of yet not completely tested techniques suitable for wider application for mosquito control, an evaluation of more conventional techniques and alternative, low-tech systems for monitoring interventions on adult mosquitoes. We also provide some simple ideas for easily applied interventions that could be implemented without delay. The situations where the techniques might be most useful are also discussed.

The history of malaria control goes through three phases, otherwise known as DDT (Fig. 1.2).

'Delight', when a recently described technique (such as IRS with insecticide) is effective and helps change the paradigm; 'Despair', when it is obvious that said technique is failing badly; and 'Think-of-something-else'. With respect to the situation at the time this goes to press, we are very much in the 'Think-of-something-else' stage. A number of these 'something else's' are described by experts in their field, in the chapters not written by the editor. The editor's chapters differ in this respect. They are more informal and describe some of the experiences of 45 years of living mostly in the tropics where they seem relevant.

In describing the scientific response to the Covid pandemic, Spiegelhalter (2020) noted that 'Those trying to defend the public good against potentially dangerous misinformation can end up denying uncertainty. But if someone is only telling you about the potential benefits of a measure, or citing arguments that support their position, it's worth asking whether there are any potential harms or evidence that points the other way'. This is the appropriate attitude to take when reading this book, or indeed anything concerned with health.

Thus, within, there are state-of-the-art academic chapters written by experts that describe many of the techniques being developed, or even becoming widespread in use, and more basic and opinionated chapters written by the editor.

Due to the editors delay, as mentioned in the Acknowledgements, Chapter 5 was originally published in Parasites and Vectors (15,473) and Chapter 6 in *Acta Tropica* (222, 106045).

References

Attenborough, D. (2022) *A Life on Our Planet: My Witness Statement and A Vision for the Future*. Grand Central Publishing, New York, p. 272 .

Behar, A. (2024) Opinion: how to get from rhetoric to reality in decolonizing development. Available at: www.devex.com/news/opinion-how-to-get-from-rhetoric-to-reality-in-decolonizing-development-107045 (accessed 23 September 2024).

Bhatt, S., Weiss, D.J., Cameron, E., Bisanzio, D., Mappin, B. *et al.* (2015) The effect of malaria control on *Plasmodium falciparum* in Africa between 2000 and 2015. *Nature* 526, 207–211.

Buller, A. (2022) *The Value of a Whale: On the Illusion of Green Capitalism*. Manchester University Press, Manchester, UK.

Caesar, L., McCarthy, G.D., Thornalley, D.J.R., Cahil, N. and Rahmstorf, S. (2021) Current atlantic meridional overturning circulation weakest in last millennium. *Nature Geoscience* 14, 118–120. DOI: 10.1038/s41561-021-00699-z.

Charlwood, J.D. (1984) Which way now for malaria control. *PNG Medical Journal* 27, 159–162.

Charlwood, J.D. (2023) A self-sustaining system to reduce migration of *Anopheles stephensi* in East Africa. *Acta Tropica* 245, 106981.

Charlwood, J.D. (2024) Uneasy bedfellows: Public-private partnerships for malaria control. *Malaria World Journal* 15, 9. DOI: 10.5281/zenodo.11046816.

Charlwood, J.D., Macia, G.A., Manhaca, M., Sousa, B., Cuamba, N. *et al.* (2013) Population dynamics and spatial structure of human-biting mosquitoes, inside and outside of houses, in the Chockwe irrigation scheme, Southern Mozambique. *Geospatial Health* 7, 309–320.

De Bono, E. (1998) *Simplicity*. Viking, p. 320 .

Ditlevsen, P. and Ditlevsen, S. (2023) Warning of a forthcoming collapse of the Atlantic meridional overturning circulation. *Nature Communications* 14, 4254. DOI: 10.1038/s41467-023-39810-w.

Dubos, R.J. (1987) *Mirage of Health: Utopias, Progress and Biological Change*. Rutgers University Press, New Brunswick, New Jersey, p. 282 .

Feachem, R.G.A., Chen, I., Akbari, O., Bertozzi-Villa, A., Bhatt, S. *et al.* (2019) Malaria eradication within a generation: Ambitious, achievable, and necessary. *The Lancet Commissions* 394(1023), 1056–1112.

Gillies, M.T. (1964) The role of secondary vectors of malaria in North-East Tanganyika. *Transactions of the Royal Society of Tropical Medicine* 58, 154–158.

Grossman, D. (2024) Scientists under arrest: The researchers taking action over climate change. *Nature* 626(8000), 710–712.

IPBES (2019) Global assessment report on biodiversity and ecosystem services of the intergovernmental science-policy platform on biodiversity and ecosystem services. In: Brondizio, E.S., Settele, J., Díaz, S. and Ngo, H.T. (eds) *IPBES Secretariat*. IBPES, Bonn, Germany, p. 1148. DOI: 10.5281/zenodo.3831673.

Kleinschmidt, I., Bradley, J., Knox, T.B., Mnzava, A.P., Kafy, H.T. *et al.* (2018) Implications of insecticide resistance for malaria vector control with long-lasting insecticidal nets: A WHO-coordinated, prospective, international, observational cohort study. *Lancet Infectious Diseases* 18, 640–649. DOI: 10.1016/S1473-3099(18)30172-5.

Latour, B. (1998) To modernize or to ecologize, that is the question. In: Castree, N. (ed.) *Remaking Reality: Nature at the Millenium*. Routledge, Oxford, UK, pp. 221–242.

Lindsay, S.W., Thomas, M.B. and Kleinschmidt, I. (2021) Threats to the effectiveness of insecticide-treated bed nets for malaria control: Thinking beyond insecticide resistance. *Lancet Global Health* 9, e1325–e1331.

Maxmen, A. (2021) Has COVID taught us anything about pandemic preparedness?. *Nature* 596, 332–335. DOI: 10.1038/d41586-021-02217-y.

Okumu, F. (2020) The fabric of life: What if mosquito nets were durable and widely available but insecticide-free?. *Malaria Journal* 19, 260.

PAHO (2024) News/PAHO calls for collective action in response to record increase in dengue cases in the Americas. Available at: https://www.paho.org/en/news/28-3-2024-paho-calls-collective-action-response-record-increase-dengue-cases-americas (accessed 13 September 2024).

Riddell, R., Ahmed, N., Maitland, A., Lawson, M. and Taneja, A. (2024) Inequality Inc. OXFAM. Available at: https://www.oxfam.org/en/research/inequality-inc (accessed 13 September 2024).

Solomon, R.C. (2003) *The Joy of Philosophy: Thinking Thin versus the Passionate Life*. Oxford University Press, Oxford, UK, p. 288 .

Spiegelhalter, D. (2020) Those who tell us what to do during the pandemic must earn our trust. *Guardian Opinion*. Available at: https://www.theguardian.com/uk/commentisfree (accessed 26 November 2020).

Stanaway, J.D., Shepard, D.S., Undurraga, E.A., Halasa, Y.A., Coffeng, L.E. *et al.* (2016) The global burden of Dengue: An analysis from the global burden of disease study 2013. *Lancet Infectious Diseases* 16, 712–723.

Swift, J. (1710) *The Examiner*, #14. November 9, 1710.

Weaving, H. (2022) Insects will struggle to keep pace with global temperature rise – which could be bad news for humans. The Conversation. Available at: https://theconversation.com/insects-will-struggle-to-keep-pace-with-global-temperature

-rise-which-could-be-bad-news-for-humans-190791 (accessed 13 September 2024).

WHO (1978) In: *Primary health care Report of the International Conference on Primary Health Care, Alma Ata, USSR, 6-12 September 1978*. World Health Organization, Geneva, Switzerland.

WHO (2014) *World malaria report 2014*. World Health Organization, Geneva, Switzerland.

WHO (2018) *Global report on insecticide resistance in malaria vectors: 2010–2016*. World Health Organization, Geneva.

Wilkes, T.J., Matola, Y.G. and Charlwood, J.D. (1996) *Anopheles rivulorum*, a vector of malaria in Africa. *Medical and Veterinary Entomology* 10, 108–110.

Wilkinson, R. and Pickett, K. (2010) *The Spirit Level: Why Equality Is Better for Everyone*. Penguin, London.

Wilkinson, R. and Pickett, K. (2024) Why the world cannot afford the rich. *Nature* 627, 268–270.

Mosquito Life Histories and Their Relevance for Control

2

J. D. Charlwood*

Global Health and Tropical Medicine, Instututo de Hygiene e Medicina Tropical, Universidade NOVA de Lisboa, Lisbon, Portugal

Abstract

In order for disease transmission to be sustained, sufficient numbers of vectors must occur in relatively close proximity to humans. The disease organism needs to be able to develop in the mosquito, and the mosquito needs to survive long enough for it to mature. Without an understanding of these factors, it is difficult to devise suitable strategies for their control. The concept of ecological niches is explored, highlighting species-specific adaptations and niche occupancy by different mosquito species. Even though many of the characteristics of their biology are shared (e.g. all mosquitoes have four larval stages when growth is possible, and all female mosquitoes are monogamous), there are wide differences in the biology of even closely related species. Urbanization's impact on malaria transmission and the potential for species replacement are also discussed.

Introduction

It may seem obvious that in order to attempt the control of any mosquito, information about their life history is mandatory. But supposing all control efforts were aimed at a single life history stage, then it is only information on that stage that is required, and if only one activity during a part of that stage that is deemed important, then it is only that aspect that is likely to be investigated. As Fred Soper describes in his book about the eradication of *Anopheles gambiae* from Brazil, 'it seemed rather more expedient to learn how to get rid of gambiae

*Corresponding author: jdcharlwood@gmail.com

 Practical Control of Mosquito Disease Vectors (ed J.D. Charlwood)
DOI: 10.1079/9781789248845.0002

by actually getting rid of gambiae' (than learning how to do so) (Soper and Wilson, 2011). This can lead to a considerable bias in sampling techniques used. For example, human-landing collection (HLC) is by far the most used technique for the sampling of anopheline mosquitoes (van de Straat *et al.*, 2021). How much such restricted sampling has hampered our understanding of the mosquito life history remains to be seen. At the end of the day, it – or a surrogate measure – remains the most useful estimate for epidemiological models. But as far as understanding mosquito behaviour, they are not enough.

Early work on the ecology of malaria vectors was largely concentrated on the larval stages, again because that is how they were controlled. Both in Brazil and Egypt, it was larval control that enabled Soper to eliminate *An. gambiae* s.l. from these countries. He did not, of course, get rid of malaria. This is because the remaining vector *Anopheles darlingi* (now called *Nyssorhynchus darlingi*) occupied a different larval habitat not amenable, or targeted, to larval control. Differences in the habitat occupied by the different species (i.e. their larval niche) resulted in Malaysia in the idea of 'species sanitation'. This was a far more sophisticated approach than that adopted after the discovery of DDT and the start of, what Harrison (1979) called the 'spray-gun war' when adult mosquitoes were the target. The use of DDT [and subsequent insecticides for indoor residual spraying (IRS)] meant that the information required of the entomologist was reduced to a small number of questions such as 'Do the mosquitoes rest inside (and, if so, where do they rest)?' and 'Does the insecticide kill the insect?' The use of IRS was in part a Pandora's box. It had huge, and well-documented, benefits, but it resulted in a continuing dependency on insecticides and an 'arms race' between the development of resistance in the mosquito and a reduction in transmission. This is acknowledged by medical entomologists, and an approach familiar to all, but still not as widely implemented as it needs to be, is integrated vector management (IVM); in other words, not using just a single technique but using two or more techniques that act synergistically to achieve control. And for this, a more detailed knowledge of the biology and ecology of the different species is mandatory (Ferguson *et al.*, 2010). A big part of this enhanced knowledge comes from monitoring the mosquito population. Indeed, the World Health Organization (WHO) (2015) now considers monitoring (= measuring) of mosquito vectors of disease an integral part of control. Disease surveillance, including entomological surveillance, serves as the basis for all control programme activities. Data should help in decisions regarding how to control the vectors. Given that it is now possible to collect and analyse geospatial data in real time, programme managers should be able to evaluate the effect of different interventions, levels of coverage or approaches to control. Mobile phones are a particularly useful tool for this.

Mosquitoes are holometabolous and undergo complete metamorphosis. All mosquitoes, therefore, have 'adventures in two worlds', initially as aquatic larvae and subsequently (if they manage to survive that) as terrestrial/aerial adults. The concept of the ecological niche is that each species occupies a unique 'hyperspace' that involves biological and environmental characteristics. Each species occupies a specific niche, which makes it avoid competition with other species, while at the same time engaging in competition with conspecifics. Indeed, evolution can be thought of as 'life endlessly probing all of the possible options' (whether morphologies or DNA and protein sequences) that enable them to occupy different hyper-dimensional spaces, or niches (Inquisitive Biologist, 2022). Species, groups of interbreeding natural populations that are reproductively isolated from other such groups, have a *fundamental* niche (i.e. the entire set of conditions under which an animal population or species can survive and reproduce itself) and a *realized* niche, the actual set of conditions used by the species after interactions with other species (predation and especially competition) have been considered.

Some of the characteristics of the niche occupied by different species may be similar, in which case, if the resource is limited, the species will be in competition for that particular resource. In general, however, natural selection will operate such that competition is avoided. Although competition may perhaps be reduced in one life stage, it may occur in another. For example, in general (and not forgetting that 'all generalizations are false'), the larval habitat of the various genera differs. *Anopheles* (from the Greek word for 'hurtful') occupy clean water, culicines may occupy water with a high organic content, while *Aedes* (aka *Stegomyia*) occupy small containers. At the same time, anophelines are surface feeders and lie parallel to the water surface, while culicines have a siphon, hang below the surface and feed deeper in any water body. Differences are even more extreme in *Mansonia* and *Coquillettidia* mosquitoes. The respiratory siphon of immature stages of these genera is shaped to pierce the root of aquatic plants for oxygen. Larvae and pupae do not need to breathe on the water surface like other mosquito larvae. Even within genera, there are differences that tend to reduce competition. African vectors (the *An. gambiae* complex and the *Anopheles funestus* group) evolved with humans. *Anopheles funestus* is the most anthropophilic of mosquitoes. Males and newly emerged females rest inside houses alongside females who had recently fed on their favourite food – humans. Blood-fed females stay indoors until ready to lay eggs. It has been argued that the first human settlements would probably have been established close to permanent water of the sort occupied by *An. funestus* larvae, meaning that this species has a longer association with humans than other mosquitoes (Charlwood *et al.*, 1995b). A key development in the evolution of *An. funestus* was the adoption of a late-night-biting cycle, which enabled the mosquito to avoid host defensive mechanisms and host density-dependent avoidance mechanisms. While *An. funestus* continued to occupy more permanent water bodies with emergent vegetation, the creation of new niches as a result of the clearing of forest areas by invasive early humans for agriculture expanded the number of small sunlit pools, 'with which every entomologist working in Africa is familiar' (Gillies and de Meillon, 1968). These pools were exploited by the mosquito that evolved into *An. gambiae*. To paraphrase Voltaire, 'if *An. gambiae* did not exist, humans would have created it'. Even within the *An. gambiae* complex, however, the different members of the complex occupy different larval habitats: it is these that largely determine their behaviour. *Anopheles merus* and *Anopheles melas* occupy brackish water but on different sides of Africa; *Anopheles bwambae* breeds in brackish water from geothermal springs (only found in Buranga hot springs in Bwamba County, Toro District, Uganda), while *An. gambiae* and *Anopheles arabiensis* occupy similar larval habitats, but as adults have diverged in their choice of mating site and blood meal host. The recently described *Anopheles fontenillei* (Barrón *et al.*, 2019) breeds in pools of the sort occupied by the other freshwater members at the edge of forest patches and close to natural marshes frequented by wild animals. It is considered to be zoophilic.

Anopheles gambiae itself is now considered to consist of two species that differ in their preferred larval habitat, which is dependent on the presence of predators. Both species (*An. gambiae* and *Anopheles coluzzii*) are largely anthropophagic; as Sherrington (1955) says, 'rich food it is, human blood'. Thus, it is through the geographical separation of the larval habitat that the species evolve. This may be happening now. In Sudan, there is evidence that *An. arabiensis* is increasing the range of habitats used by larvae to include water with high organic content (Azrag and Mohammed, 2018). This has yet to be accompanied by a genetic change, but should this happen, and should this also influence mating behaviour, then a new 'urban' species may evolve (see Box 2.1).

Box 2.1. What's in a name? – The importance of mosquito identification.

'A superior man, in regard to what he does not know, shows a cautious reserve. If names be not correct, language is not in accordance with the truth of things. If language be not in accordance with the truth of things, affairs cannot be carried on to success.' Confucius Analects, Book XIII, Chapter 3.

It is now recognized that many mosquitoes belong to species complexes (in which there are no morphological characteristics that can be used to separate them) or closely related groups (in which the differences are slight and often difficult to discern). The different members may differ in their propensity to transmit disease. The possibility that certain morphologically similar species might actually be different was postulated because, in Europe, there were areas where the malaria vector *Anopheles maculipennis* occurred but where there was no malaria transmission. This led to the concept of anophelism sans malaria. In some regions, the mosquito preferred to feed on animals, while in others it was anthropophilic. Differences in mating behaviour were also observed. In 1926 Falleroni described distinct morphological differences in the eggs of certain populations (Hackett, 1934). Twelve members of the complex, which can still largely be distinguished according to the morphology of their eggs are recognized today, although genetic techniques, are now predominantly used. Almost anywhere one looks one finds a species complex be it among the malaria vectors in Africa or the dengue vectors in South America. It is beyond the scope of this book to describe these in any detail but a knowledge of which mosquitoes are not actually vectors can sometimes lead to considerable savings since they do not need to be controlled.

On the other hand, similar niches may be occupied by different species in different places. Thus, *Anopheles punctulatus* breeds in sunlit pools in Papua New Guinea similar to those occupied by *An. gambiae* in Africa. Forest-dwelling *Anopheles dirus* in South-east Asia and *Ny. darlingi* in South America face similar challenges to each other, and so they too have similar ecologies. Should the species occupy the same larval habitat (as can occur with *An. gambiae* and *An. arabiensis*), then competition between the adults may be reduced – in this case by the insects tending to feed on different hosts; *An. gambiae* largely feeds on humans, while *An. arabiensis* has more catholic feeding habits, as does another member of the complex, *Anopheles quadriannulatus*, which largely feeds on animals (and so is not involved in malaria transmission). Similarly, even though both *Aedes aegypti* and *Aedes albopictus* larvae are to be found in small containers, one species occupies more urban habitats, and the other is found in more rural or forested habitats.

When a dominant member of a closely related group of mosquitoes is reduced as a result of a control campaign, then the usually less successful competitor may occupy the vacated larval niche, and so species replacement occurs. Thus, *Anopheles rivulorum* may replace *An. funestus*, and *An. arabiensis* may replace *An. gambiae*. These species, although capable of transmitting malaria (Wilkes *et al.*, 1996), are much less-efficient vectors. This is in large part because of their tendency to be less anthropophagic and to have naturally occurring higher mortality rates than the primary vectors. Under these circumstances, reducing man-mosquito contact rather than killing the mosquito may be a more suitable way of reducing transmission, since this will enable the less-efficient larval competitor to maintain itself and so help prevent the re-establishment of the primary vector once the effect of control (such as the efficacy of an insecticide over time) is reduced.

Invasive Species

Invasive species pose a particular problem. They are species that have not evolved in the particular environment or niche that they are invading, so the usual checks and balances that apply to local species may not apply, potentially allowing them to proliferate unchecked. They may occupy a niche or habitat that has previously been left unoccupied.

By far the most destructive invasive species is, of course, *Homo sapiens* – 'that monster man(un)kind'. The arrival of humans to previously uncolonized habitats has meant the elimination of a wide variety of species including, but not limited to, all the larger mammals. Invasive mosquito vectors of disease, often abetted by humans providing the means of long-distance movement, have also resulted in major epidemics of disease. In the 1930s, *An. arabiensis* arrived in Brazil with the '*aviso*' mail boats (boats that made the journey in 3 days) from Dakar. Its presence was recognized after an outbreak of malaria in Natal. It was controlled in the city but – perhaps as a result of this success and a lack of political will – spread unchecked through much of the north-east of the country. An initial suggestion to flood breeding sites near the coast with salt water, which may have alleviated the problem very early on, was refused. The years 1932–1938 are considered the 'silent' years during which time the mosquito spread but did not cause large epidemics of the disease. This was perhaps due to a relative lack of rainfall that inhibited the mosquito population. This changed in 1938 when there was a major epidemic and many deaths. The mosquito was eventually eradicated from Brazil using a veritable army of people applying the larvicide Paris Green to breeding sites. Fred Soper, the person responsible for the effort was known as 'the commander' because of his military-style approach to the problem. A similar approach and result was obtained when *An. gambiae* invaded Egypt in the 1940s (Shousha, 1948). Other invaders, also aided and abetted by humans, have been more successful and remain a problem.

There is a major industry in the transportation and storage of used motor-car tyres around the world. In 1990, there were a billion such tyres in the USA, many of which came from countries where *Ae. albopictus* was endemic. After rain, these tyres provide a suitable habitat for the drought-resistant eggs. In this way, the mosquito has invaded a number of countries, and once it has arrived, it is very difficult to control or eradicate.

Although an inefficient vector of lymphatic filariasis, *Culex quinquefasciatus* is an efficient vector of avian malaria (*Plasmodium relictum*) (Warner, 1968). It entered Hawaii between 1826 and 1830 and has played, and continues to play, an important role in the decline and extinction of many native forest birds from the island (Fig. 2.1). Although Hawaii is well known for its unique scenic beauty and its fascinating native flora, fauna, bird and marine life, it is also called the extinction capital of the world. The islands' 77 bird species and subspecies extinctions account for approximately 15% of global bird extinctions during the last 700 years.

On some islands, over 80% of the original land bird species are now extinct. It is malaria that is responsible. The expansion of *Cx. quinquefasciatus* into forest habitats is facilitated by the spread of another invasive species. Feral pigs, the descendants of domestic pigs introduced into the islands by early settlers, knock over tree ferns, tear through the fibrous and woody outer layers, and eat the starchy pith leaving behind a bowl- or trough-shaped, nutrient-rich cavity that fills up with rainwater. In forests, where tree ferns and pigs occur together, these cavities are the most abundant and productive habitat for larval mosquitoes. Feral pigs increase the numbers of mosquitoes found in the forest and so exacerbate malaria transmission to the native birds.

As pointed out by Charlwood *et al.* (2024), 'This also illustrates how an understanding of the ecology of the vector and its interactions with the environment can produce an unexpected 'best' method of control'. In this case, kill the feral pigs (Liao *et al.*, 2017). This provides a win-win, since it reduces transmission, provides food, helps preserve the forest and surely could be marketed

Fig. 2.1. An ʻIʻiwi (*Drepanis coccinea*), or scarlet honeycreeper, from Hawaii. It is apparently still relatively common, but 8 of 10 species of bird declared extinct in the USA in 2023 were from Hawaii. Extinction was driven by avian malaria transmitted by the invasive *Culex quinquefasciatus,* aided and abetted by feral pigs. (Photo, HarmonyonPlanetEarth, CC BY 2.0 <https://creativecommons.org/licenses/by/2.0>, via Wikimedia Commons.).

as a sport to adventure-seeking tourists. Invasive species can be viewed as an external component of the ecosystem and can be eliminated for the benefit of the local fauna and flora.

Urbanization and Mosquito Vectors of Disease

Urbanization is continuing apace in Africa, with 43% of Africans now living in cities and with the number of megacities in sub-Saharan Africa rising from 10 in 1990 to 28 in 2014. An estimated 62% of the population live in slums. Urbanization in general reduces malaria transmission (while at the same time possibly enhancing the transmission of dengue and other arboviruses). This is because the typical habitat for the larvae of the usual vectors is not as readily available as it is in more rural areas. New niches for larvae may, however, be produced as a result of urbanization. As described above, there is evidence that in some areas of Africa, members of the *An. gambiae* complex are adapting to this expansion by occupying polluted water with high organic content (Azrag and Mohammed, 2018). A further threat to malaria elimination in urban areas is *Anopheles stephensi*. In India, it is one of the principal malaria vectors and is a newly arrived invader to Africa. It has to date been found in Djibouti and Sudan and also appears to be well-established in Ethiopia (Sinka *et al.*, 2020). It breeds in water tanks and smaller bodies of clean water in towns. Should *An. stephensi* find its way (perhaps again with human assistance) to larger centres of population in sub-Saharan Africa, especially those where the domestic water supply is irregular and uncertain so that people store water in large containers or tanks when it is available, then the consequences could be catastrophic. This would be exacerbated by the tendency of the mosquito to bite outdoors at dusk rather than indoors at night (as is the case for the principal African malaria vectors), which would render methods of control, such as mosquito nets treated with insecticide, ineffective.

The Mosquito Life Cycle

It is perhaps worth describing the basics of the mosquito life cycle, indicating why some species might be vectors and where some of the different control techniques might operate depending on the behaviour of the mosquito.

All animals can potentially produce more offspring than actually survive to reproduce in the next generation. There are a number of factors that will influence the population dynamics of animals in general. It has been estimated by Klowden (2007) that should all the eggs laid by a single *Ae. aegypti* at the start of the summer survive and her offspring likewise survive and reproduce, then by the end of the summer, there would be 7×10^{21} mosquitoes with a biomass about 33,000 times the weight of the total human population. It is therefore a good thing that constraints to population growth operate. Density-independent factors, especially meteorological factors, act on the population irrespective of population density while density-dependent factors, such as predation, will have an increasing effect as numbers increase. How much these affect the overall population dynamics depends on the stability of the environment.

All mosquitoes start their life as eggs, and as Samuel Butler once said, 'A hen is only the egg's way of making another egg'. In other words, each stage in the life history of a mosquito is as important as all the others. Each species will develop a series of trade-offs to ensure their survival. This may mean that females lay a large number of eggs per egg batch but only lay a few batches (meaning that there is a low adult survival) or may lay a few eggs per batch but many batches (which implies a high survival rate and therefore a higher probability of the insect being a vector of disease). Alternatively, there may be a high survival rate among larvae but relatively a low survival of adults. Given that high adult survival is a critical factor for a mosquito to transmit disease, such mosquitoes are also unlikely to be vectors.

For the egg, it is important that it is laid in a suitable place for the larvae to develop. Oviposition site selection by female mosquitoes is the primary determining factor in the habitat distribution of mosquito larvae, but the various cues used for oviposition remain only partially known. Colour, moisture and volatile chemical stimulants appear to play a role in choice in most species. Eggs of anophelines are characterized by having floats on the side and do not survive desiccation. Those of culicines are laid on the water surface together and form a 'raft' visible to the naked eye. Eggs of *Aedes* are laid separately, and those of species like *Ae. aegypti* can withstand desiccation for long periods (although it is not the egg but rather the unhatched first instar larva that is inside the egg at these times). This means that in areas where control of adults has been complete, *Aedes* vectors of dengue can reappear months later without there being reinvasion from outside. Since the viral diseases that they transmit can pass from the adult female to her eggs, this also means that diseases such as dengue can reappear even in the face of control.

Mosquito larvae pass through four instars. These are the only times when the mosquito can grow. Growth is continuous throughout each instar, as seen in the progressive increase in size of the body regions that are covered with thin cuticle. The time required for development of the larval stages depends on several factors, the most important of which is water temperature. The quantity of reserves present at metamorphosis is largely dependent on the availability of nutrients during the fourth instar. Adult females are usually heavier than males. Perhaps because of this, they emerge later than the males. *Anopheles* generally inhabit groundwater and exhibit rapid development and little ability to adjust their development time to accommodate variations in availability of nutrients. *Aedes* larvae in small containers can, however, have a protracted development, depending on food availability.

Adult mosquitoes only encounter insecticides for short periods, and it is often only the females that do so. On the other hand, if the insecticide is in the water in which they live, both female and male larvae swim in it throughout their life. It is perhaps in the larval stage that resistance to insecticides is most likely to develop in pool breeders, largely as a result of low-level exposure to the runoff of the same products from agricultural use. Thus, resistance to pyrethroids (which were not widely used in the early days of insecticide-based vector control) was first recorded in the cotton growing areas of the Ivory Coast, where it was used extensively against the bollworm – *Earias insulana* (actually a moth) – attacking the cotton.

The control of larvae is one area that has often been relatively overlooked in recent years. One reason for this is that control of larvae only reduces the number of adults and not their survival, which, as recognized by MacDonald (1957), is a key reason why the control of adult malaria vectors is more effective than larval control, unless it is extremely well done. An example of extremely well done control of larval malaria vectors was the elimination of invasive *An. gambiae* s.l. in Brazil in the 1940s (Soper and Wilson, 2011). Presently, the WHO considers that larval control of malaria vectors should only be attempted if the breeding sites are 'Few, Fixed and Findable'. The control of immature stages of *Ae. aegypti* and *Ae. albopictus* and removal of potential oviposition sites is, however, a widely used technique.

The Origin of Species

Although his book was called *The Origin of Species*, Darwin did not actually describe how they might arise. Coluzzi *et al.* (1979, 1985) described the ecological theory of species origin. Basically, in this model, speciation occurs when some habitats at the edge of a species distribution (likely to be marginal habitats), or newly created habitats (as occurred with the development of agriculture in forested areas) are occupied, often as a group of populations that are separated by space (in other words, as a metapopulation). These spatially separated populations interact as individual members, move from one population to another but are liable to extinction and recolonization. Eventually, the populations occupying these habitats (niches) adapt to them so that they become the new optimal habitat (niche). Should the adaptation include modifications to the mating behaviour, then they will become separate species. Thus, understanding the different mating behaviours of species is important. The process may be enhanced because, at the edge of a species range, there is often less genetic variability than at the centre of its distribution.

Mating in *Aedes* and *Anopheles*

Aedes, like *Ae. aegypti*, are stenogamic, meaning that they can mate in small spaces. Individual male *Aedes* search out and mate with individual females. Male *Aedes* will encounter females of other species and previously mated females of their own species (who will not mate again). This means that *Aedes* males need to have a repertoire of techniques, such as contact chemical recognition, to recognize conspecific females that have already mated and members of other species that initially appear to be the same species and have the same flight tone.

Mating in *Anopheles* is a group phenomenon. *Anopheles* mosquitoes are eurygamic, meaning that they mate in large open spaces in swarms at dusk, 'Pegasuses of the dew' that 'wheel in the still air', according to the Spanish poet Lorca. Swarms are aggregations of males that generally form over distinct areas of contrast, called swarm markers, be that a bush, a distinctly coloured patch of earth or some other area of contrast. These markers are not always easy to find, and it may be that vertical markers also play a role in swarm location. Swarming, like lekking in birds, probably has several advantages that contribute to its evolution. When two sexes are separated in space and searching for each other, it makes sense for one of them to remain in one place while the other does the searching. Virgin females are also attracted to such locations and enter swarms, where they are mated. Short-range attraction of males to females is modulated by the female flight tone. Males hear and are attracted to the female flight tone, which is 200 Hz lower than their own, when the fibrillae on their antennae are erect, which only happens for a short time around dusk (Charlwood and Jones, 1979; Somers *et al.*, 2022). Differences in the location of swarms resulting from different responses to markers are the primary mate-recognition system (Paterson, 1985) that maintains isolation between members of the *An. gambiae* complex. Males act like 'boids' (Charlwood *et al.*, 2002). 'Boids' is an agent-based computer programme developed by Craig Reynolds (Reynolds, 1987) that simulates the flocking of birds or the shoaling of fish. A general feature of the agents in simple agent-based models is that it can give rise to remarkably complex and surprising aggregate outcomes. Boids only need to follow three rules for a swarm or flock to occur. Each boid acts: (i) to move towards the perceived centre of mass of the boids in its neighbourhood; (ii) to match velocities with boids in its neighbourhood; and (iii) to maintain a minimum distance from other objects in the environment, including other boids (Stereiny and Griffiths, 1999). Virgin female *An. coluzzii* in many ways act as surrogate males and surrogate boids. They will 'investigate' potential markers by making turning movements over them similar to the males but of much shorter duration. During this flight, after she locates the swarm site, she attracts a male who, matching her speed as per boid rules, without altering her behaviour, attaches itself and mates with her. As the sex ratio becomes more male biased, chance (often considered to be 'noise' in the study of mating systems) contributes relatively more to the overall mating pattern because an increasing number of males remain unmated, even if they have a trait preferred by the females (Klug and Stone, 2021). In addition to sperm, male members of the *An. gambiae* complex donate a 'nuptial gift', a gelatinous mating plug, to the female (Fig. 2.2). The plug is absorbed after 24 hour and, hormonally, induces monogamy in the females.

The cost of these gifts means that males cannot mate many times in succession. Females only mate once in their life. The sperm from that mating are stored in a spermatheca (the number of which varies according to genera). There remains some debate about the potential for females to choose males (in other words, for sexual selection to occur). Proponents of theories of sexual selection do not know how this might work but suggest that because the female

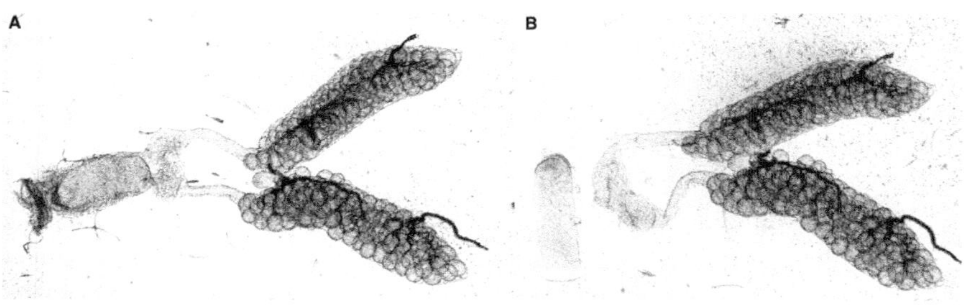

Fig. 2.2. A mating plug *in situ* in an *Anopheles arabiensis* from Tanzania, B the plug removed. If male *Homo sapiens* had to sacrifice an equivalent amount of body mass each time they had sex, they might be less promiscuous. (Photo: J.D. Charlwood.).

only mates once, it behoves her to choose the best mate possible. But, because of the very high male-female ratio in swarms and the trivial advantage for a female even if she chooses the perfect male (Williams, 1992), this is perhaps unlikely. As pointed out by Charlwood (2023), just being in a swarm may be sufficient for the male to have passed any female 'test'.

Male *Anopheles* are only likely to meet virgin females of their own species (who come to them to mate). Male *Anopheles*, therefore, do not need to have the refinements that are required by *Aedes*, and, in the laboratory, will attempt to mate with anything that makes the right sound.

Monogamy is induced by hormones from the male accessory glands transferred to the female by the male during mating. In *Aedes*, the effect is immediate. By mating with a sterile male or one with incompatible sperm, but with suitable accessory gland hormones, females will lay sterile eggs, and the population will decrease. This is the rationale behind the use of *Wolbachia* (Chapter 6) or radioactively sterilized males (SIT). As Charlwood pointed out in 2003:

> *Anopheles gambiae* shares many characteristics of other insects that undergo the oogenesis-or-flight syndrome (Rankin *et al.*, 1986): males form stationary swarms where mating takes place and during mating males provide a substantial mating plug which may have a nutrient function; breeding sites are temporary, and there is a high mortality among larvae.

Blood Feeding and Its Epidemiological Importance

In addition, newly emerged females have undeveloped ovaries and do not feed to repletion but rather take an incomplete blood meal used for adult maintenance rather than egg development.

At other times, anopheline females, and the majority of culicine females, feed to repletion, each blood meal giving rise to a complete egg batch. In other words, the insects are gonotrophically concordant, the number of egg batches laid equalling the number of blood meals taken.

The act of feeding was beautifully described by Sherrington (1955): 'The so called 'bite' happens like this. The gnat when she alights on the skin tests the place with her labellum. Then steadying her head against the skin, she stabs by styles with dagger points and saw-like edges. Swaying her head as she uses her mouthparts she works these through the skin. They go through and among the blood vessels and carry with them a tiny tube like a hollow needle, close behind the stabbing style. It leads from a poison gland it injects a droplet of juice into the stabbed wound. This makes the blood vessel flush; they bring more blood to the stabbed spot. Also, the juice delays the time of clotting of the blood which might baulk the gnat of her full meal by cutting short the yield of blood from the tiny wound'.

D H Lawrence also described the act of feeding in a memorable poem :

'The Mosquito'

When did you start your tricks Monsieur?

What do you stand on such high legs for?

Why this length of shredded shank

You exaltation?

Is it so that you shall lift your centre of gravity upwards

And weigh no more than air as you alight upon me,

Stand upon me weightless, you phantom?

I heard a woman call you the Winged Victory

In sluggish Venice.

You turn your head towards your tail, and smile.

How can you put so much devilry

Into that translucent phantom shred

Of a frail corpus?

Queer, with your thin wings and your streaming legs

How you sail like a heron, or a dull clot of air,

A nothingness.

Yet what an aura surrounds you;

Your evil little aura, prowling and casting a numbness on my mind.

That is your trick, your bit of filthy magic:

Invisibility, and the anæsthetic power

To deaden my attention in your direction.

But I know your game now, streaky sorcerer.

Queer, how you stalk and prowl the air

In circles and evasions, enveloping me,

Ghoul on wings

Winged Victory.

Settle, and stand on long thin shanks

Eyeing me sideways, and cunningly conscious that I am aware,

You speck.

I hate the way you lurch off side-ways into air

Having read my thoughts against you.

Come then, let us play at unawares,

And see who wins in this sly game of bluff.

Man or mosquito.

You don't know that I exist, and I don't know that you exist.

Now then!

It is your trump

It is your hateful little trump

You pointed fiend,

Which shakes my sudden blood to hatred of you:

It is your small, high, hateful bugle in my ear.

Why do you do it?

Surely it is bad policy.

They say you can't help it.

If that is so, then I believe a little in
Providence protecting the innocent.

But it sounds so amazingly like a
slogan

A yell of triumph as you snatch my
scalp.

Blood, red blood

Super-magical

Forbidden liquor.

I behold you stand

For a second enspasmed in oblivion,

Obscenely ecstasied

Sucking live blood

My blood.

Such silence, such suspended
transport,

Such gorging,

Such obscenity of trespass.

You stagger

As well as you may.

Only your accursed hairy frailty

Your own imponderable
weightlessness

Saves you, wafts you away on the
very draught my anger makes in its
snatching.

Away with a pæan of derision

You winged blood-drop.

Can I not overtake you?

Are you one too many for me

Winged Victory?

Am I not mosquito enough to out-
mosquito you?

Queer, what a big stain my sucked
blood makes

Beside the infinitesimal faint smear
of you!

Queer, what a dim dark smudge you
have disappeared into!

But he got the sex wrong! He could be forgiven, since even today, mistakes are often made, if not by sex then by genera! (Fig. 2.3).

Feeding *Aedes* (now *Ochlerotatus*) *cantans* in the UK and feeding *Ae. (Stegomyia) albopictus* in Cambodia appeared to attract other mosquitoes to the feeding site (Charlwood *et al.*, 1995a, 2014). Why this should be so is difficult to explain. Video evidence of this would be useful.

If feeding is a regular process, then this also provides information on the longevity of the mosquito. Exceptions are the species like *An. gambiae*, which may take two blood meals during the first gonotrophic cycle (and go through a pre-gravid phase), and those which aestivate as a way to avoid hot dry conditions and which may, at such times, take occasional small meals for maintenance. In general, though, the life of a female mosquito, once mated, becomes one of searching for blood, finding a suitable place to rest while her eggs develop, finding a suitable oviposition site and returning to feed. It is the different behaviours and characteristics of the mosquito during this part of the life cycle that determines whether she is likely to become a vector. Four factors are important: (i) abundance; (ii) contact with man (or the host that will

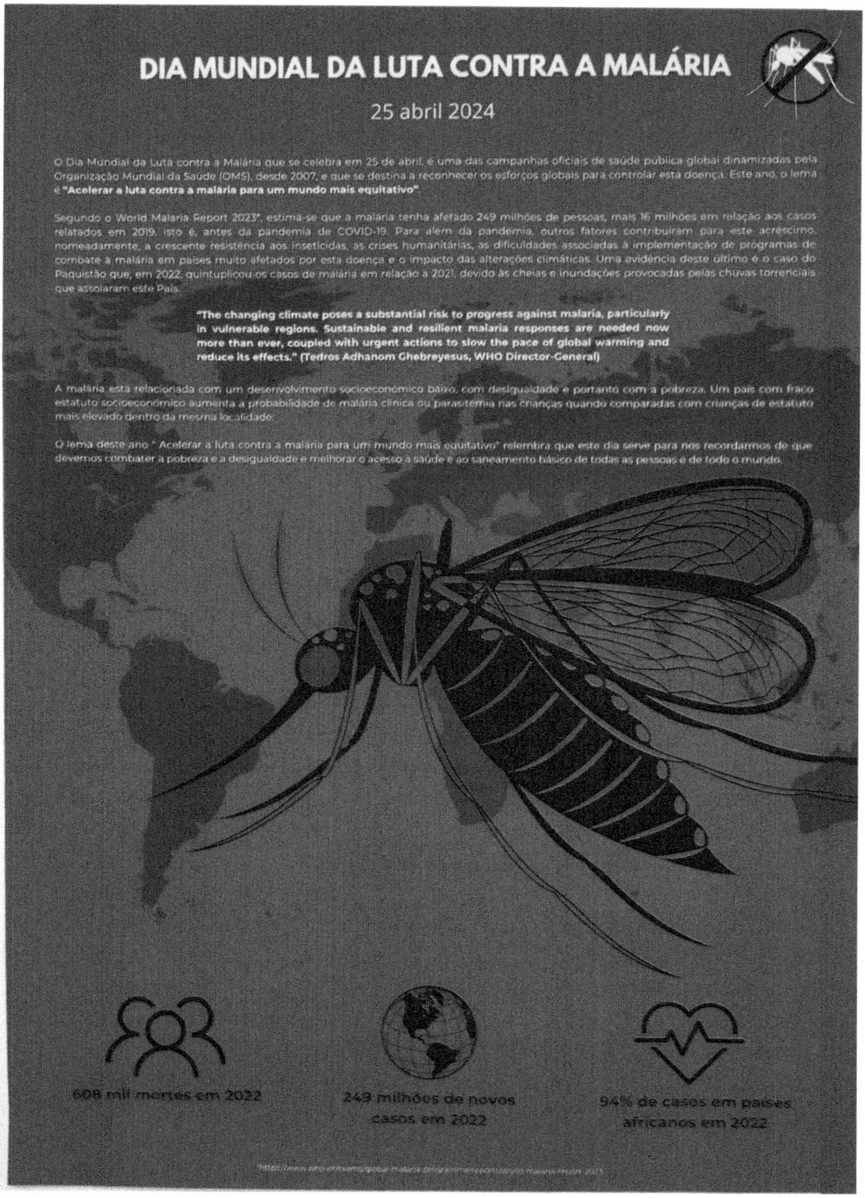

Fig. 2.3. A photograph of a physical poster for World Malaria Day by the World Health Organization (the 25th of April, the anniversary of the Aboudja declaration) clearly shows an *Aedes* mosquito (which do not transmit human malaria) and not an *Anopheles* (which do!). (Photo J.D. Charlwood.).

suffer from the disease); (iii) susceptibility to infection; and (iv) longevity. For human diseases, like malaria, factors that influence host-mosquito contact, such as the propensity to enter houses, the host preference and behaviour after feeding are especially important. House construction (Chapter 8) is therefore an important determinant of exposure to malaria. In general, houses built on stilts reduce the number of anophelines entering them (although *An. dirus*, which may normally feed on monkeys in the tree canopy, is apparently not inhibited in this way).

In order for disease transmission to be sustained, sufficient numbers of vectors must occur in relatively close proximity to humans. The disease organism needs to be able to develop in the mosquito (so *Culex* are not vectors of human malaria but, as noted above, are important vectors of avian malaria), and the mosquito needs to survive long enough for the parasite to develop. The mosquito needs to feed at least twice on the host that suffers from the disease, the first feed being one in which the insect acquires the parasite (in which case it is infected), and the second after the parasite has completed its development when it transmits the parasite (in which case it is infectious).

Host Location

Among night-biting insects many species will approach the odour plumes that emanate from the eaves and other openings of houses. Should there be openings close to ground level, then a variety of species will enter, but if, as is usually the case, the openings are those between the roof and walls (either at the eaves or at the gable ends of houses), then it is only the vectors that enter. In other words, when hitting the wall of the house, non-vectors turn sideways while vectors turn upwards (Snow, 1987; Lindsay and Snow, 1988).

The season of peak abundance of the different vectors will depend on the kind of water body used by the larvae. Mosquitoes that breed in temporary pools (such as the members of the *An. gambiae* complex) are most abundant during the wet season (and tend to be 'boom or bust' species that show rapid changes in abundance according to rainfall), while those occupying more permanent sites, such as *An. funestus*, are less affected by the rain (Charlwood and Braganca, 2012) and tend to peak towards the end of the wet season. This means that where they occur together, the effective transmission season for malaria is extended. Other species (such as *Cx. quinquefasciatus*) that breed in man-made structures may also peak once the rains have stopped.

Fig. 2.4 (from Kiware *et al.*, 2017, used with license CC BY 4.0) describes a typical life cycle of a malaria vector with possible methods of control according to the different activities. Within this overall pattern there are a number of variations, which render the insect vulnerable to different control techniques.

Major differences include biting time and place and resting site. Many vectors of human disease rest inside houses (that is they are endophilic). It is known that the material used in the construction of the house, the number of openings and amount of smoke in the house may affect the time spent by the

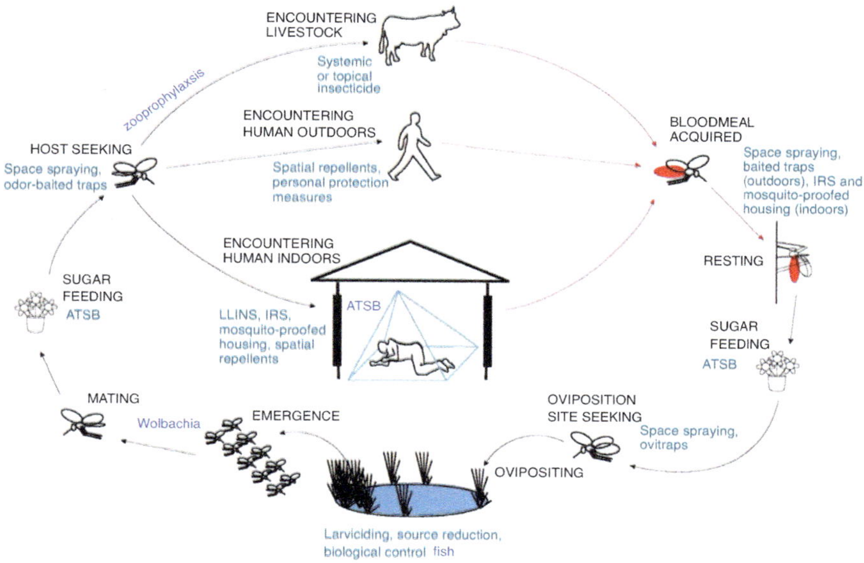

Fig. 2.4. Opportunities for existing and novel vector control tools that can be used to target mosquitoes both indoors and outdoors and at all stages of the mosquito life and feeding cycles. ATSB – Attractive Toxic Sugar Baits. (From Kiware *et al.*, 2017.).

mosquito inside and the house resting site. Traditional houses provide a more equitable environment inside than equivalent outdoor sites. Such houses are warmer in the cool season and cooler in the hot season than outdoor sites. This may reduce mortality in the mosquito and thus enhance vectorial capacity of the insect. One factor that influences relative endophily is the availability of suitable alternative resting sites outdoors.

Changes in biting time as a result of interventions such as IRS, which have been recorded in a number of species, may depend on the number of alternative hosts in the area. As pointed out by Gillies (1988), where alternative hosts are few and far between, changes in activity patterns and a tendency to bite outdoors may develop. Whether such changes result in the development of behavioural resistance (in which case, the change is the result of genetic changes in the mosquito) or of local ecological factors will affect whether the change becomes 'fixed' and continues, even in the absence of the intervention that caused it in the first place.

Sugar Feeding

Both males and females feed on nectar. This sugar feeding is a relatively unstudied behaviour in mosquitoes. The use of Attractive Toxic Sugar Baits (ATBS) is one avenue that is being explored as a supplementary method of control (Chapter 10). Males do not feed on blood and depend on suitable sugar sources for their survival, but females will also feed on suitable sources of sugar when they are available. Flower-visiting mosquitoes may pick up pollen when they feed and so may act as incidental pollinators (Fig. 2.5). How important they might be is unknown, although as long ago as 1913, there was evidence that they may be important in the pollination of some orchids (Dexter, 1913), however even there, a small moth was apparently the more important pollinator (Foster, 2024). Foster goes on to describe some of the factors that, in fact, make mosquitoes inefficient pollinators at best.

As pointed out by Feachem *et al.* (2019), any product that is used outdoors must be carefully evaluated for its effect on the ecosystem, as reductions to biodiversity can result in unintended consequences to human and environmental health.

Females feed on blood. The Goddess Athene, in *The Iliad*, gave the warrior Menelaus the courage equivalent to the 'persistent daring of the fly' (presumably a mosquito, or perhaps *Stomoxys*) which 'finds human blood so sweet it keeps attacking however often it is flicked away'. Different species will show different levels of this daring. *Aedes aegypti* will continue to attack even a defensive host, each time taking small amounts of blood. It has been calculated that they will feed two or more times per cycle, with smaller females feeding more often (Farjana and Tuno, 2013).

Female mosquitoes locate their hosts using a combination of olfactory and visual cues, with perhaps a heat signature when in close proximity to the host. Feeding at different times of the diel has its advantages and disadvantages, as described in Table 2.1 (adapted from Gibson and Torr, 1999).

The number of bites, and more especially the number of infective bites, that an 'average' person receives in a given time is a key parameter in the epidemiology of vector-borne disease. This is difficult to determine directly.

Fig. 2.5. Male mosquitoes only feed on nectar and so may act as pollinators. (Photo Beatriz Moisset. USDA forest service.).

Table 2.1. Advantages and disadvantages of feeding during the day or night (adapted from Gibson and Torr, 1999).

	Day	Night	Crepuscular
Disadvantages	1. Greater risk of desiccation	1. Poor visual clues (especially colour)	1. Short window of opportunity to feed
	2. Greater wind turbulence	2. Low wind speed and hence poor directional clues in host-odour plumes	2. Predators (especially dragonflies) still active
	3. Greater risk from predators	3. Greater background levels of atmospheric carbon dioxide	3. Sit-and-wait strategies not feasible
	4. Host mobile (disadvantage for odour-responding insects?)	4. Host less mobile, so sit-and-wait strategies less feasible	
	5. Greater risk from defensive behaviour of active host	5. Cold temperatures at high altitudes	
	6. Hot and low humidity in low tropical areas		
Advantages	1. Good visual cues	1. Less risk of desiccation	1. Visual cues available
	2. Higher wind speeds providing good directional cues in odour plumes	2. Host more likely to be at rest so reduced risk from host defensive behaviour	2. Hosts may congregate
	3. Reduced background levels of atmospheric carbon dioxide	3. Less risk from predators	3. Wind speed and humidity optimal
	4. Host mobile, making a sit-and-wait strategy feasible	4. Less atmospheric turbulence and hence more continuous odour plumes	4. Temperature optimal

In the past, the method in general use was the 'landing collection' in which sentinel individuals sat with exposed lower legs and collected insects as they attempted to feed on them. While this technique is likely to be the sample that best reflects biting exposure (especially outdoors), there are a number of problems inherent to it. Human-landing catches require a considerable amount of supervision, are expensive to run, and, most significantly, they often expose the collector to pathogen transmission (Achee *et al.*, 2015). These considerations have recently led the WHO to prioritize the search for substitutes to HLC (World Health Organisation, 2017). Where the collector would otherwise be outside, ignoring the mosquitoes, then collecting them at this time by HLC would perhaps reduce their exposure. For nocturnally active endophagic mosquitoes, there are also a number of proxies available (Chapter 4). Light traps (notably the CDC miniature light trap) are widely used indoors, and a range of options (notably tent traps) are available outdoors (Charlwood *et al.*, 2013). The ratio of insects biting indoors and outdoors can change according to local weather conditions. When it rains, insects are more likely to enter houses to feed (Charlwood *et al.*, 2012). It is therefore useful to collect both indoor biting and outdoor biting insects simultaneously. The possibility of there being separate indoor and outdoor biting populations within a species has been examined on a number of occasions, but so far, none have been found (Charlwood *et al.*, 1986; Killeen *et al.*, 2016). The collection of day-biting *Aedes* and mosquitoes active at dusk is more problematical. They are much more difficult to catch.

Although many models of vector-borne disease use an 'average' person's exposure for their calculations, exposure is rarely homogeneous. Exposure tends to follow the so-called 80/20 rule, or Pareto principle, in that 80% of exposure falls on 20% of the shoulders (Woolhouse *et al.*, 1997). Estimates are strongly biased in heterogeneous environments, and such heterogeneous biting has important implications for the dynamics and control of mosquito-borne disease. This localization of transmission means that once the reason for the concentrated distribution of the vectors is known, or understood, then control applied only to that 'focal' area might be possible. This should reduce costs and make the system more efficient. If an area approaches elimination of a disease, then the focus shifts to controlling areas where new cases occur rather than straightforward entomological monitoring.

Dispersal

Dispersal, or the flight range, of mosquitoes is another factor that will influence the efficacy of interventions. Dispersal is likely to be influenced by several extrinsic and intrinsic factors, including among others: meteorological conditions, local topography, physiological status of mosquitoes, body size, population density and availability of oviposition and resting sites. Dispersal is commonly thought to indicate flights at random, with air movement being important. Most dispersal is leptokurtic; in other words, many more individuals move smaller distances than would be expected at random. Insects with large geographic ranges are more likely to be divided by geomorphology and climate than species with smaller ranges. They are more likely to develop clines and subspecies, especially where the habitat is fragmented, as it often is at the edge of a species range. The average range of adult mosquitoes is more an effect of the environment than the inherent flight ability of the insect. In *Ae. aegypti*, this may only be a few hundred metres (Trpis and Hausermann, 1986), and for most malaria vectors, where suitable oviposition sites and feeding sites are in close proximity, dispersal will be considerably less than in areas where they are widely separated. Becker *et al.* (2010) classified mosquitoes into (i) weak flyers, which live close to their hosts' habitats; (ii) weak to moderate flyers, which are found in woodlands; (iii) moderate to strong flyers, which disperse along field and forest edges (as do *Cx. quinquefasciatus* in Hawaii or a variety of anophelines in the Amazon); (iv) and strong flyers, which are found in open areas and disperse over long distances. Indeed, the range of flight distance in different mosquito species has been calculated to range from 50m to 50 km (Verdonschot and Besse-Lototskya, 2014). Dispersal range affects such things as the spread of insecticide resistance as well as the probability of importation of disease within a controlled area. Resistance is, however, not an easy metric to measure. By requiring very little equipment and being simple to perform a recently described measure may help in increasing sampling (Charlwood and Kampango, 2021) (see Chapter 4, Sampling).

In conclusion, despite more than a century of study, there remain a large number of things that we do not know about mosquito vectors of disease. In the early years of the present century, the main vector of malaria in East Africa, *An. gambiae*, suffered a population decline that could not be associated with any particular intervention or environmental factor. Something must have been responsible. How much infection affects mosquitoes in the field is also moot. There is some evidence that a mosquito infected with malaria has a longer gonotrophic cycle than uninfected insects (thus increasing the likelihood that she will survive to become infectious) (Charlwood and Tomás, 2011), while at the same time, an infectious mosquito is more likely to attempt to feed on multiple hosts (due to blocked salivary ducts) during any one feeding cycle. This may reduce their survival, but it enhances transmission (Anderson *et al.*, 2000). Although this is a side effect of having numerous sporozoites in the salivary glands, the mosquito inadvertently becomes what Dawkins called the 'extended phenotype' of the parasite (Dawkins, 1982).

Basic research (i.e. research not primarily designed to evaluate a control scheme), and research into radical new approaches, can occasionally be transformative, although the likelihood that they will be is necessarily unknown.

References

Achee, N.L., Youngblood, L., Bangs, M.J., Lavery, J.V. and James, S. (2015) Considerations for the use of human participants in vector biology research: A tool for investigators and regulators. *Vector Borne Zoonotic Disease* 15, 89–102.

Anderson, R.A., Knols, B. and Koella, J. (2000) *Plasmodium falciparum* sporozoites increase feeding-associated mortality of their mosquito hosts, *Anopheles gambiae* s.l. *Parasitology* 120, 329–333.

Azrag, R.S. and Mohammed, B.H. (2018) *Anopheles arabiensis* in Sudan: A noticeable tolerance to urban polluted larval habitats associated with resistance to Temephos. *Malaria Journal* 17, 204.

Barrón, M.G., Paupy, C., Rahola, N., Akone-Ella, O., Ngangue, M.F. *et al.* (2019) A new species in the major malaria vector complex sheds light on reticulated species evolution. *Scientific Reports* 9, 14753. DOI: 10.1038/s41598-019-49065-5.

Becker, N., Petric, D., Zgomba, M., Boase, C., Madon, M. *et al.* (2010) *Mosquitoes and Their Control*. Springer, Heidelberg, Dordrecht, New York.

Charlwood, J.D. (2003) Chapter 5: May the force be with you: measuring mosquito fitness in the field. In: Scott, T. and Takken, W. (eds) *Ecological Aspects for Application of Genetically Modified Mosquitoes*. Kluwer Academic Publishers, Dordrecht, The Netherlands, p. 243.

Charlwood, J.D. (2023) Swarming and mate selection in *Anopheles gambiae* mosquitoes (Diptera: Culicidae). *Journal of Medical Entomology* 60, 857–864.

Charlwood, J.D. and Braganca, M. (2012) The effect of rainstorms on adult *Anopheles funestus* behavior and survival. *Journal of Vector Ecology* 37, 1–5.

Charlwood, J.D. and Jones, M.D.R. (1979) Mating behaviour in the mosquito *Anopheles gambiae* s.l. I. Close range and contact behaviour. *Physiological Entomology* 4, 111–120.

Charlwood, J.D. and Kampango, A. (2021) Knockdown times in a simple assay determination of insecticide resistance in malaria vectors. *Transactions of the Royal Society of Tropical Medicine and Hygiene* 115, 1348–1349.

Charlwood, J.D. and Tomás, E.V.E. (2011) Do developing malaria parasites manipulate their mosquito host? – Evidence from infected *Anopheles funestus* (Giles) from Mozambique. *Transactions of the Royal Society of Tropical Medicine and Hygiene* 105, 352–354.

Charlwood, J.D., Billingsley, P.F. and Hoc, T.Q. (1995a) Mosquito-mediated attraction of female European but not African mosquitoes to hosts. *Annals of Tropical Medicine and Parasitology* 89, 327–329.

Charlwood, J.D., Birley, M.H. and Graves, P.M. (1986) Capture-recapture studies of females of the *Anopheles punctulatus* group of mosquitoes (Diptera, Culicidae) from Papua New Guinea. *Bulletin of Entomological Research* 76, 211–227.

Charlwood, J.D., Kampango, A., Omari, M. and Okumu, F. (2024) Simple techniques for a complex problem: Sampling malaria vectors in Africa. *Acta Tropica* 251, 107104.

Charlwood, J.D., Macia, G.A., Manhaca, M., Sousa, B., Cuamba, N. *et al.* (2013) Population dynamics and spatial structure of human-biting mosquitoes, inside and outside of houses, in the Chockwe irrigation scheme, Southern Mozambique. *Geospatial Health* 7, 309–320.

Charlwood, J.D., Pinto, J., Sousa, C.A., Ferreira, C. and Rosário, V.E. (2002) The swarming and mating behaviour of *Anopheles gambiae* (Diptera: Culicidae) from São Tomé Island. *Journal of Vector Ecology* 27, 178–183.

Charlwood, J.D., Smith, T., Kihonda, J., Billingsley, P.F. and Takken, W. (1995b) Density independent feeding success of malaria vectors in Tanzania. *Bulletin of Entomological Research* 85, 29–35.

Charlwood, J.D., Tomás, E.V.E., Egyir-Yawson, A., Kampango, A. and Pitts, R.J. (2012) Feeding frequency and survival of *Anopheles gambiae* from a rice growing area of ghana. *Medical and Veterinary Entomology* 26, 263–270.

Charlwood, J.D., Tomás, E.V.E., Kelly-Hope, L. and Briët, O.J.T. (2014) Evidence of an 'invitation' effect in feeding sylvatic *Stegomyia albopicta* from Cambodia. *Parasites and Vectors* 7, 324.

Coluzzi, M., Petrarca, V. and Deco, M.A. (1985) Chromosomal inversion intergradation and incipient speciation in *Anopheles gambiae. Bollettino di Zoologia* 52, 45–63.

Coluzzi, M., Sabatini, A., Petrarca, V. and Deco, M.A. (1979) Chromosomal differentiation and adaptation to human environments in the *Anopheles gambiae* complex.?. *Transactions of the Royal Society of Tropical Medicine and Hygiene* 73, 483–497.

Dawkins, R. (1982) *The Extended Phenotype: The Long Reach of the Gene*. Oxford University Press, Oxford, UK.

Dexter, J.S. (1913) Mosquitoes pollinating orchids. *Science* 37, 867.

Farjana, T. and Tuno, N. (2013) Multiple blood feeding and host-seeking behavior in *Aedes aegypti* and *Aedes albopictus* (Diptera: Culicidae). *Journal of Medical Entomology* 50, 838–846.

Feachem, R.G.A., Chen, I., Akbari, O., Bertozzi-Villa, A., Bhatt, S. *et al.* (2019) Malaria eradication within a generation: Ambitious, achievable, and necessary. *The Lancet Commissions* 394(1023), 1056–1112.

Ferguson, H.M., Dornhaus, A., Beeche, A., Borgemeister, C., Gottlieb, M. *et al.* (2010) Ecology: A prerequisite for malaria elimination and eradication. *PLoS Medicine* 7(8), e1000303.

Foster, W.A. (2024) Pollination of plants by disease vectors: An assessment. Genen Convene. Available at: https://www.geneconvenevi.org/pollination-of-plants-by-disease-vectors-a-risk-assessment/ (accessed 14 February 2024).

Gibson, G. and Torr, S.J. (1999) Visual and olfactory responses of hematophagous Diptera to host stimuli. *Medical and Veterinary Entomology* 13, 2–23.

Gillies, M.T. (1988) Chapter 16: Anopheline mosquitoes: Vector behaviour and bionomics. In: Wernsdorfer, I.W.H. and McGregor, I. (eds) *Malaria, Principles and Practice of Malariology*. Churchill Livingstone, Edinburgh, UK, pp. 453–485.

Gillies, M.T. and de Meillon, B. (1968) *The Anophelinae of Africa South of Sahara (Ethiopian Zoogeographical Region)*. The South Africa Institute for Medical Research, Johannesburg.

Hackett, L.W. (1934) The present status of our knowledge of the sub-species of *Anopheles maculipennis. Transactions of the Royal Society of Tropical Medicine and Hygiene* 28, 109–114.

Harrison, G. (1979) *Mosquitoes, Malaria and Man: A History of the Hostilities since 1880.* John Murray, London.

Inquisitive Biologist (2022). Available at: https://inquisitivebiologist.com/2022/09/17/book-review-from-extraterrestrials-to-animal-minds-six-myths-of-evolution/(accessed (accessed 23 September 2024).

Killeen, G.F., Govella, N.J., Lwetoijera, D.W. and Okumu, F.O. (2016) Most outdoor malaria transmission by behaviourally-resistant *Anopheles arabiensis* is mediated by mosquitoes that have previously been inside houses. *Malaria Journal* 15, 225. DOI: 10.1186/s12936-016-1280-z.

Kiware, S.S., Chitnis, N., Tatarsky, A., Wu, S., Castellanos, H.M.S. *et al.* (2017) Attacking the mosquito on multiple fronts: Insights from the Vector Control Optimization Model (VCOM) for malaria elimination. *PLoS ONE* 12(12), e0187680.

Klowden, M. (2007) *Physiological Systems in Insects.* Elsevier, Amsterdam, The Netherlands.

Klug, H. and Stone, L. (2021) More than just noise: Chance, mating success, and sexual selection. *Ecology and Evolution* 11, 6326–6340.

Liao, W., Atkinson, C.T., LaPointe, D.A. and Samuel, M.D. (2017) Mitigating future avian malaria threats to Hawaiian forest birds from climate change. *PLoS ONE* 12(1), e0168880.

Lindsay, S.W. and Snow, R.W. (1988) The trouble with eaves; house entry by vectors of malaria?. *Transactions of the Royal Society of Tropical Medicine and Hygiene* 82, 645–646.

MacDonald, G. (1957) *The Epidemiology and Control of Malaria.* Oxford University Press, London.

Paterson, H.E. (1985) The recognition concept of species. In: Vrba, E.S. (ed.) *Species and Speciation.* Transvaal Museum Monograph, No. 4. Transvaal Museum, Transvaal, South Africa, pp. 21–29.

Rankin, M.A., McAnelly, M.L. and Bodenhamer, J.E. (1986) The oogenesis-flight syndrome revisited. In: Danthanarayana, W. (ed.) *Insect Flight: Dispersal and Migration.* Springer, Berlin, Germany, pp. 27–48.

Reynolds, C. (1987) In: *SIGGRAPH '87: Proceedings of the 14th Annual Conference on Computer Graphics and Interactive Techniques*, Association for Computing Machinery, New York, pp. 25–34.

Sherrington, C. (1955) *Man on His Nature.* Pelican Books, Harmondsworth, UK.

Shousha, A.T. (1948) Species-eradication. The eradication of *Anopheles gambiae* from upper Egypt, 1942–1945. *Bulletin of the World Health Organization* 1, 309–353.

Sinka, M.E., Pironon, S., Massey, N.C., Longbottom, J., Hemingway, J. *et al.* (2020) A new malaria vector in Africa: Predicting the expansion range of *Anopheles stephensi* and identifying the urban populations at risk. *Proceedings of the National Academy of Sciences of the United States of America* 117(40), 24900–24908. DOI: 10.1073/pnas.2003976117.

Snow, W.F. (1987) Studies of house-entering habits of mosquitoes in The Gambia, West Africa: Experiments with prefabricated huts with varied wall apertures. *Medical and Veterinary Entomology* 1, 9–21.

Somers, J., Georgiades, M., Su, M.P., Bagi, J., Andrés, M. *et al.* (2022) Hitting the right note at the right time: Circadian control of audibility in *Anopheles* mosquito mating swarms is mediated by flight tones. *Science Advances* 8(2), eabl4844. DOI: 10.1126/sciadv.abl4844.

Soper, F.L. and Wilson, B.D. (2011) *Anopheles Gambiae No Brasil: 1930–1940.* Editora Fiocruz, Rio de Janeiro, Brazil.

Stereiny, K. and Griffiths, P.E. (1999) *Sex and Death: An Introduction to Philosophy of Biology.* University of Chicago Press, Chicago, Illinois.

Trpis, M. and Hausermann, W. (1986) Dispersal and other population parameters of *Aedes aegypti* in an African village and their possible significance in epidemiology of vector-borne diseases. *American Journal of Tropical Medicine and Hygiene* 35(6), 1263–1279.

van de Straat, B., Russell, T.L., Staunton, K.M., Sinka, M.E. and Burkot, T.R. (2021) A global assessment of surveillance methods for dominant malaria vectors. *Science Reports* 11, 15337.

Verdonschot, P.F. and Besse-Lototskya, A.A. (2014) Flight distance of mosquitoes (Culicidae): A metadata analysis to support the management of barrier zones around rewetted and newly constructed wetlands. *Limnologia* 45, 69–79.

Warner, R.E. (1968) The role of introduced diseases in the extinction of the endemic *Hawaiian avifauna. Condor* 70, 101–120.

Wilkes, T.J., Matola, Y.G. and Charlwood, J.D. (1996) *Anopheles rivulorum*, a vector of human malaria in Africa. *Medical and Veterinary Entomology* 10, 108–110.

Williams, G.C. (1992) *Natural Selection: Domains, Levels and Challenges*. Oxford University Press, Oxford, UK.

Woolhouse, M.E.J., Dye, C., Etard, J.F., Smith, T., Charlwood, J.D. *et al.* (1997) Heterogeneities in the transmission of infectious agents: Implications for the design of control programs. *Proceedings of the National Academy of Sciences of the United States of America* 94(1), 338–342.

World Health Organization (2015) *Global Technical Strategy for Malaria 2016–2030*. World Health Organization, Geneva, Switzerland.

World Health Organization (2017) *Ethical Issues Associated with Vector-Borne Diseases*. World Health Organization, Geneva, Switzerland.

A Short History of Malaria Control

3

Carla A. Sousa* and João Pinto

Global Health and Tropical Medicine, GHTM, Associate Laboratory in Translation and Innovation Towards Global Health, LA-REAL, Instituto de Higiene e Medicina Tropical, IHMT, Universidade NOVA de Lisboa, UNL, Rua da Junqueira 100, 1349-008 Lisboa, Portugal

Malaria leaves 2 M hectares of land uncultivated in Italy; poisons 2 M inhabitants and kills ca. 15 thousand each year. There is not another health problem that so profoundly affects the prosperity of our country. (G. Fortunato and L. Franchetti, Italian Parliament, 1898)

Malaria is the main disease and a primary cause of poverty in Africa. Every day 3,000 children die from malaria. Each year there are 500 million cases among children and adults. (G.H. Brundtland, WHO General Director, 1998)

Abstract

Malaria, caused by *Plasmodium* parasites, has plagued humanity for millennia, and historical records from ancient civilizations describe malaria-like symptoms. Significant advancements against the disease began in the 17th century, with Jesuit missionaries from Peru introducing quinine as a treatment. European colonization spread malaria but also initiated control measures. Identification of the parasite and of its mosquito vectors was a key discovery, leading to targeted control strategies. The 20th century saw further progress with DDT and chloroquine. Based on these efficient tools, the World Health Organization (WHO) launched the Global Malaria Eradication Program in 1955. This Program was successful in eradicating malaria from temperate areas but faced challenges in tropical areas. The Roll Back Malaria Partnership was launched in

*Corresponding author: casousa@ihmt.unl.pt

DOI: 10.1079/9781789248845.0003

1998, emphasizing prevention and treatment, supported by recent advances such as LLINs, ACTs and RDTs. The most recent WHO 2016–2030 strategy aims at a 90% reduction in malaria by 2030. The history of malaria control is a testimonial to scientific innovation, international collaboration and persistent challenges. From ancient herbal remedies to modern genomic technologies, the fight against malaria continues to progress, driven by the ultimate goal of eradication.

Graphical support:

Box 3.1. Phases of the Global Programme for Malaria Eradication (1955–1969) (figure author's own).

Box 3.2. Differences between eradication programmes and control programmes (figure author's own).

Box 3.3. Schematic representation of the Global Technical Strategy for Malaria 2016-2030 (adapted from The malERA Refresh Consultative Panel on Combination Interventions and Modelling, 2017).

From Ancient Times to the First Malaria Control Tool

The first report describing malaria as a human disease is a controversial issue. The first written references to a malaria-like disease appear in the Hippocratic Corpus, in Greece (4th and 5th centuries BC), in ancient Indian texts (with no ascertained date) and in Chinese literature of the I millennium BC (Sallares *et al.*, 2004). There is little doubt that benign tertian and quartan fevers (caused respectively by *Plasmodium vivax* and *Plasmodium malariae*) were endemic diseases in the Old World around 500 BC. The presence of *Plasmodium falciparum* in ancient Europe is more problematic to establish; however, palaeopathological data, including manuscripts of Hippocrates, support the hypothesis that *falciparum* malaria was present in Greece since the Mesolithic.

Literary sources also indicate that around the 5th century BC, the disease was already active in Sicily and in southern Italy, where Greek colonies were established since the 8th–6th centuries BC. It spread to western central Italy during the 1st and 2nd centuries BC but reached the northern areas only during the Medieval period (Jones, 1907 quoted in Grmek, 1994; Sallares *et al.*, 2004). Molecular evidence confirms the presence of *P. falciparum* malaria in the south of Italy in 1st–2nd centuries AD (Marciniak *et al.*, 2016).

Malaria spread throughout Europe during the Roman Empire by the continuous migration of soldiers, slaves, merchants and administrators (Sallares *et al.*, 2004). However, the process was also driven by the dispersal of certain mosquito species (Bruce-Chwatt and Zulueta, 1980), like the expansion of *Anopheles labranchiae* and *Anopheles sacharovi* after the extensive deforestation that occurred in the Hellenistic and Roman periods (Jones, 1907 quoted in Grmek, 1994).

In the 17th and 18th centuries AD, malaria was common in southern Europe as well as in some northern areas, including the Netherlands, the North Sea coast of Germany, Sweden and Finland (Bruce-Chwatt, 1988; Huldén *et al.*, 2005). However, without reliable descriptions of epidemics, the disease must also have been common in eastern Europe since the 15th century, as several references to malaria symptoms are present in Polish and Russian folklores (Bruce-Chwatt, 1988).

It was at the beginning of the 17th century that one of the most important events of malaria history took place: the discovery of therapeutic effects of the Peruvian bark, cinchona, against intermittent fevers. How this new remedy and the knowledge of its benefits reached Europe is a more or less obscure tale (Boyd, 1949). The most probable hypothesis is that it was brought by Jesuit priests to Spain, where it was used in treatment of Miguel de Barreda, a Professor of Theology at Alcala de Henares, in 1639 (Guerra, 1977 quoted in Bruce-Chwatt, 1988). With the discovery of cinchona, humankind had access to the first tool for malaria control. However, it was only in 1820 that the main alkaloids of the bark were finally isolated and the chemical process of obtaining quinine and its salts was described by Pelletier and Caventou, two French pharmacists (Bruce-Chwatt, 1988).

Early Successes

In the middle of the 18th century, malaria started to decline in northern Europe. Environmental changes due to the drainage of marshy areas, novel practices in agriculture and stock breeding, as well as better treatment of cases with quinine, are the most common explanations for the natural disappearance of malaria from regions such as England (Bruce-Chwatt, 1988; Dobson, 1994). Improvements in social condition with benefits in nutritional status, hygiene and child care provided the human population with better weapons to fight the disease. The introduction of new crops to provide cattle with winter forage, the increased size and health of the herds, the stabling of cattle and pigs and improvements in house construction have all contributed to an increasing dissociation between humans and mosquitoes (Dobson, 1994).

It was only at the end of the 19th century that the aetiological agent of malaria and its life cycle were uncovered. In 1880, Alphonse Laveran first described the malaria parasites in human blood samples, a discovery that contributed to his Nobel Prize in Physiology or Medicine awarded in 1907. In 1898, the life cycle of *Plasmodium* was elucidated by Sir Ronald Ross, based on work conducted with avian malaria, and confirmed by Italian malariologists Giovanni Battista Grassi, Amico Bignami and Giuseppe Bastianelli, demonstrating the transmission of malaria by mosquitoes (Dobson, 1999). In the 20th century, although declining in northern Europe without any specific control measures, malaria was still an endemic disease in large areas of the southern and eastern parts of the continent. In areas where changes in crop production and stock rearing were absent, malaria did not decline. In Russia, from the Black Sea to Siberia, malaria remained a major public health problem for the entire first half of the 20th century, in contrast with other countries at similar latitudes (Reiter, 2001).

It was the impact of malaria during the First World War and the difficulties in assuring a continuous supply of quinine that stimulated German scientists in the pursuit of new drugs for malaria treatment. In 1924, Schulemann and his colleagues produced the first synthetic anti-malarial compound, named Plasmochin (pamaquine), and before the Second World War, several new synthetic drugs were available for malaria therapy (Bruce-Chwatt, 1988). It was also due to the war efforts – but this time during the Second World War – that the anti-malarial effects of two other compounds were discovered: chloroquine and amodiaquine. These 4-aminoquinoline derivates proved to be two of the most effective anti-malarial compounds and remained therapeutic and suppressive drugs of choice for over 25 years (Bruce-Chwatt, 1988).

At the beginning of the Second World War, another major discovery took place in Switzerland. In 1939, Muller and Wiesmann have testified the remarkable insecticide proprieties of a synthetic compound, dichlorodiphenyltrichloroethane, to which was given the abbreviated name of DDT (Bruce-Chwatt, 1988). This compound with its four extraordinary characteristics (cheap, long persistence, highly toxic to insects by contact but low toxicity to humans) revolutionized malaria control. Before 1944, it was obvious that methods of malaria control by source reduction using larvicides such as kerosene and Paris Green dust were only feasible in urban centres or limited areas. In contrast, this new insecticide could be applied to vaster regions. In 1944, the first trial with DDT spraying was conducted in the region of Rome, Italy (Bruce-Chwatt, 1988). For the first time, residual indoor spraying (IRS), a simple, economically sustainable mosquito control method could be used, especially in rural areas.

The Golden Age – Global Programme for Malaria Eradication

In the middle of the 20th century, several countries carried out successful national malaria control programmes. In Albania, Italy, Greece and Portugal, these programmes received major stimulus and generous financial support from the Rockefeller Foundation. In the course of these and other successful post–Second World War campaigns, it became obvious that the complete elimination of the mosquito vectors was not a *sine qua non* condition for the cessation of malaria transmission. Thus, the idea of eradication starts to build up and in 1955 this concept was adopted by WHO as the ultimate goal of the fight against malaria. In the *Global Programme for Malaria Eradication* (1955–1969) approved on the 8th United Nations General Assemblage for Health, (see Box 3.1) carried out in Mexico city, May 1955 (WHO, 1956) the malaria eradication concept is defined as 'the ending of transmission and of the reservoir of infective cases in a campaign limited in time, carried to a such degree of perfection that when it comes to an end there is no resumption of transmission' (WHO, 1956).

Box 3.1. Phases of the Global Programme for Malaria Eradication (1955-1969) (figure author's own).

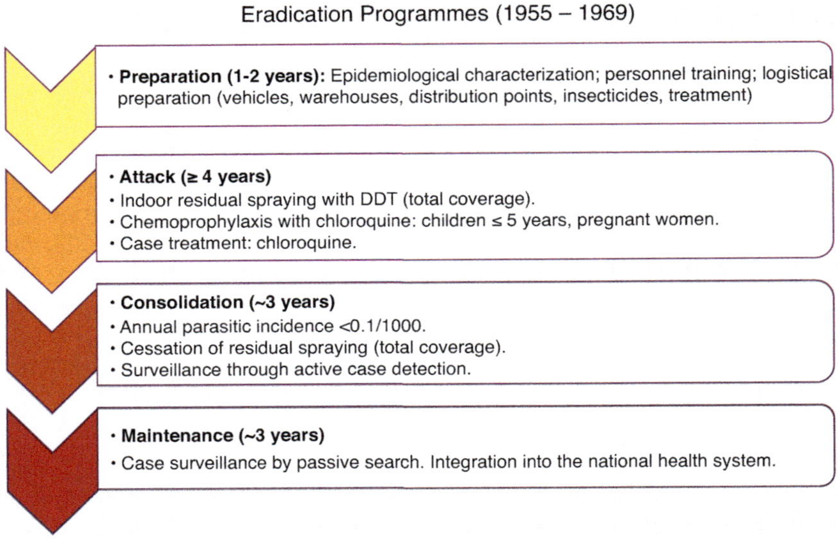

Behind the eradication notion is the need for total coverage and perfectly implemented actions (Nájera *et al.*, 2011). Regarding implementation, there is the need to address with equal commitment the fight against the parasite and the vector. As to the vector, actions were focused on three main aspects: (i) reduce human-vector contact, (ii) reduce/eliminate breeding sites and (iii) eliminate adult vectors, which are the life stage that transmit the parasite. By this time, most of the European malaria control programmes had adopted eradication strategies and, by 1975, Europe was free from indigenous malaria (WHO, 2006). For the first time in history, endemic malaria disappeared from a worldwide region.

Encouraged by the European success, along with the eradication of the African malaria vector *Anophele gambiae* introduced in Brazil around 1930 and eliminated in 1940 (Parmakelis *et al.*, 2008), control efforts started to be intensified in tropical regions. It was also when the golden history of malaria eradication started a path of disillusions and half-achievements. In spite of the efforts, from the 1960s onwards, there was a resurgence of malaria cases. In addition to the ecological conditions of tropical regions being more favourable to malaria transmission, it is possible to identify the major causes that led to the decline of eradication programmes in tropical regions, namely:

- Difficulties in sustaining the eradication programmes, mainly in maintaining full coverage.

- Financial disinvestment by countries that contributed the most to the WHO Special Account for the Eradication of Malaria, namely the USA that ceased its contribution in 1963.
- Failures of the surveillance systems, which were not sensitive enough to detect the presence of problematic areas where local transmission persisted.
- The evolution of resistance to chloroquine and DDT by parasite and vector populations, respectively.

From Eradication to Control

It is generally accepted that the Global Malaria Eradication Programme was never truly implemented in Africa, due to the recognition of the weaknesses of the health structures in this continent, which did not allow the implementation of the principles of total coverage and operational perfection (Teklehaimanot and Bosman, 1999). In 1969, the 22nd General Health Assembly revised the objective of eradicating malaria, reverting it in favour of control (see Box 3.2), recognizing that there were countries where eradication was not feasible in the short term, and it was imperative to adapt control strategies to different eco-epidemiological situations and taking into account the available economic and administrative resources (WHO, 1973).

In the following years, malaria prevalence continued to increase. An 'Africanization' of this disease was observed, with the vast majority of cases being concentrated in this continent. Several aspects may have contributed to this development:

- The increase in financial constraints, aggravated by the 1970 oil crisis, which amplified problems with logistics and human resources.
- Epidemics associated with the expansion of primary sector activities in developing countries, in particular, mining and deforestation.

Box 3.2. Differences between eradication programmes and control programmes (figure author's own).

Eradication *versus* control

	Eradication	Control
Objective	Interruption of transmission, elimination of foci	Reduction of mortality and incidence to acceptable levels
Duration	Limited in time	Undefined
Intervention area	All transmission areas	Selected areas (e.g. > mortality)
Total coverage	Indispensable	Not necessary
Implementation	Perfect	Good
Funding	Capital investment	Recurrent
Health System (HS)	Independent of HS	Implemented by HS

- The spread of drug and insecticide resistance, particularly in the African continent where malaria was predominantly caused by the most lethal species, *P. falciparum*, being transmitted by the most efficient malaria vectors, *Anopheles gambiae s.l.* and *Anopheles funestus s.l.*
- The prohibition of the use of DDT, first in the control of agricultural pests and later in public health, a process involving controversy, given the importance of this insecticide and the apparent fragility of the scientific evidence that led to this prohibition (Edwards, 2004).

It was also a period with a strong disinvestment in scientific research for developing alternative control methods. Based on the recognition of this weakness, the WHO launched, in 1974, the TDR Programme – Special Programme for Research and Training in Tropical Diseases. Nevertheless, the global increase in malaria prevalence continued until the end of the 1990s. According to the World Health Report 1998, malaria was the 8th cause of death in the world, with *ca.* 1.5–2.7 million annual deaths, and the 2nd cause of morbidity, with *ca.* 300–500 million new cases annually (WHO, 1998). By the end of the 20th century, the unbearable impact of malaria on global health and socio-economic development brought this disease back on to the international political agenda.

In 1998, the *Roll Back Malaria* (RBM) partnership was launched, a global platform for resource mobilization and coordination of malaria control actions (Nabarro and Tayler, 1998). This partnership would bring together more than 500 stakeholders, including malaria-endemic countries and their bilateral and multilateral partners, the private sector, non-governmental organizations, foundations and academic institutions. The RBM initiative will centralize global efforts to control malaria, promoting partnerships among stakeholders and ensuring resource optimization. At the RBM African summit on 25 April 2000, the leaders of 44 African countries signed the Abuja Declaration, which establishes the principles of political convergence for the fight against malaria in Africa over the next 10 years. From this day on, 25 April is celebrated as World Malaria Day (RBM, 2003).

In the subsequent years, there was a significant increase in investment in scientific research. Programmes such as the multilateral initiative against malaria (MIM), the INCO-DC (Scientific cooperation with developing countries) and INCO-DEV (European research cooperation for development) of the European Commission (Ntoumi and Priebe, 2010), followed by the 6th and 7th European Framework Programmes, which financed large malaria research consortia (Holtel *et al.*, 2011), stood out as main contributors. The main North American funding agencies for health and development (e.g. National Institutes of Health/National Institute of Allergy and Infectious Diseases, USA Agency for International Development) also increased funding for scientific research in malaria. Established in 1994, the Bill & Melinda Gates Foundation became the world's largest private foundation dedicated to funding scientific research, not only for malaria but also other diseases of poverty and neglected tropical diseases, in the context of global health (McCoy *et al.*, 2009). This greater availability of financial resources led not only to important scientific advances at the level of basic research on the biology of parasites and vectors and parasite-host interactions but also to important translational achievements.

Resistance to anti-malarial drugs and insecticides were subjects that showed the greatest progress. The mechanisms underlying resistance to the most used drugs during this period were elucidated and the identification of genetic markers associated with resistance made it possible to map the geographic distribution and study the epidemiology of resistance to antimalarials (Cui *et al.*, 2015). Likewise, insecticide resistance mechanisms and their genetic basis were identified in the main malaria vectors (Corbel and N'Guessan, 2013). The identification of mutations associated with knockdown resistance and the development of efficient genotyping methods allowed geographic mapping and a better understanding of the evolutionary processes that led to the dissemination of this type of target-site resistance (Pinto *et al.*, 2007; Santolamazza *et al.*, 2008).

It was based on all the research carried out until late 1990s that new control tools were developed and integrated into national malaria control programmes. Namely:

- Rapid diagnostic tests (RDTs). The adoption of these immunochromatographic assays by control programmes allowed easier access to the parasitological diagnosis of malaria, especially in remote areas with scarce technical and human resources, thus avoiding unnecessary consumption of anti-malarial drugs and, therefore, reducing waste and risk of resistance emergence (Mouatcho and Goldring, 2013).
- Artemisinin-based combination therapies (ACT). The use of ACT as the first line of treatment for uncomplicated *P. falciparum* malaria has been adopted by most endemic countries since the beginning of the 21st century (White, 1999).
- Preventive intermittent therapy. This measure has been adopted by control programmes as a way of protecting target groups most susceptible to developing severe malaria, in particular, pregnant women, newborns and children up to 5 years of age (WHO, 2014).
- Long-lasting insecticide-treated mosquito nets. These nets, more durable, have solved one of the main problems that control programmes faced, the periodic treatment of the nets, in order to guarantee their insecticidal effect over time (Dhiman and Veer, 2014).

In September 2000, the United Nations General Assembly unanimously adopted the United Nations Millennium Declaration, launching a global partnership to reduce extreme poverty by establishing eight development goals, each with a set of targets to meet by 2015. These objectives became known as the United Nations Millennium Goals (United Nations, 2000). Goal 6, 'Combat HIV/AIDS, malaria and other diseases', proposed as Goal 6C: 'By 2015, stop and begin to reverse the incidence of malaria and other serious diseases'. In the years that followed, the World Malaria Reports of the WHO reported increases in funding and in the availability of control tools in endemic areas. In sub-Saharan Africa, the number of insecticide-treated mosquito nets delivered by manufacturers to national control programmes increased from *ca.* 5 million in 2004 to close to 88 million in 2009 (WHO, 2010). Rapid diagnostic tests that were distributed by ministries of health increased from less than 200 thousand to *ca.* 30 million between 2005 and 2009 and the global number of treatments with the ACT artemether +lumefantrine doubled in the same period.

In 2006, WHO reintroduced DDT in the malaria control agenda due to its cost-effectiveness (WHO, 2011). This decision sparked controversy (Weissmann, 2006), reigniting debates on DDT's environmental impact and leaving a mark of apparent regression to the paths taken in the past.

Control efforts led to a decrease in global malaria levels throughout the first decade of the 21st century. In 2006, three African countries and one territory had recorded a 50% reduction in mortality and/or malaria cases: Eritrea, Rwanda, São Tomé and Príncipe, and the island of Zanzibar (Tanzania) (WHO, 2008). Outside of Africa, malaria cases had declined by 50% or more in 22 countries between 2000 and 2007. In 2007, of the 109 countries/territories affected by malaria, 82 were in the control phase, 11 in the pre-elimination phase, 10 in elimination and six in the reintroduction prevention phase (WHO, 2008). In January 2007, the United Arab Emirates was the first country to be certified by the WHO as 'malaria free', since Australia and Singapore in the 1980s (WHO, 2007).

From Control to Eradication, Again!

In October, during the Malaria Forum organized by the Bill & Melinda Gates Foundation, an old idea was rescued from the past (Bill & Melinda Gates Foundation, 2007). The foundation's patrons launched the appeal that, with sufficient international commitment to guarantee the intensification of control interventions, malaria can be eradicated. This statement was supported by the Secretary General of the WHO Dr Margareth Chan in her speech at the same Forum. The paradigm of malaria control would therefore be changed to include elimination and, as the ultimate objective, eradication.

In response to the change in the paradigm from control to eradication, the RBM launched the *Global Malaria Action Plan* (GMAP) in 2008 (RBM, 2008). The GMAP proposes a strategy based on three components: (i) control malaria to reduce the impact of the disease; (ii) eliminate malaria country-by-country over time; and (iii) investigate new tools that support global control and elimination efforts. With respect to control, GMAP defines a strategy based on two stages: (i) scaling up to achieve universal coverage of all populations at risk, with locally appropriate and supported control interventions; and (ii) sustain control to maintain universal coverage of interventions until countries enter the elimination phase. This will require strong political commitment, particularly when malaria levels become low, to avoid re-establishment of transmission.

The GMAP promoted the elimination of malaria in all countries that have the conditions to do so based on: (i) a robust epidemiological surveillance with efficient case detection and management system, which allows for the timely identification and treatment of all confirmed cases of malaria, including asymptomatic ones, for the detection of outbreaks; (ii) differentiated vector control strategies targeted at active outbreaks; (iii) education and communication programmes to tackle community fatigue regarding interventions; and (iv) cross-border cooperation initiatives to minimize the risk of reintroduction.

The GMAP recommended increased investment in scientific research to support the control and elimination of malaria, particularly in regions of high transmission. Three key areas were identified: (i) research for the development of new interventions or improvement of existing ones; (ii) research into informed health policies, which will define the type of interventions that best adapt to different contexts; and (iii) operational and implementation research, to allow a better understanding of the use and effectiveness of interventions in the field.

The GMAP ends in 2015 with its goals only partially achieved (WHO, 2015a, 2016a). Of the 106 endemic countries in 2000, 57 have reduced the incidence of malaria by more than 75%. However, malaria remained an African problem, with around 88% of cases and 90% of deaths recorded in 2015 occurring on that continent. By December 2015, 20 countries were in the elimination or pre-elimination phase, but only two (Cape Verde and Swaziland) were African. One of the biggest achievements of GMAP was a reduction of almost 50% in children mortality rate, from 158 deaths/1000 live births in 2000 to 82 deaths/1000 live births in 2015. Malaria was no longer the main cause of child mortality in sub-Saharan Africa (WHO, 2015a).

Adopted in May 2015 at the 68th World Health Assembly, the *Global Technical Strategy for Malaria* (GTSM) defined the guidelines to be implemented in the fight against malaria over the next 15 years (WHO, 2015b). Aiming to a malaria-free world, an ambitious set of objectives and targets have been set to reduce malaria cases in 90% by 2030 (WHO, 2015b).

Five principles underlie the GTSM (WHO, 2015b): (i) all countries must implement intensive efforts to eliminate malaria through interventions adapted to local contexts; (ii) accelerated progress must be attained through a multisectoral approach and a strong national leadership; (iii) interventions must be optimized based on improved surveillance; (iv) there must be equity in access to health services, particularly for vulnerable populations in remote areas; and (v) innovation in control strategies must allow countries to maximize the chances of eliminating malaria. It is based on these principles that three pillars and two support elements are defined for global efforts to eliminate malaria (WHO, 2015b):

- Pillar n°1: 'Ensure universal access to prevention, diagnosis and treatment of malaria'. The WHO recommends the implementation of two core interventions, in a complementary manner: (i) prevention strategies, based on vector control and, in certain contexts and population groups, chemoprophylaxis; and (ii) universal diagnosis and timely treatment of malaria.
- Pillar n°2: 'Accelerate efforts towards elimination and attainment of malaria-free status' by intensifying detection and control of outbreaks in low transmission areas through active surveillance and rapid response. Develop and implement innovative solutions to combat resistance to insecticides and drugs, residual transmission and *P. vivax* hypnozoite reservoirs.
- Pillar n°3: 'Transform malaria surveillance into a nuclear intervention'. Countries with endemic or potentially re-emerging malaria need efficient health information systems to support national control programmes. These systems help allocate resources to affected populations, identify coverage gaps, detect outbreaks and evaluate intervention impacts and tool effectiveness.
- Support element n°1: 'Harnessing innovation and expanding research'. Basic research is essential for a better understanding of parasites and their vectors, to develop more effective diagnostic tests and drugs, innovative vector control methods and other tools (e.g. vaccines). Implementation research is the key to optimizing impact and cost-effectiveness of new methodologies that are paramount to engage communities in new malaria control tools.
- Support element n°2: 'Strengthen the enabling environment'. Strong political commitment, robust and stable funding, and greater multisectoral collaboration are key factors to progress towards malaria elimination. To optimize national responses against malaria, strengthening health systems and improving the surrounding environment are crucial. Empowering communities, building capacity and providing supportive supervision are key elements to achieving the strategy's vision, goals and milestones (see Box 3.3).

Struggling Towards Eradication

Due to the success of the rapid decline in malaria mortality from 2000 to 2015, WHO issues the 'strategy and better estimate of the likely duration of effort to be maintained over the last mile' for malaria eradication (WHO, 2020). Based on past examples of diseases that were eradicated, this document continues the consensus produced over the last decade with the collaborative effort of the Malaria Eradication Research Agenda (malERA, 2011). In the document 'Malaria eradication: benefits, future scenarios

Box 3.3. Schematic representation of the Global Technical Strategy for Malaria 2016-2030 (adapted from The malERA Refresh Consultative Panel on Combination Interventions and Modelling, 2017).

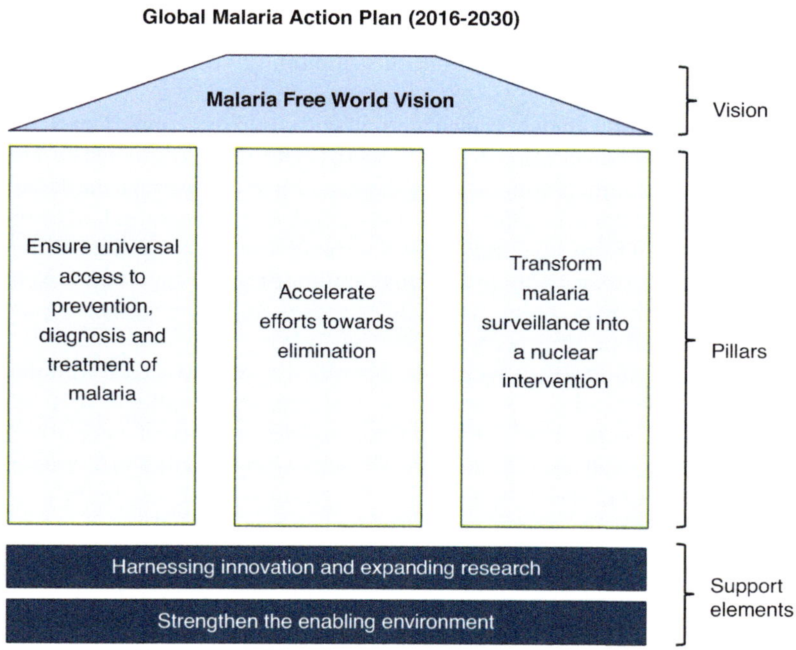

and feasibility', WHO recommends a proactive, strategic and humanitarian approach to address malaria elimination, emphasizing the need for R&D and implementation science, contributing to the universal health coverage agenda to strengthen health systems, building strategies on solid evidence and accurate surveillance and actively engaging affected communities in control and elimination efforts.

By 2021, WHO published an update on the *Global Technical Strategy for Malaria* (GTSM 2021 update). Recognizing shortcomings in the 2020 milestones, namely the 40% reduction of malaria mortality rates (which have only reached up to 20%) and the decrease of 40% of case incidence (which showed a very shy reduction of 3%), the GTSM 2021 update added a sixth principle, stating 'a resilient health system underpins the overall success of malaria response' (WHO, 2021). This principle reflects the impact of the Covid epidemics, reinforcing the idea that health systems have not only to provide quality malaria services but also have to adapt and respond to disruptive events such as epidemics and natural disasters. Acknowledging that the progress in reducing malaria mortality and morbidity has slowed, stopped or even reversed in some endemic countries, the GTSM 2021 update refers to population growth, with the consequent decrease in investment per person, and the substandard quality of services, which are unable to deal with high-impact health emergencies, as some of the major threats to the progress of malaria eradication. Climate change, urbanization, economic development and deforestation are recognized game changers, which reinforce the need for optimizing coverage of interventions, including access to quality health services.

The GTSM 2021 update is also aligned with the Sustainable Development Goals (SDGs, United Nations, 2015). It seems clear that progresses in malaria control and elimination supported by universal health coverage contribute to this development agenda. Efforts to prevent and control malaria both contribute to and benefit from sustainable development. Key factors include malaria's role in the poverty cycle, its prevalence among disadvantaged and excluded populations with poor access to health services and its negative impact on children's education, thus perpetuating the cycle.

As part of Supporting Elements described in GTSM 2021 update, the role of new insecticides, the impact of strategies for exophagic vectors, the development of genetically modified mosquitoes as part of vector control programmes as well as the success of vaccine candidates are considered important additions to the arsenal of future interventions. In July 2013, the World Health Organization Pesticide Scheme (WHOPES) presented a collection of results evaluating the safety and efficacy of chlorfenapyr, an insecticide from the halogenated pyrrole chemical group (WHOPES, 2013). Chlorfenapyr presents a mode of action that is quite different from the pyrethroids not presenting cross-resistance with neurotoxic insecticides (N'Guessan *et al.*, 2007; Raghavendra *et al.*, 2011). In January 2016, the WHO recommended for the first time the large-scale implementation of an anti-malarial vaccine, RTS,S/AS01 (WHO, 2016b). By that time 25 new malaria vaccine candidates were in clinical trials (Hemingway *et al.*, 2016). Integrating operational research carried out in the last decade, in 2023, WHO issued a recommendation for the deployment of pyrethroid-chlorfenapyr insecticide-treated mosquito nets (ITNs) in comparison with pyrethroid-only ITNs and a conditional recommendation for the deployment of these nets instead of pyrethroid–piperonyl butoxide (PBO) nets. Since 2019, Ghana, Kenya and Malawi have been delivering the malaria vaccine RTS,S/AS01 (RTS,S), and by 2021, at least 28 African countries have expressed interest in introducing this vaccine have part of their control programmes. In October 2023, the R21/Matrix-M (R21) malaria vaccine became the second vaccine recommended by WHO for the prevention of *P. falciparum* malaria in children living in malaria-endemic regions, especially in areas of moderate to high transmission (WHO, 2023).

Vector control strategies are moving forward with large investments on strategies based on genetically modified mosquitoes. Currently, two major projects are carrying out pilot studies in West African countries: Target Malaria

(Uganda and Mali: https://targetmalaria.org/), aiming to suppress vector populations, and University of California Malaria Initiative (São Tomé and Príncipe: https://stopmalaria.org/) in which *Plasmodium*-refractory genetically modified mosquitoes will be released to replace the natural *Anopheles coluzzii* population as a means to interrupt malaria transmission. These control strategies based on either suppression or replacement of natural vector populations through the use of genetically modified mosquitoes are aligned with WHO's new *Global Malaria Programme Operational Strategy* (GMPOS). In 2024, WHO included as 'preferred product characteristics' genetically modified mosquitoes and products such as non-invasive diagnostic tests, multi-pathogen tests at point of care, second-generation vaccines and monoclonal antibodies (WHO, 2024). This last report presenting comprehensive operational plans and detailing specific activities will be developed and sustained to achieve GTSM goals (WHO, 2015b, 2021). It is also proposed that the progress of these activities must be monitored in accordance with the theory of change (acknowledging that the global health ecosystem is evolving), tracked using rigorous monitoring and evaluation tools, and reported to donors and partners.

This new operational strategic programme (GMPOS), while reinforcing the main messages of GTSM 2021 update, relies on four strategic functional objectives: '(i) drawing up norms and standards; (ii) accelerate the development of new tools and innovations; (iii) promote the use of strategic information for impact; and (iv) provide technical leadership of the global response'; and cross-cutting pillar, 'context-based country support' (WHO, 2024). Designed to meet the GTSM 2030's targets, it also admits that based on current trends, the 2030's milestones for malaria mortality and morbidity are currently out of reach.

Conclusion

Despite the success in recent years, there are still several obstacles threatening the ultimate goal of eradicating malaria. Many of the threats have been known since the Global Malaria Eradication Programme of the 1950s. The weakening of control programmes, the increase in transmission potential and the occurrence of resistance to anti-malarial drugs and insecticides appear as main causes of the resurgence of malaria.

Insufficient funding has been pinpointed as one of the main factors weakening control programmes, but other factors were also identified. These include intrinsic factors, such as weak implementation and execution capacity, difficulties in the distribution chain, inefficient epidemiological surveillance and quality control systems (e.g. counterfeit drugs); and extrinsic issues, such as armed conflicts, natural disasters and weak community cooperation. It is feared that, as in the past, we will see a gradual reduction in political commitment and an associated decrease in funding for malaria control as the incidence of the disease reduces.

If the emergence of resistance to chloroquine was considered one of the main obstacles to the eradication of malaria in the 1950s–1960s, it is with the same concern that we see the first reports of resistance to ACTs. The first cases of declining effectiveness were recorded in 2008, on the border between Thailand and Cambodia (Wongsrichanalai and Meshnick, 2008). Resistance to ACTs is no longer confined to South-east Asia (Lu *et al.*, 2017). Much of the success of elimination will depend not only on the development of new drugs but also on the ability to prolong the therapeutic efficacy of ACTs.

Likewise, resistance to insecticides is seen as a serious threat to the sustainability of malaria vector control. Currently, resistance to pyrethroids has developed in virtually all primary malaria vectors, although with varying levels and geographic distributions (Quiñones *et al.*, 2015).

As the prevalence and distribution of malaria regress, challenges related to case detection and management increase, particularly in remote regions, where transmission foci from asymptomatic individuals with low parasitaemia may persist. These asymptomatic reservoirs may be the basis for the occurrence of epidemics in regions where elimination has been achieved through phenomena that increase the transmission potential, such as population mobility, environmental changes, extreme events, armed conflicts and other health emergencies, such as the arboviral (dengue, chikungunya, Zika) or COVID epidemics. Another aspect to take into account is that the elimination of malaria has been primarily aimed at eliminating *P. falciparum*. In regions where there is the (co)existence of *P. vivax* and *Plasmodium ovale*, capable of developing hypnozoites, the elimination of malaria will be more difficult due to limitations in the detection and treatment of these dormant forms, which could lead to relapses of malaria, consequently promoting the re-establishment of transmission. Finally, the transmission of zoonotic malaria parasites such as *Plasmodium knowlesi* represents a new challenge to malaria control, particularly in regions bordering forest biomes, where the increasing urbanization of deforested areas will tend to increase the risk of transmission (Shearer *et al.*, 2016).

All these possible setbacks and potential threats to the progress achieved in the last three decades in malaria control are well recognized by WHO in both their GTSM (WHO, 2015b, 2021) and Global Vector Control Response, 2017–2030 (WHO, 2017). However, these challenges can be overcome through global cooperation across communities, countries, languages and cultures to implement context-specific strategies in diverse and changing epidemiological, environmental and sociological settings. National ownership of malaria control efforts is fundamental; countries must lead their own initiatives while receiving support and encouragement from WHO and partners to advance towards reduction and, ultimately, malaria eradication, bearing always in mind that 'Grassi's giant with clay feet' is still standing!

References

Bill & Melinda Gates Foundation; (2007) *Malaria Forum: Collaboration, Innovation, Impact*. Final Report, Bill & Melinda Gates Foundation, Washington, DC, 20 pp. Available at: https://docs.gatesfoundation.org/Documents/MalariaForumReport.pdf (accessed 13 September 2024).

Boyd, M.F. (1949) Historical review. In: Boyd, M.F. (ed.) *Malariology*. W.S. Saunders Company, Philadelphia, Pennsylvania and London, pp. 3–25.

Bruce-Chwatt, L.J. and Zulueta, J. (1980) *The Rise and Fall of Malaria in Europe. A Historico-epidemiological Study*. Oxford University Press, Oxford, UK, p. 240 .

Bruce-Chwatt, L.J. (1988) History of malaria from prehistory to eradication. In: Wernsdorfer, W. and McGregor, I. (eds) *Malaria. Principles of Malariology*. Churchill Livingstone Inc, London, pp. 1–60.

Corbel, V. and N'Guessan, R. (2013) Distribution, mechanisms, impact and management of insecticide resistance in malaria vectors: A pragmatic review. In: Manguin, S. (ed.) *Anopheles Mosquitoes – New Insights into Malaria Vectors*. InTech, Rijeka, Croatia (e-book), pp. 579–633.

Cui, L., Mharakurwa, S., Ndiaye, D., Rathod, P.K. and Rosenthal, P.J. (2015) Antimalarial drug resistance: Literature review and activities and findings of the ICEMR network. *American Journal of Tropical Medicine and Hygiene* 93(Suppl. 3), 57–68.

Dhiman, S. and Veer, V. (2014) Culminating anti-malaria efforts at long lasting insecticidal net?. *Journal of Infection and Public Health* 7, 457–464.

Dobson, M.J. (1994) Malaria in England: A geographical and historical perspective. *Parassitologia* 36, 35–60.

Dobson, M.J. (1999) The malariology centenary. *Parassitologia* 41, 21–32.

Edwards, J.G. (2004) DDT: A case study in scientific fraud. *Journal of American Physicians and Surgeons* 9, 83–88.

Grmek, M.D. (1994) Malaria in the Eastern Mediterranean in prehistory and antiquity. *Parassitologia* 36, 1–6.

Hemingway, J., Shretta, R., Wells, T.N.C., David, B., Djimdé, A.A. *et al.* (2016) Tools and strategies for malaria control and elimination: What do we need to achieve a grand convergence in malaria?. *PLoS Biology* 14, e1002380.

Holtel, A., Troye-Blomberg, M. and Penas-Jimenez, I. (2011) EU-funded malaria research under the 6th and 7th framework programmes for research and technological development. *Malaria Journal* 10, 11.

Huldén, L., Heliovaara, K. and Heliövaara, K. (2005) Endemic malaria: An 'indoor' disease in Northern Europe. Historical data analysed. *Malaria Journal* 4, 19.

Lu, F., Culleton, R., Zhang, M., Ramaprasad, A., von Seidlein, L. *et al.* (2017) Emergence of indigenous artemisinin-resistant *Plasmodium falciparum* in Africa. *New England Journal of Medicine* 376, 991–993.

malERA (2011) malERA - A Research Agenda for Malaria Eradication. PLoS Collections. Available at: http://collections.plos.org/malera (accessed 22 January 2024).

Marciniak, S., Prowse, T.L., Herring, D.A., Klunk, J., Kuch, M. *et al.* (2016) *Plasmodium falciparum* malaria in 1st-2nd century CE Southern Italy. *Current Biology* 26, R1220–R1222.

McCoy, D., Kembhavi, G., Patel, J. and Luintel, A. (2009) The Bill & Melinda gates foundation's grant-making programme for global health. *Lancet* 373, 1645–1653.

Mouatcho, J.C. and Goldring, J.P. (2013) Malaria rapid diagnostic tests: Challenges and prospects. *Journal of Medical Microbiology* 62, 1491–1505.

Nabarro, D.N. and Tayler, E.M. (1998) The “roll back malaria” campaign. *Science* 280, 2067–2068.

Nájera, J.A., González-Silva, M. and Alonso, P.L. (2011) Some lessons for the future from the Global Malaria Eradication Programme (1955–1969). *PLoS Medicine* 8, e1000412.

N'Guessan, R., Boko, P., Odjo, A., Akogbéto, M., Yates, A. *et al.* (2007) Chlorfenapyr: A pyrrole insecticide for the control of pyrethroid or DDT resistant *Anopheles gambiae* (Diptera: Culicidae) mosquitoes. *Acta Tropica* 102, 69–78.

Ntoumi, F. and Priebe, G. (2010) Africanizing scientific knowledge: The multilateral initiative on Malaria as a model? *Malaria Journal* 9(Supp. 3), S7.

Parmakelis, A., Russello, M.A., Caccone, A., Marcondes, C.B., Costa, J. *et al.* (2008) Historical analysis of a near disaster: *Anopheles gambiae* in Brazil. *American Journal of Tropical Medicine and Hygiene* 78, 176–178.

Pinto, J., Lynd, A., Vicente, J.L., Santolamazza, F., Randle, N.P. *et al.* (2007) Multiple origins of knockdown resistance mutations in the Afrotropical mosquito vector *Anopheles gambiae*. *PLoS ONE* 11, e1243.

Quiñones, M.L., Norris, D.E., Conn, J.E., Moreno, M., Burkot, T.R. *et al.* (2015) Insecticide resistance in areas under investigation by the international centers of excellence for malaria research: A challenge for malaria control and elimination. *American Journal of Tropical Medicine and Hygiene* 93(Suppl. 3), 69–78.

Raghavendra, K., Barik, T.K., Sharma, P., Bhatt, R.M., Srivastava, H.C. *et al.* (2011) Chlorfenapyr: A new insecticide with novel mode of action can control pyrethroid resistant malaria vectors. *Malaria Journal* 10, 16.

RBM (2003) *The Abuja Declaration and Plan of Action*. Roll Back Malaria/World Health Organization, Geneva, Switzerland (WHO/CDS/RBM/2003.46), p. 11.

RBM (2008) *Global Malaria Action Plan. For a Malaria-Free World*. Roll Back Malaria Partnership, Geneva, Switzerland, p. 271.

Reiter, P. (2001) Climate change and mosquito-borne disease. *Environmental Health Perspectives* 109(Supp. 1), 141–161.

Sallares, R., Bouwman, A. and Anderung, C. (2004) The spread of malaria to Southern Europe in antiquity: New approaches to old problems. *Medical History* 48, 311–328.

Santolamazza, F., Calzetta, M., Etang, J., Barrese, E., Dia, I. *et al.* (2008) Distribution of knock-down resistance mutations in *Anopheles gambiae* (Diptera: Culicidae) molecular forms in West and West-Central Africa. *Malaria Journal* 7, 74.

Shearer, F.M., Huang, Z., Weiss, D.J., Wiebe, A., Gibson, H.S. *et al.* (2016) Estimating geographical variation in the risk of zoonotic *Plasmodium knowlesi* infection in countries eliminating malaria. *PLoS Neglected Tropical Diseases* 10, e0004915.

Teklehaimanot, A. and Bosman, A. (1999) Opportunities, problems and perspectives for malaria control in Sub-Saharan Africa. *Parassitologia* 41, 335–338.

The malERA Refresh Consultative Panel on Combination Interventions and Modelling (2017) MalERA: An updated research agenda for combination interventions and modelling in malaria elimination and eradication. *PLoS Medicine* 14, e100245.

United Nations (2000) Transforming Our World: The 2030. Agenda for Sustainable Development, United Nations General Assembly, New York, 9 pp. Available at:

https://sdgs.un.org/sites/default/files/publications/21252030%20Agenda%20for%20Sustainable%20Development%20web.pdf (accessed 14 December 2023).

United Nations (2015) United Nations Millennium Declaration. A/RES/55/2, United Nations, New York, 41 pp. Available at: http://www.USAID.org/millennium/declaration/ares552e.pdf (accessed 18 January 2024).

Weissmann, G. (2006) DDT is back: Let us spray!. *FASEB Journal* 20, 2427–2429.

White, N. (1999) Antimalarial drug resistance and combination chemotherapy. *Philosophical Transactions of the Royal Society B* 354, 739–749.

WHO (1956) *6th Report of the Expert Committee on Malaria.* WHO/MAL/180. World Health Organization, Geneva, Switzerland, p. 93.

WHO (1973) *Handbook of Resolutions and Decisions of the World Health Assembly and the Executive Board.* Volume I, 1948-1972, 1st to 25th WHA and 1st to 50th EB, World Health Organization, Geneva, Switzerland, pp. 66–81.

WHO (1998) *World health report, 1988.* World Health Organization, Geneva, Switzerland, 232 pp.

WHO (2006) *Regional strategy: From Malaria Control to Elimination in the WHO European Region, 2006–2015.* World Health Organization, Regional Office for Europe, Copenhagen, Denmark, 50 pp.

WHO (2007) *Malaria Elimination: A Field Manual for Low and Moderate Endemic Countries.* World Health Organization, Geneva, Switzerland, 85 pp.

WHO (2008) *World Malaria Report, 2008.* World Health Organization, Geneva, Switzerland, 190 pp.

WHO (2010) *World Malaria Report, 2010.* World Health Organization, Geneva, Switzerland, 204 pp.

WHO (2011) *The Use of DDT in Malaria Vector Control: WHO Position Statement.* Global Malaria Programme, World Health Organization, Geneva, Switzerland, 10 pp.

WHO (2014) *WHO Policy Brief for the Implementation of Intermittent Preventive Treatment of Malaria in Pregnancy Using Sulfadoxine-pyrimethamine (IPTp-SP).* Global Malaria Programme, World Health Organization, Geneva, Switzerland, 13 pp.

WHO (2015a) *World malaria report, 2015.* Global Malaria Programme, World Health Organization, Geneva, Switzerland, 243 pp.

WHO (2015b) *Global technical strategy for malaria 2016–2030.* Global Malaria Programme, World Health Organization, Geneva, Switzerland, 29 pp.

WHO (2016a) *World malaria report, 2016.* Global Malaria Programme, World Health Organization, Geneva, Switzerland, 148 pp.

WHO (2016b) Malaria vaccine: WHO position paper – January 2016. *Weekly Epidemiological Report* 91, 33–52.

WHO (2017) *Global Vector Control Response 2017–2030.* World Health Organization, Geneva, Switzerland, 51 pp.

WHO (2020) *Malaria Eradication: Benefits, Future Scenarios and Feasibility.* A Report of the Strategic Advisory Group on Malaria Eradication. World Health Organization, Geneva, Switzerland, 116 pp.

WHO (2021) *Global Technical Strategy for Malaria 2016–2030, 2021 Update.* World Health Organization, Geneva, Switzerland, 30 pp.

WHO (2023) *World Malaria Report, 2023.* Global Malaria Programme, World Health Organization, Geneva, Switzerland, 283 pp.

WHO (2024) *Global Malaria Programme Operational Strategy 2024–2030.* World Health Organization, Geneva, Switzerland, 68 pp.

WHOPES (2013) *Report of the Sixteenth WHOPES Working Group Meeting*. World Health Organization Pesticide Evaluation Scheme, Geneva, Switzerland, 183 pp.

Wongsrichanalai, C. and Meshnick, S.R. (2008) Declining artesunate-mefloquine efficacy against *falciparum* malaria on the Cambodia-Thailand border. *Emerging Infectious Diseases* 14, 716–719.

Sampling of Mosquito Vectors of Disease

4

J. D. Charlwood*

Global Health and Tropical Medicine, Instituto de Higiene e Medicina Tropical, Universidade NOVA de Lisboa, Lisbon, Portugal

Abstract

Monitoring of vectors is a cornerstone of control. Without an assessment of risk, it is impossible to determine what the most suitable intervention is and, once applied, whether or not it has any effect. Effective monitoring requires effective sampling. Mosquito behaviour can be roughly divided into four 'compartments', each of which has a number of 'best' simple methods of collection. Here, some recent techniques are described for the collection of larvae and adults, with an emphasis on those that can be performed with minimal equipment or training.

> The effectiveness of a baited trap for a particular species of fly primarily depends on its responses to the trapping device in the presence of the attractant stimuli used.
>
> (Gillies, 1974)

> Entomological surveillance should utilise multiple complementary collection methods across different micro-habitats to sample different behaviours.
>
> (van de Straat *et al.*, 2021)

Introduction

A number of books have already been written on sampling techniques for mosquitoes (in particular, Silver, 2008), and it is not our intention to replicate

*Corresponding author: jdcharlwood@gmail.com

(ed J.D. Charlwood)
DOI: 10.1079/9781789248845.0004

these in any detail. Rather, we describe techniques that we have found useful, or developed, and describe some notes on how best to set them up.

Monitoring of the vectors is now recognized by the WHO as one of the pillars leading to disease elimination. Without it, you are blind to what is going on. It is likely that in the future, there will be fewer resources available for this. Even with the simplest of tools, it is, however, possible to obtain most of the measures required for an assessment of adult vector populations. Sampling strategies depend on the information required. The amount and intensity of sampling effort depends, to a large extent, on budget, logistics and, of course, the question being asked. The most basic questions perhaps being, 'What is the mosquito responsible for disease transmission?', 'How many bites does the 'average' person (or a proxy) suffer, both indoors and outdoors?' and 'How many of these bites potentially infected them with sporozoites or virus?' Other questions that might be asked are likely to depend on resources and logistics as much as anything else.

Sampling will never be free and is unlikely to ever be so comprehensive that every question is adequately answered. As pointed out by Foley *et al.* (2009), 'Accurately geo-referenced collection data are crucial for understanding mosquito biogeography, ecology, and the impact of environmental changes, as well as for species distribution modelling, planning mosquito surveys, and for determining disease risk'. It may involve the establishment of widespread sentinel houses in which the householder does the collection, using trapping methods that require at most a torch as a source of energy, as described below, or it may be more intensive – and so less biodiversity friendly – random sampling. The data collected can be uploaded on to a site/platform that acts as a central repository. Obviously, the aim is to develop an understanding of vector dynamics, vectorial capacity and overall ecology of the mosquitoes, both as larvae and adults, and then to use this information to devise strategies to reduce disease transmission.

Part of the difficulty faced by many entomologists may be knowing how many samples of what sort are needed in order to be able to produce a meaningful data set suitable for epidemiological analysis. One of the important measures in vector-borne disease, especially malaria, is the Entomological Inoculation Rate (EIR). This is calculated by multiplying the sporozoite rate in the mosquito by the number of bites an 'average' host receives. The aim of interventions is to reduce this figure, mainly by reducing the sporozoite rate, which may be done by reducing the survival rate of the mosquito.

Vector Behaviour and Sampling

Mosquito behaviour can be roughly divided into four 'compartments' each of which has a number of 'best' simple methods of collection (Charlwood *et al.*, 2024).

1. *Endophagic, endophilic* mosquitoes feed and rest inside houses.
2. *Exophagic endophilic* mosquitoes may feed outside but rest inside houses.
3. *Endophagic exophilic* mosquitoes may feed inside houses but rest outdoors.
4. *Exophagic, exophilic* mosquitoes both feed and rest outdoors.

These are not hard and fast distinctions. For example, mosquitoes such as *An. arabiensis* may demonstrate all four behaviours, in varying degrees, in different populations or at different times (Charlwood *et al.*, 1995, 1999; Charlwood and Edoh, 1996). Mosquitoes are likely to occur in a metapopulation patchwork of semi-isolated subpopulations, whose ecology is defined as much by the environment as by intrinsic characteristics (Verdonschot and Besse-Lototoskaya, 2014). Given the substantial geographic heterogeneities in disease burden, assessment of the bionomics of local mosquitoes is likely to require a range of bespoke interventions for different populations.

A summary of possible collection methods for malaria vectors according to their different behaviours is shown in Table 4.1. Here, we first describe 'energy-free techniques' that can be performed by anyone with the simplest equipment before describing one or two of the more sophisticated methods used by the professional entomologist.

Endophagic endophilic

Both male and female endophilic mosquitoes, such as *An. funestus*, spend much of their adult lives inside houses. Such mosquitoes, including males as well as females, tend to leave houses at dusk when they are positively phototactic, the males to swarm (and, if they are lucky, to mate), newly emerged females to swarm and mate, and gravid females to oviposit. They can easily be collected by placing a piece of white mosquito netting over the open door of a house (which makes it the lightest part of the house and so

Table 4.1. Summary of the different behaviour patterns among mosquitoes and the method described in the text to sample them.

	Endophilic	Exophilic
Endophagic	exit resting light-trap	window trap outdoor resting light-trap
Exophagic	exit landing single net trap resting tent-trap	landing single net trap outdoor resting tent-trap

attracts the mosquitoes) and having someone inside the house aspirating mosquitoes when they land on the netting in their attempt to exit. This kind of collection only lasts around 25–30 min and does not need a power source. The mosquitoes leaving do not attempt to bite, and so the collection poses no risk to the collector. Another advantage of such collections is that the females can easily be separated into different age/abdominal classes. The relative ratio of the different abdominal stages provides information on any exophilic tendency of the insects and on the relative survival of both males and females (Charlwood *et al.*, 2023). The system can be seen at https://youtu.be/SL8FeIuY1GM. It can also be modified for outdoor collection as performed in Eritrea with *An. arabiensis* (Charlwood *et al.*, 2021). This collection can be used with any endophilic species including those that feed outside but then rest inside (or in man-made structures).

Exophagic endophilic

In addition to exit collections, another collection that can provide useful data on endophilic mosquitoes is the indoor resting collection. A minimum requirement for resting collections is an aspirator, torch and holding cup. More sophisticated methods of resting collection are described below. In Eritrea, a single round of collection (around 20 min) was sufficient to sample between a quarter and two-thirds of the total in the house or shelter depending on the method used (Charlwood *et al.*, 2021), and in Muleba, Tanzania, approximately 40% of the total collected from five rounds of collection were collected during the first round (Charlwood *et al.*, 2018).

There is often a significant difference in the number of mosquitoes found in different houses (Charlwood *et al.*, 1995). It is therefore better to collect from a different house than having a second look in the same house. Such collections provide information on longevity and contact with man, as well as density.

Endophagic exophilic

Insects, which feed indoors but leave shortly after feeding, can perhaps be best collected using window traps.

Window traps

This is another kind of exit collection that also does not require an energy source. The most commonly used window exit/entrance traps are the Muirhead-Thomson (1947) traps, and their various modifications, which are fitted over bedroom windows to catch exiting mosquitoes. These traps are basically cages that on the side facing the interior of the house consist of a netting funnel that allows mosquitoes to enter but not to return inside

Fig. 4.1. Window trap in Muleba, Tanzania. The inverted cone or 'lobster-pot' design allows mosquitoes to enter but not leave. Two or three dead mosquitoes were left every day on the paper. Their non-removal indicated that ants were not affecting the sample. (Photo: J.D. Charlwood.).

(Fig. 4.1). These traps are relatively bulky and difficult to transport but, once in place, can be used for long periods before they might need repair or replacement.

Ambient light (especially moonlight) may influence window-trap collections. The window is the only source of illumination at night in many houses. On nights when moonlight is present, the window stands out, while at other times, bedrooms may be pitch black so that there is little to guide the mosquito into the trap.

Glue net trap

Glue net traps (GNTs) are an alternative to conventional window traps (Müller *et al.*, 2017). They are assembled from rigid, dark green plastic garden fencing with 0.2 cm wide net separated by 0.8 cm square holes. Frames can be constructed by overlapping two layers of 1.0 cm thick plywood fixed with nails and wood glue. The netting is sandwiched between the two wooden frames and secured with screws. The net is painted with glue (Tanglefoot® Glue, Tanglefoot Company, Grand Rapids, MI, USA) to capture insects trying to land or pass into or out of the house (Fig. 4.2). Mounting one GNT outside the window and one inside, both entering and exiting mosquitoes can be collected separately. In field trials, the trap was particularly effective at collecting insects attempting to enter bedrooms (Müller *et al.*, 2017). Traps can be mounted over the windows

Fig. 4.2. Close-up of two *An. gambiae* s.l. caught in the GNT. Mosquitoes trying to pass the netting get caught by the glue. Note the body almost never touches the glue. (From Müller *et al.*, 2017.).

in a variety of ways. In the original field study, traps were suspended from rope secured by a nail. One advantage of the relatively large mesh size used in the GNT is that it hardly affects air circulation, meaning that the room remains comfortable for the occupants.

Exophagic exophilic

For exophagic mosquitoes (that do or do not rest inside houses), the simplest way to determine outdoor densities is to undertake outdoor human landing collections (HLC). They only require an aspirator, torch and holding cup, and they are easily learnt. Landing collections might expose the collector to pathogens (Achee *et al.*, 2015; World Health Organization, 2017), but they are the best way to determine the exposure of people sitting outside in the evening, and since sitting outside is a 'normal' activity, it need not involve greater exposure than normal. They also only need to be undertaken until people normally go indoors.

An alternative to the standard HLC is for the collector to sit under a double-sized mosquito net raised around 10 cm off the ground. Collectors might sit for 40 min and then collect from the walls of the net. The collector may be protected from mosquitoes (by wearing appropriate clothing, including wellington boots) or can undertake a standard landing collection when not sampling from the walls. Unlike double-net traps, it does not need any special tailoring to work, and unlike double net traps, which, when not in use, are just stored somewhere, the net is just a normal net and so goes over someone's bed when not in use.

The wide-scale use of LLINs for vector control may result in changing biting patterns among previously endophagic vectors and result in low levels of residual transmission outdoors. These mosquitoes may best be collected by HLC or outdoor resting collection. Anywhere that the mosquitoes rest, but do not feed, can be considered an outdoor site. If the males of a species do not rest inside houses (but the females do), the collection of males also indicates that a site can be considered to be 'outdoors', even though it might be a structure like a latrine (Charlwood *et al.*, 2018).

Mosquitoes that rest in long grass can be collected using a net on a frame. In this case, the frame is placed over the area to be sampled and a collector enters under the net and beats the grass so that insects fly up from it. They can then be collected off the walls of the net. A similar type of collection was undertaken in Madang, Papua New Guinea (PNG), but in this case, the frame was covered with black plastic. A 30 × 30 cm exit trap, in the style of a window trap, made of white netting was attached to one side. In coastal areas of PNG, the local malaria vector, *Anopheles farauti*, rests among discarded coconut husks in bomb craters (i.e. where bombs exploded during the Second World War). The frame was placed over such discarded husks. As the temperature increased, the mosquitoes were observed to crawl further into the pile of husks. At sunset, however, like the endophilic *An. funestus* from Mozambique, they left their resting site and flew into the window trap on the side of the frame.

The same principle can be applied to insects resting in clay pots that are otherwise difficult to extract. In this case, an exit trap is placed over the pot before sunset and removed once it is dark. Thus, the pot is open during the night and in the early hours of the morning, which allows the insects searching for a daytime resting site to enter. The main caveat to this procedure is that the collector needs to be in the village/sampling site after dark. An alternative is to use sticky netting inside the pot to collect insects that enter to rest. One advantage of this method is that mosquitoes collected over several days can be sampled together.

There are a number of conflicting tensions when it comes to sampling malaria and dengue vectors in a routine way over long periods. On the one hand, a replicable system of collection, such that a meta-database can be developed that encompasses sampling throughout the vectors' geographical range, is a good thing. On the other hand, the local ecological characteristics of a vector may require novel sampling techniques that may not work elsewhere. The inquisitive scientist does not want to be straightjacketed into a system that does not allow for individual innovation, especially if the mosquito in question appears to be behaving in a way different from the accepted norm. Some sort of mean should be attempted. We also, therefore, indicate how novel experiments could be performed within a potential sampling scheme.

Alternative Trapping Techniques

Endophagic host-seeking insects

Light-traps provide a low-cost way of achieving high sampling intensities and provide valuable entomological measures of the impact of vector control on human exposure (Huho *et al.*, 2015). The most common replacement for HLC indoors are CDC light-traps, so their use in a standardized fashion enables meta-analysis to be undertaken. Nevertheless, within a 'universal' sampling frame, other experiments might be performed. For example, it is generally recommended that light-traps are placed circa 1.5m off the ground at the end of the bed in which someone is sleeping under a mosquito net, but how the placement of the trap influences the catch might vary according to house type. In general, light-traps collect very few male mosquitoes, but this may depend on the time that the trap is switched on. Males that are resting indoors exit houses at dusk by flying towards the light. Should the trap be switched on before dusk in a darkened house, it may collect these males. There is a positive relationship between males inside houses in *An. funestus* and young, virgin, females resting inside. Should a light-trap collect many males, it is likely that the trap would be catching these as well. Such females, who are not there to feed on blood but only to rest before going out to mate, are difficult to distinguish from other older, more dangerous females. This would tend to result in an overestimate of risk. The effect on the number of male mosquitoes in the trap might be determined according to different start times. Ideally, of course, one would not want there to be any difference in catches either according to placement or start times (thus enabling an uncomplicated meta-analysis).

Standard CDC light-traps come with a metal mesh grid installed at the entrance to the trap and a standard plastic collection cup. The grid is designed to exclude larger insects, like moths, from entering the trap. Such large insects are almost non-existent inside houses. Mosquitoes are battered when they pass through this grid. Our advice would be to remove it. (Again, it would be an easy test to determine if this has an effect, either on the number of mosquitoes collected or their survival rate.)

The plastic container used for standard collections almost always results in 100% mortality among the insects. Alternative collection bags are available. At the Ifakara Health Institute (IHI) in Tanzania, they make large, long bespoke bags. These may enhance survival of collected mosquitoes in light-traps. Live mosquitoes are useful because they can be dissected for parity determination and because a number of analyses require mosquitoes to be killed immediately into preservative, such as mRNA.

If live mosquitoes are not required, then there are advantages to killing the mosquitoes at the time of collection. For example, supposing a mosquito enters the trap at 1900 hour and survives until 0900 hour (when it is killed by the entomologist), then it will be 14 hours older than it was at the time it was collected, a proxy for the time that it attempted to feed. This might be sufficient for sporozoites from a burst oocyst to travel to, and enter, the salivary glands and would mean that, had the insect survived, it would be considered to be an infectious one rather than merely being an infected one. This would bias epidemiological estimates (but it would at least be a bias in the right direction).

Silver bullet trap

A recently developed alternative to the standard CDC light-trap is the 'Silver bullet' trap (Mbare *et al.*, 2023). The trap is made of polyvinyl chloride (PVC) and aluminium. A polycrystalline 12-Watt solar panel at the top of the trap charges the trap's lithium-ion battery pack when placed in the sun. The trap also has a plug-in mains adapter that can be used to charge it. Three light-emitting diodes (LEDs) that emit UV, white and red lights are fitted on the trap, and a programmable microcontroller controls the LED output. In addition, the traps have a USB port charger, which can be used for charging mobile phones and tablets (Fig. 4.3). The trap is made in South Africa and is cheaper than the standard CDC light-trap.

In semifield and field trials, the trap collected an equivalent sample to CDC traps. It has a number of advantages over conventional CDC light-traps. The solar panel means that it does not depend on a supply of charged batteries for

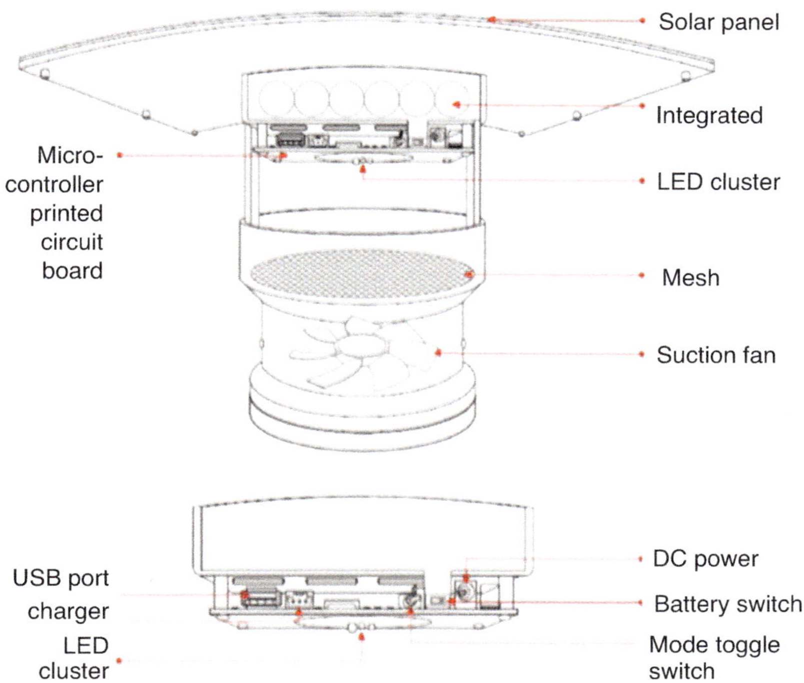

Fig. 4.3. Details of the Silver Bullet light-trap. (From Mbare *et al.*, 2023.).

it to work. It can be programmed to produce different types of light (including UV light), can charge a mobile phone and is robust. Traps emitting light in the UV spectrum trapped twice as many mosquitoes of all genera, and both females and males, as traps emitting white LED or incandescent light (Mbare *et al.*, 2023). The main disadvantage is that it is difficult to transport because it cannot be dismantled. This means that it is most usefully established in sentinel houses rather than in a changing set of random houses.

Exophagic Host-Seeking Insects

Tent-traps

While operationally practical, CDC light-traps (or Silver Bullet traps) used outdoors do not adequately sample the outdoor biting fraction of malaria vectors (Fornadel *et al.*, 2010). Thus, there was no significant correlation between *Anopheles gambiae* s.l., HLC outdoors and outdoor CDC light-traps reported by Costantini *et al.* (1998). Similarly, Cooke *et al.* (2015) concluded that a CDC light-trap hung close to an occupied, open-sided rain shelter made from a domed one-man tent was a 'limitation' of their study. Other traps, notably the MM-X trap (Njiru *et al.*, 2006), and more recently the 'Suna' trap (Homan *et al.*, 2016), have been developed to catch outdoor biting mosquitoes. Nevertheless, they do not provide an easily quantifiable estimate of exposure, nor are they cheap or easily available. Tent-traps are the simplest alternative solution. Among available designs, the Furvela tent-trap (Charlwood *et al.*, 2017) is cheap, uses 'off the shelf' components, is easy to set up and provides a sample that is directly equivalent to a CDC light-trap (Fig. 4.4). The host is not exposed to mosquito bites and is able to get a good night's sleep, thus being able to work normally the following day.

The anthropophilic members of the *An. gambiae* complex are 'indifferent' to the location of the host, with indoor and outdoor biting ratios in human landing collections generally being similar (e.g. Kabbale *et al.*, 2013; Dambach *et al.*, 2018). This similarity is shown when Furvela tent-traps have been compared to CDC light-traps (Fig. 4.5). This has led some authors to suggest that 'tent-traps such as the Furvela tent-trap are not truly outdoor trapping tools but mimics of an indoor environment where the trapped mosquito species exhibit house entry traits' (Abong'o *et al.*, 2021). However, it does not require an 'entry behaviour' on the part of the mosquito, and we consider that it really does sample the outdoor biting fraction of the mosquito fauna. Thus, species known to be exophagic are collected in high numbers in the tent-trap but may be absent from indoor CDC light-trap collections (Charlwood *et al.*, 2013).

We do not, however, know how the trap actually works. Conditioned air, produced by the host inside the tent, is sucked out from the opening by the CDC fan. This air presumably hits the back of the collection bag. Conical bags (as supplied by John Hock company) have been used by the present authors.

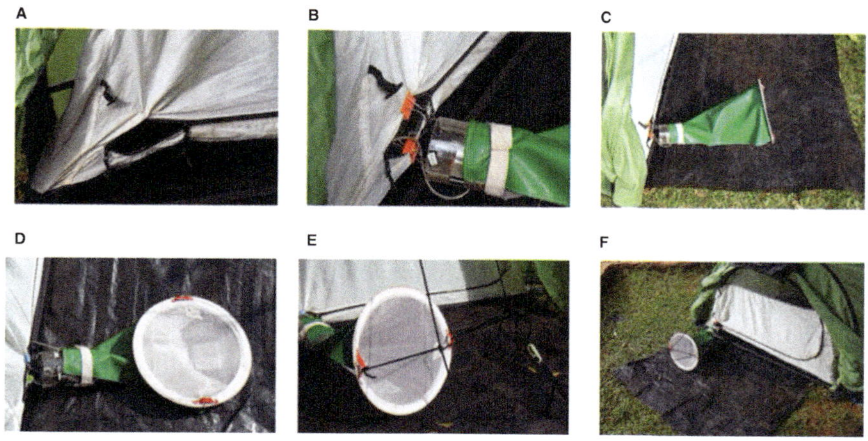

Fig. 4.4. The Furvela tent-trap. Installation procedure. (i) Sew open the door of the tent to be approximately the diameter of the standard CDC light-trap; (ii) attach a CDC trap (without light or lid) to the tent approximately 4 cm from the opening, pass the lead to the trap into the tent where it can be switched on by the sleeper; (iii) attach a conical netting collection bag (with optional rain cover) to the trap; (iv) eyelets are sewn into the end of the bag; (v) these attach to guy ropes to keep the trap suspended horizontally in front of the opening; (vi) the complete trap ready for use. (From Charlwood *et al.*, 2017.).

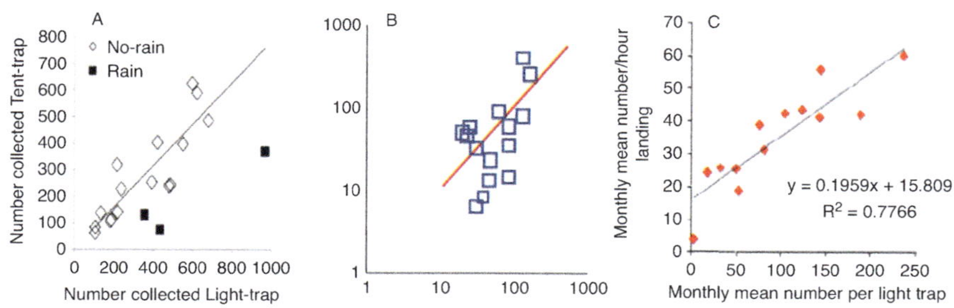

Fig. 4.5. A) Relationship between numbers of *An. coluzzii* collected in a Furvela tent-trap and a CDC light-trap in Okyereko, Ghana, the black squares represent nights when it rained; B) Relationship between numbers of *A. gambiae* in Furvela tent-traps and light-traps in Tanzania; C) Relationship between numbers of *A. coluzzii* in light-traps and landing collections from São Tomé.

The conditioned air (with a higher than normal CO_2 concentration) doubtless spreads from the back of the bag to the outside, and it is this that attracts the mosquitoes. However, how they find their way into the trap opening instead of accumulating on the outside of the bag is not really known. The bag as supplied is designed for use with a standard CDC light-trap, so when used in a light-trap, the attachment of the bag to the body of the trap only needs to support its own weight. A greater amount of tension on the attachment is present when the bag is used in a Furvela tent-trap (because it also needs to resist the pull from the string that keeps everything horizontal), and the usual Velcro attachment of these bags does not easily support this. A strong rubber band of the appropriate size and strength (the sort that may fit on a person's wrist; mine says 'Beat Heartbreak forever') ensures that the bag stays on the trap.

Presently, we do not know how much of an effect (if any) different types or styles of collection bag have on the functioning of the trap. The long Ifakara CDC collection bag may be useful when hung vertically with a CDC trap, but if the above theory of how the trap works is correct, it may be less useful when used with the tent-trap. The trap also puts a greater strain on the body of the trap. Traps made with thick Perspex for the trap body are better than ones made with thinner Perspex.

In areas where they might be a problem, the distribution of an anti-ant powder spread around the pole that supports the trap and the trap opening is recommended. Although the Furvela tent-trap can be used with almost any tent, there is likely to be a delay between the time that the bait (the person sleeping in the tent) enters the tent and the time that it starts working efficiently. This is because there is likely to be a delay before the CO_2 concentration inside the tent is high enough for it to attract mosquitoes.

Mosquitoes are attracted to large objects (Bidlingmayer, 1994). Tents may in themselves attract mosquitoes, and larger tents may attract greater numbers. But this has also not yet been examined.

Given that, with the Furvela trap, collection bags can be easily substituted, they can be changed when villagers go indoors so that a more accurate estimate of outdoor exposure can be determined. This was done in the village of Kyamyorwa in northern Tanzania. In this case, a villager was given a tally counter, a watch with an alarm and a handheld Global Positioning System (a Garmin eTrex). Every hour after the alarm sounded, they followed a specific route and, with the tally counter, counted the number of residents that were outside their houses. The route encompassed 26 houses with 193 residents. The villager reported to a field assistant sleeping in a tent-trap at the end point of his route. The track on the GPS was checked, the number of people counted noted down and the counter and alarm reset. This enabled the time at which people went indoors to be determined. When all residents were inside, the counting stopped. At this time, the cages on the tent-traps were changed and the mosquitoes in a window trap were collected. By 21:00 hour, the majority of people had gone inside (Fig. 4.6).

In the collections undertaken prior to an intervention, when the population was sufficiently large to make these collections worthwhile, only 9.7% of the 2117 *A. gambiae* collected during the night were caught before people went to bed (i.e. in the first 3 hour period of collection) (Fig. 4.6). This is a proportion similar to that described from an earlier study elsewhere in Tanzania (Killeen

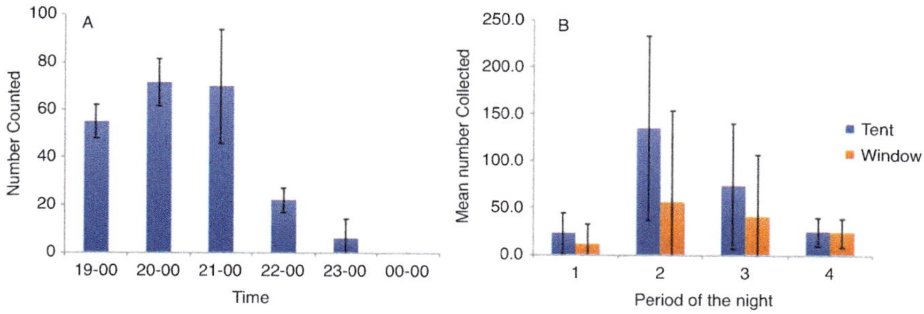

Fig. 4.6. Number of residents from 26 houses outside their houses by hour of the night, B) Mean number of *A. gambiae* collected from tent and window traps, Kyamyorwa, Tanzania.

et al., 2006). Hence, outdoor exposure was only a fraction of the exposure indoors, and an intervention of IRS and LLINs, used together, effectively controlled malaria in the village.

We are advocates of sequential, daily sampling, especially when the models of Birley and colleagues can be applied to the data (Birley and Rajagopalan, 1981; Birley and Boorman, 1982; Charlwood *et al.*, 1985, 2023). Unfortunately, most of the efforts to apply such time-series analysis of host-seeking females collected sequentially for 30 days or more have been only partially successful. One of the premises of the analysis is that a constant proportion of the total population is collected every night. During most of the time-series collections, only indoor sampling was undertaken. There is some reason to believe, however, that insects normally attracted to host outdoors are more likely to enter houses and less likely to enter tent-traps on rainy nights (Fig. 4.5). This increases the trap's efficiency and upsets this premise (Charlwood *et al.*, 2011). A time series of daily indoor and outdoor collection made for at least 3 weeks is the most suitable sample for determining the size of the host-seeking population of mosquitoes. Two pairs of traps are likely to be sufficient for survival rate determination.

The necessity of having an outdoor collection is less important for endophilic mosquitoes like *An. funestus* since the great majority of the whole population can be collected in exit collections independent of weather conditions. More recently, the models of Birley and colleagues have been updated by Tom Smith so that missing data can be estimated and so that estimates of the duration of the oviposition (gonotrophic) cycle do not need to be an integer (Charlwood *et al.*, 2023).

Survival rate determination – traditionally done by dissection but now with a number of alternative systems in the pipeline, which enables the potential number of vectors, compared to the actual number, to be estimated – is the best alternative to time-series data.

Resting Insects

Given that mosquitoes spend a large part of their adult life 'resting', where they may be especially vulnerable to control, knowing where they do so is important. The collection of resting insects, in particular, females, was especially important when IRS was the main means of control. Under these circumstances, it is important to know what proportion of the population rests inside houses, what proportion rests on the roof or walls and at what height they do so on the walls. Such information can help the focus of IRS. In traditional African houses, where the roofs are made with thatch, the insects can often be found there. This changes when other roof types, in particular, corrugated iron, are used.

Two methods are traditionally used to collect mosquitoes resting inside houses: Pyrethrum Spray Catches (PSC) (Silver, 2008), or the collection of resting adults, using either mechanical or manual aspirators. Mosquitoes collected by PSC are killed by the insecticide, and the location of their resting site within the house cannot be determined. Presently, the method of choice for the collection of live insects that can be separated into the location of their resting site is the mechanical Prokopack collector (Vazquez-Prokopec, 2009), which uses the same collection cup as the earlier backpack aspirator developed by CDC.

The Prokopack can be made on a DIY (do-it-yourself) basis (with instructions given in the original publication), although a lighter weight and more streamlined version is available for purchase from the John W. Hock Company (https://www.johnwhock.com/). The main caveat to the use of the Prokopack and the backpack aspirator, if live mosquitoes are required, is that the collection cups should be changed every 5 min or less. It has the advantage over manual aspiration in that it does not depend on the visual ability of the collector and over backpack aspirators in that it is much lighter and easier to use (Maia *et al.*, 2011; Charlwood *et al.*, 2018). The addition of an extension rod also means that roofs can easily be sampled.

What proportion of the resting population is collected by each method is also important. For PSC, it was estimated to be approximately 70% (Silver, 2008). Estimates of the number of insects not collected can be obtained through the technique of removal sampling in which sequential samples are made from the same site.

With the exception that an exit collection is not valid the evening after a resting collection (although it might provide an estimate of sampling efficiency of the resting collection), the sampling techniques described here are not generally mutually exclusive. For example, an outdoor landing catch may not affect the numbers collected in exit traps.

Collection of Gravid Insects

A number of traps have been developed to sample gravid mosquitoes, especially those that oviposit in water with a high organic content such as pit latrines and septic tanks (Silver, 2008).

In Madang, Papua New Guinea, a combined exit and entrance trap for the collection of *Armigeres milnensis* that was breeding in a septic tank in the grounds of the Papua New Guinea Institute of Medical Research was developed (Fig. 4.7).

Newly emerged insects and females, which had recently oviposited, and gravid females attempting to enter the septic tank were sampled using the trap illustrated in the figure. This consisted of a wooden box (50 cm on a side) with 14-cm-diameter holes on the top and on one side, into which tin cylinders had been fixed (Fig. 4.8). A standard netting cage, 30 cm on a side, was placed over the top tin. A removable brass mesh cone was attached to the side tin at its end closest to the box. The other end of the tin was covered with brass mesh. A similar tin, used as an entrance trap, was attached to the first tin by a locking collar. A small 6 V motor and fan assembly was installed over the vent to the septic tank in line with the side exit tin. The fan, powered by a regulated supply at a nominal 4 V setting, blew 'conditioned air' (i.e. air containing odours emanating from the septic tank vent) out through the side tins. During collections, the trap was operated for 30 min/h. At other times, the trap was removed so that free ingress and egress were possible, enabling population replacement to continue.

Interestingly, the mosquito became active during a partial eclipse of the sun on 11 June 1983.

Sampling of Larvae and Pupae

Mosquito larvae can be found in a wide variety of water bodies. The nature of the water body largely determines the kind of sampling technique used. The small containers inhabited by dengue vectors can be difficult to find.

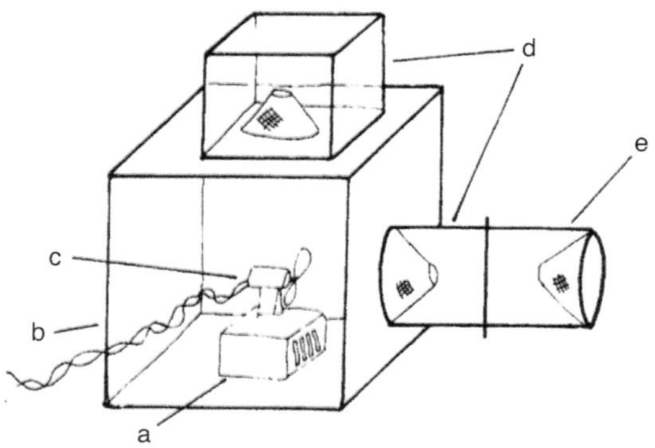

Fig. 4.7. Diagram of the entrance-exit trap used for the collection of *Armigeres milnensis* in Papua New Guinea: (a) septic tank vent; (b) wooden box 50 cm on a side; (c) tan and motor assembly; (d) exit traps with wire mesh cones; (e) entrance trap. (From Charlwood and Galgal, 1985.).

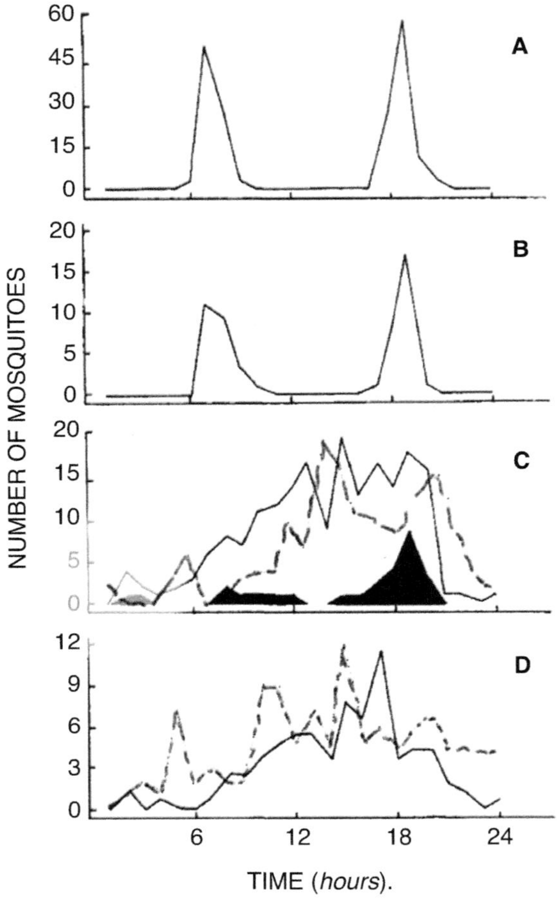

Fig. 4.8. Williams mean activity patterns of *Armigeres milnensis* from Madang, Papua New Guinea – A) man biting; B) Number collected in entrance trap; C) Number of females collected in the exit trap (newly emerged solid line) recently oviposited (solid graph) and emerged in the laboratory (dotted line); D) Number of males collected in the exit trap (solid line) and emerging in the laboratory (dotted line). (From Charlwood and Galgal, 1985.).

The dipper is the almost universal method of collecting anopheles larvae. Densities may be recorded as number per dip. In some situations, alternative methods might be more useful. In large bodies of water, there may be predators, and in order to avoid these, larvae can rest very close to the edge in very shallow water. They move only when they have to. A simple way to estimate density of *An. arabiensis* at a Masai camp close to the town of Ifakara, Tanzania, was to count the number in a given length of edge. Today, a more accurate way to do this would be to take a digital photograph. This can either be counted at leisure, and by more than a single person, or can be interpreted by AI software.

Another alternative for small-pool (puddle) breeding insects is to use a metal cylinder (made from an old tin) that can be pushed into the mud. Water is added to the cylinder, and using a tea strainer, larvae can be removed. One technique is to stir the water in the cylinder when it is half-full and pass the tea strainer through it so that larvae are collected in the mesh. They are then transferred to a tray with clean water and removed with a pipette. Using this method, all stages of larvae are collected. In order to determine survival rates, once counted, the larvae are returned to the puddle, and a collection is made on the following day. One problem with such a sampling technique is that it may artificially prolong the life of the puddle by adding water to it when the cylinder is removed. The technique was adopted in São Tomé.

In some instances, merely locating the habitat is difficult. The use of drones for remote sensing is increasing (see Chapter 5).

Monitoring Resistance to Insecticides

The problem of insecticide resistance and/or provision of a net type that fails to kill mosquitoes is (i) how to distinguish between the two; (ii) how to measure it; and (iii) how to establish a system for retreatment if required

The failure of a net to kill local mosquitoes in bioassay tests may generally be considered to be demonstrations of resistance, but this need not be so. The following cautionary tale from Papua New Guinea was recently recorded by the business magazine *Bloomberg* with the heading 'One business decision made the World's losing battle with malaria even tougher' (Edney and Fay Cortez, 2024). Briefly, a sharp rise in malaria cases from the year 2012 prompted an investigation worthy of a detective novel. The project leader of the Rotarians, an organization that had delivered 20 million nets to all areas of the country, thought that the nets delivered from 2012 were not as effective as earlier nets. Unopened nets from 2007, kept by some villagers, were tested and found to remain effective. Something changed in 2012. The findings were published in *Nature* (Vinit *et al.*, 2020), but it was left at that. A chance meeting at a conference prompted the question: was the binding changed? And lo and behold, in 2012, it was. Since it had not been required information, it was not divulged by the manufacturers. They are a capitalist firm, even if they are doing 'good'. They are driven by the profit motive. Competition is fierce. It is obvious that the company would opt for a cheaper alternative to the 'forever chemical' used for the original binder (and 'pat on the back' you are no longer using a 'forever chemical' in their production). Part of the problem is in the interpretation of the WHOPES guidelines. According to the manufacturer, since it passed the tunnel test, it is WHOPES approved. Nets that are repellent can be tested with the tunnel test (i.e. will mosquitoes fly away). A net may pass the tunnel test

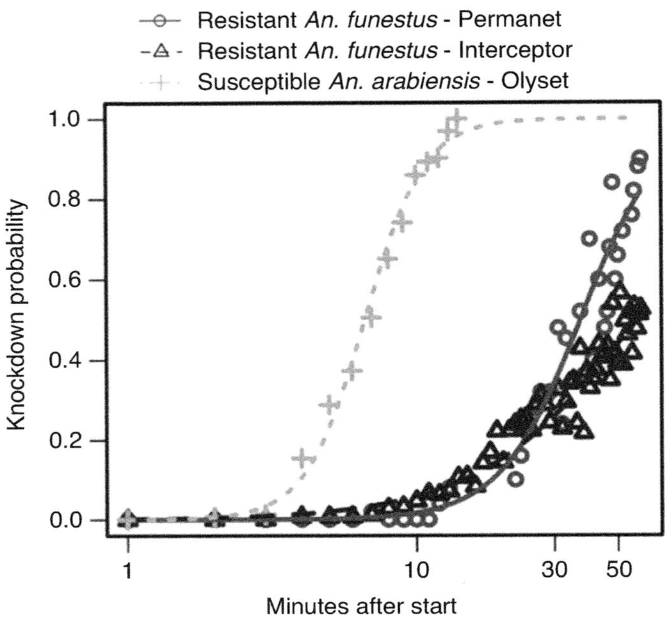

Fig. 4.9. Wireframe cage knockdown times assay of susceptible *An. arabiensis* from Muleba, Tanzania, exposed to a new Olyset net and resistant *An. funestus* from Furvela, Mozambique, exposed to used interceptor nets, which killed susceptible *An. arabiensis,* or a new 'Permanet'. Resistance status based on standard World Health Organization assays. (After Charlwood and Kampango, 2021.).

but fail a standard bioassay (as would seem to be the case in PNG). Standard bioassays themselves have their own problems; the mosquitoes can rest on the plastic cone and so avoid contact with the insecticide, giving a false impression of survival. Both kinds of tests require a considerable degree of expertise. They both have their faults. We propose that the kind of assay described by Charlwood and Kampango (2021) be more widely adopted.

Briefly, a cage is made using a new LLIN over a metal frame and a rubber band. Mosquitoes collected the night before (using one or other of the collection techniques) described by Charlwood *et al.* (2024) are introduced into the cage in batches of 20 (with five as a minimum). Numbers knocked down per minute are recorded for up to an hour (or even an hour and a half), as shown in Fig. 4.9.

There are several advantages associated with this assay over both cone and tunnel tests. There is no plastic to hide on, but there is enough space to indulge in free flight if the sides are repellent. They are performed in the morning using mosquitoes collected the night before. It does not require any particular expertise greater than the ability to count to 20. It is a system that can easily be established over a wide area, with individual householders undertaking such tests as part of a routine entomological surveillance. It provides evidence of increasing resistance levels. It eliminates the controversy of tunnel vs cone bioassay tests. As Nelson Goodman said, 'We aim for simplicity and hope for truth'.

Supposing entomological collections are being made by villagers as described above, then once a month, they might perform an assay to determine if there was a change in resistance status of the mosquito. On the other days of the month villagers might bring their own nets to be assayed (for a fee). An electronic certificate of the result would be issued. Should the net fail (but the mosquito not be resistant), then the net could be retreated – at a women's group. The net would be washed, repaired and retreated. It could then be retested (as part of the retreatment package). The system could run on the cryptocurrency ADA, as described in the chapter The role of Politics in Vector Control (Chapter 14).

With a phone that made a 5 s video every minute that records the sound and provides images of the cage walls data can be sent to a central database, where analysis is performed by computer. This would make it an even more streamlined system. High-risk areas could be flagged by AI.

References

Abong'o, B., Gimnig, J., Longman, B., Odongo, T., Wekesa, C. *et al.* (2021) Comparison of four outdoor mosquito trapping methods as potential replacements for human landing catches in western Kenya. *Parasites and Vectors* 14, 320.

Achee, N.L., Youngblood, L., Bangs, M.J., Lavery, J.V. and James, S. (2015) Considerations for the use of human participants in vector biology research: A tool for investigators and regulators. *Vector Borne Zoonotic Disease* 15, 89–102.

Bidlingmayer, W.L. (1994) How mosquitoes see traps: Role of visual responses. *Journal of the American Mosquito Control Association* 10, 272–279.

Birley, M.H. and Rajagopalan, P.K. (1981) Estimation of the survival and biting rates of *Culex quinquefasciatus* (Diptera, Culicidae). *Journal of Medical Entomology* 18(3), 181–186.

Birley, M.H. and Boorman, J.P.T. (1982) Estimating the survival and biting rates of hematophagous insects, with particular reference to the *culicoides-obsoletus* group (Diptera, Ceratopogonidae) in southern England. *Journal of Animal Ecology* 51(1), 135–148.

Charlwood, J.D. and Galgal, K. (1985) Observations on the biology behaviour of *Armigeres milnensis* lee (Diptera, Culicidade) in Papua New Guinea. *Australian Journal of Entomology* 24, 313–320.

Charlwood, J.D. and Edoh, D. (1996) Larval distribution of members of the *Anopheles gambiae* Giles complex (Diptera,Culicidae), identified by the PCR reaction, in the environs of Ifakara, Tanzania. *Journal of Medical Entomology* 33, 202–204.

Charlwood, J.D. and Kampango, A. (2021) Knockdown times in a simple assay determination of insecticide resistance in malaria vectors. *Transactions of the Royal Society of Tropical Medicine and Hygiene* 115, 1348–1349.

Charlwood, J.D., Birley, M.H., Dagoro, H., Paru, R. and Holmes, P.R. (1985) Assessing survival rates of *Anopheles farauti* (Diptera, culicidae) from Papua New Guinea. *Journal of Animal Ecology* 54, 1003–1016.

Charlwood, J.D., Smith, T., Kihonda, J., Billingsley, P.F. and Takken, W. (1995) Density independent feeding success of malaria vectors in Tanzania. *Bulletin of Entomological Research* 85, 29–35.

Charlwood, J.D., Vij, R. and Billingsley, P.F. (1999) Dry season refugia of malaria-transmitting mosquitoes in a dry savannah zone of East Africa. *American Journal of Tropical Medicine and Hygiene* 62, 726–732.

Charlwood, J.D., Tomás, E.V.E., Salgueiro, P., Egyir-Yawson, A., Pitts, R.J. *et al.* (2011) Studies on the behaviour of peridomestic and endophagic M form *Anopheles gambiae* from a rice growing area of Ghana. *Bulletin of Entomological Research* 101, 533–539.

Charlwood, J.D., Macia, G.A., Manhaca, M., Sousa, B. and Cuamba, N. (2013) Population dynamics and spatial structure of human-biting mosquitoes, inside and outside of houses, in the Chockwe irrigation scheme, southern Mozambique. *Geospatial Health* 7, 309–320.

Charlwood, J.D., Rowland, M., Protopopoff, N. and LeClair, C. (2017) The Furvela tent-trap Mk 1.1 for the collection of, outdoor biting mosquitoes. *PeerJ* 5, e3848. DOI: 10.7717/peerj.3848.

Charlwood, J.D., Protopopoff, N., Rowland, M. and LeClair, C. (2018) Studies on the resting and feeding behaviour of *Anopheles gambiae* and An. *Arabiensis* from Muleba District, North-eastern Tanzania. *Medical and Veterinary Entomology* 32, 263–270. DOI: 10.1111/mve.12299.

Charlwood, J.D., Kidane, A., Asfaha, Y.E., Weldu, L.T., Petros, F. *et al.* (2021) Novel sampling methods for monitoring *Anopheles arabiensis* from Eritrea. PeerJ 9:e11497. DOI: 10.7717/peerj.11497.

Charlwood, J.D., Smith, T.A., Kampango, A., Tomas, E.V.E. and Chitnis, N. (2023) Time series analysis of survival and oviposition cycle duration of *Anopheles funestus* (Giles) in Mozambique. *PeerJ* 11, e15230. DOI: 10.7717/peerj.15230.

Charlwood, J.D., Kampango, A., Omari, M. and Okumu, F. (2024) Simple techniques for a complex problem: Sampling malaria vectors in Africa. *Acta Tropica* 251, 107104.

Cooke, M.K., Kahindi, S.C., Oriango, R.M., Owaga C. *et al.* (2015) 'A bite before bed': exposure to malaria vectors outside the times of net use in the highlands of western Kenya. *Malaria Journal* 14, 1.

Costantini, C., Sagnon, N.F., Sanogo, E., Merzagora, L. and Coluzzi, M. (1998) Relationship to human biting collections and influence of light and bednet in CDC light-trap catches of West African malaria vectors. *Bulletin of Entomological Research* 88(5), 503–511. DOI: 10.1017/S000748530002602X.

Dambach, P., Schleicher, M., Korir, P., Ouedraogo, S., Dambach, J. *et al.* (2018) Nightly biting cycles of *Anopheles* species in rural Northwestern Burkina Faso. *Journal of Medical Entomology* 55(4), 1027–1034. DOI: 10.1093/jme/tjy043.

Edney, A. and Fay Cortez, M. (2024) One Business Decision Made the World's Losing Battle With Malaria Even Tougher. Bloomberg. Available at: https://www.bloomberg.com/news/features/2024-02-22/papua-new-guinea-s-losing-fight-against-malaria-is-a-lesson-for-the-world (accessed 23 September 2024).

Foley, D.H., Wilkerson, R.C. and Rueda, L.M. (2009) Importance of the "what," "when," and "where" of mosquito collection events. *Journal of Medical Entomology* 46(4), 717–722. DOI: 10.1603/033.046.0401.

Fornadel, C.M., Norris, L.C. and Norris, D.E. (2010) Centers for disease control light-traps for monitoring *Anopheles arabiensis* human biting rates in an area with low vector density and high insecticide-treated bed net use. *American Journal of Tropical Medicine and Hygiene* 83, 838–842.

Gillies, M.T. (1974) Methods for assessing the density and survival of blood-sucking Diptera. *Annual Review of Entomology* 19, 345–362.

Homan, T., Hiscox, A., Mweresa, C.K., Masiga, D., Mukabana, W.R. *et al.* (2016) The effect of mass mosquito trapping on malaria transmission and disease burden (solarmal): A stepped-wedge cluster-randomised trial. *Lancet* 388, 1193–1201.

Huho, B.J., Gimnig, J.E., Bayoh, N., Seyoum, A., Sikalaa, C.H. *et al.* (2015) Applications and limitations of centers for disease control and prevention miniature light-traps for measuring biting densities of African malaria vector populations: A pooled-analysis of 13 comparisons with human landing catches. *Malaria Journal* 14, 247.

Kabbale, F.G., Akol, A.M., Kaddu, J.B. and Onapa, A.W. (2013) Biting patterns and seasonality of *Anopheles gambiae* sensu lato and *Anopheles funestus* mosquitoes in Kamuli district, Uganda. *Parasites Vectors* 6, 340.

Killeen, G.F., Kihonda, J., Lyimo, E., Oketch, F.R., Kotas, M.E. *et al.* (2006) Quantifying behavioural interactions between humans and mosquitoes: Evaluating the protective efficacy of insecticidal nets against malaria transmission in rural Tanzania. *BMC Infectious Diseases* 6, 161.

Maia, M.F., Robinson, A., John, A., Mgando, J., Simfukwe, E. *et al.* (2011) Comparison of the CDC backpack aspirator and the prokopack aspirator for sampling indoor- and outdoor-resting mosquitoes in southern Tanzania. *Parasites and Vectors* 4, 124.

Mbare, O., Njoroge, M.M., Ong'wen, F., Bukhari, T. and Fillinger, U. (2023) Evaluation of the solar-powered Silver bullet 2.1 (lumin 8) light-trap for sampling malaria vectors in western Kenya. *Malaria Journal* 22, 277. DOI: 10.1186/s12936-023-04707-y.

Muirhead-Thomson, R.C. (1947) The effects of house spraying with pyrethrum and with DDT on *Anopheles gambiae* and *A. melas* in west Africa bull. *Entomological Research* 38, 449–464.

Müller, G.C., Junnila, A., Traore, M.M., Revay, E.E., Traore, S.F. *et al.* (2017) A novel window entry/exit trap for the study of endophilic behavior of mosquitoes. *Acta Tropica* 167, 137–141.

Njiru, B.N., Mukabana, W.R., Takken, W. and Knols, B.G.J. (2006) Trapping of the malaria vector *Anopheles gambiae* with odour-baited MM-X traps in semi-field conditions in western Kenya. *Malaria Journal* 5, 39.

Silver, J.B. (2008) *Mosquito Ecology*, 3rd edn. Springer, Dordrecht, Netherlands, p. 1477.

van de Straat, B., Russell, T.L., Staunton, K.M., Sinka, M.E. and Burkot, T.R. (2021) A global assessment of surveillance methods for dominant malaria vectors. *Scientific Reports* 11, 15337. DOI: 10.1038/s41598-021-94656-w.

Vazquez-Prokopec, G. (2009) A new, cost-effective, battery-powered aspirator for adult mosquito collections. *Journal of Medical Entomology* 46(6), 1256–1259.

Verdonschot, P.F.M. and Besse-Lototoskaya, A.A. (2014) Flight distance of mosquitoes (Culicidae): A metadata analysis to support the management of barrier zones around rewetted and newly constructed wetlands. *Limnologica* 45, 69–79.

Vinit, R., Timinao, L., Bubun, N., Katusele, M., Robinson, L.J. *et al.* (2020) Decreased bioefficacy of long-lasting insecticidal nets and the resurgence of malaria in Papua New Guinea. *Nature Communications* 11, 3646.

World Health Organization (2017) *Ethical Issues Associated with Vector-Borne Diseases*. World Health Organization, Geneva, Switzerland.

5

The Use of Drones for Mosquito Surveillance and Control

GABRIEL CARRASCO-ESCOBAR[1,2*], MARTA MORENO[3], KIMBERLY FORNACE[3], MANUELA HERRERA-VARELA[4], EDGAR MANRIQUE[1] AND JAN E. CONN[5,6]

[1]InnovaLab, Institute of Tropical Medicine 'Alexander von Humboldt', Universidad Peruana Cayetano Heredia, Lima, Peru
[2]School of Public Health, University of California San Diego, San Diego, CA, USA
[3]Faculty of Infectious and Tropical Diseases and Centre for Climate Change and Planetary Health, London School of Hygiene and Tropical Medicine, London, UK
[4]Grupo de Investigación en Entomología, Facultad de Medicina, Universidad Nacional de Colombia, Bogotá, Colombia
[5]The Wadsworth Center, New York State Department of Health, Albany, New York, USA
[6]Department of Biomedical Sciences, School of Public Health, State University of New York at Albany, Albany, NY, USA

General Background

New tools for mosquito control and rationale

One of the main drivers of the need for new, more effective, mosquito control is insecticide resistance across multiple vector species of *Aedes*, *Anopheles* and *Culex*. For malaria, an increase in residual (exophagic) transmission together with ecological heterogeneity in everything from weather to local human migration and housing to mosquito species' behaviours presents many challenges (Sherrard-Smith *et al.*, 2019). Higher insecticide resistance (Ranson

*Corresponding author: gabriel.carrasco@upch.pe

DOI: 10.1079/9781789248845.0005

and Lissenden, 2016) is particularly ironic given that a major cause in many countries has been the intensive roll-out of Long-Lasting Insecticidal Nets (LLINs), sometimes combined with *indoor residual spraying (IRS)* that reduced malaria cases dramatically from 2000 to 2015 (Bhatt *et al.*, 2015; World Health Organization, 2020).

In recent years, global health security has been threatened by the geographical expansion of arboviruses such as dengue, Yellow Fever and emergent viruses, such as Zika and chikungunya (Brady and Hay, 2020). The increased range of competent *Aedes* mosquitoes worldwide has been linked to global travel, urbanization modifications and environmental changes (Bhatt *et al.*, 2015; Messina *et al.*, 2015; Kraemer *et al.*, 2019). Resistance in *Aedes* (Moyes *et al.*, 2017; Valle *et al.*, 2019) and *Culex* is prevalent in insecticides employed both for larvae and adult control, compromising effective control of dengue and other arboviruses (Grisales *et al.*, 2013) and lymphatic filariasis (Yadouléton *et al.*, 2015; Lopes *et al.*, 2019). The global landscape of mosquito-borne diseases and disease dynamics fluctuates frequently, and there could be serious public health consequences in the success of mosquito surveillance and control programmes if the issues of insecticide resistance and outdoor transmission are not adequately addressed.

A major complication of residual transmission in malaria-endemic regions is locating the species and populations of questing and resting vectors that are outside houses (Durnez and Coosemans, 2013). Those that feed and/or rest in the peridomestic area, including in and around animal enclosures, may be vulnerable to interventions, such as insecticide-treated barrier screens (with the obvious caveat of local resistance), spatial repellents or endectocide treatment of domestic animals (Killeen *et al.*, 2017). Depending on local/regional ecology and human behaviour, it may be effective to detect and target aquatic stages. For controlling both aquatic and adult stages, *remote-sensing* data have been used together with predictive modelling for risk, incidence and detection of transmission hotspots and landscape profiles in relation to mosquito-borne pathogens for several years (Palaniyandi, 2012; Viana *et al.*, 2017). Treatment of transmission hotspots can be effective in reducing, possibly in eliminating arthropod-borne pathogens, but in low transmission settings, targeting the household level (the smallest unit), or even the community, may not translate to nearby non-targeted areas (Stresman *et al.*, 2019).

This chapter outlines opportunities and challenges for integrating drones into vector surveillance and intervention strategies across the mosquito life cycle (Fig. 5.1). We present a five-step systematic environmental mapping strategy that we recommend be undertaken in locations where a drone is expected to be used, outline the key considerations for incorporating drone or other Earth Observation data into vector surveillance and provide two case studies of the advantages of using drones equipped with multispectral cameras: (i) to locate anopheline vector breeding sites in Amazonian Peru; and (ii) to map fine-scale deforestation or other environmental modification that affect wildlife movements.

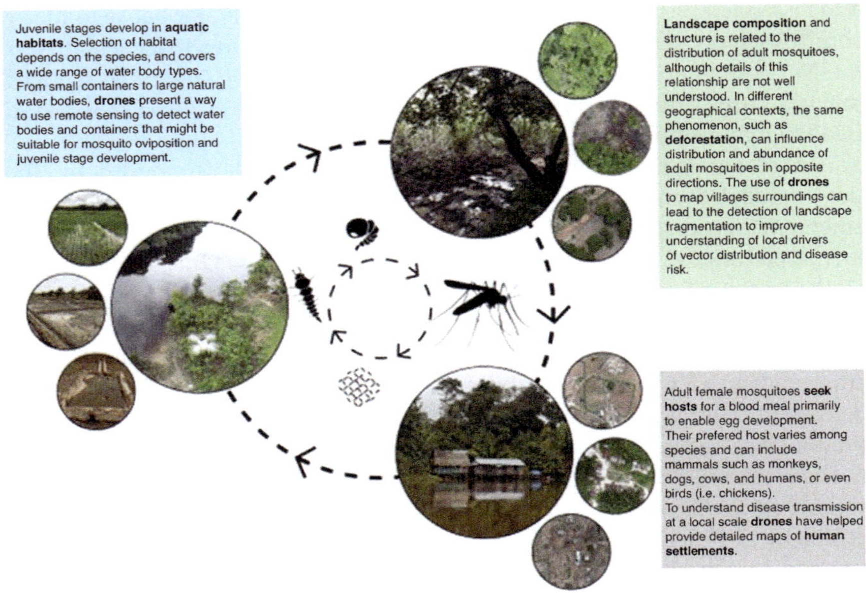

Fig. 5.1. To understand the various stages at which drones can be used in surveillance and control of vector-borne pathogens, it is important to follow the vector life cycle and relate it to the environment wherein mosquitoes develop. Juvenile stages develop in aquatic habitats; in contrast, adult mosquitoes are broadly dispersed in the landscape. Drones are being used to produce high-resolution maps of these landscapes and to assist surveillance and control activities in the field.

We describe in some detail a series of pre-flight considerations that include local and regional drone regulations, privacy, safety, community engagement and feasibility. To illustrate the range of practical applications, we discuss examples of mapping juvenile aedine breeding containers and preliminary ongoing intervention studies to enable the use of drones to disperse biological control agents, chemical larvicides, *Wolbachia*-carrying mosquitoes or sterile males. Included at the end of the chapter is a glossary. We encourage serious consideration of high-resolution drone imagery data using specific sensors that can be incorporated into predictive distribution models and environmental suitability analysis to assess the risk of establishment and spread of mosquito-borne pathogens.

Water bodies and breeding site detection

Aquatic stages of mosquitoes are amenable to the use of *drones (also known as Unmanned aerial vehicle – UAV)* for identification, surveillance and treatment. Generally, container breeders such as *Aedes aegypti* and *Aedes albopictus* oviposit in many kinds of water-filled receptacles (Russell *et al.*, 2001; Fonseca *et al.*, 2013), whereas non-container breeders (e.g. *Culex* species) are associated with stagnant water rich in organic matter (Lopes *et al.*, 2019), and *Anopheles* choose natural (ponds, puddles, river edges, swamps) or anthropogenic water bodies such as fish ponds, irrigated rice fields and mining ponds. Preferred aquatic site profiles have been established for many individual vector species (e.g. *Anopheles dirus*, *Culex quinquefasciatus* and *Ae. albopictus*) that are essential for both *ground truthing* and training remote-sensing tools to enable detection of specific types of breeding sites.

Commercial drones have been used for surveying potential breeding sites in contrasting environments (rice fields, ponds and urban and peri-urban areas in Zanzibar, Tanzania), as demonstrated for malaria vectors by Hardy *et al.* (2017). In tidal marsh in California, USA, Haas-Stapleton *et al.* (2019) used a drone fitted with a multispectral camera to map accumulated surface water with the objective of prioritizing sites for larval stage mosquito inspection and to improve drainage ditches to reduce mosquito oviposition. This study also tested the capacity of a drone with a high-magnification camera to visualize immature mosquito stages of *Culex pipiens* from varying heights and colour-contrasted containers, demonstrating its potential to estimate mosquito abundance that would reduce work effort in mosquito habitat ground surveys. In Amazonian Peru, water bodies positive for the presence of the malaria vector *Nyssorhynchus darlingi* mosquito larvae were identified with an overall classification accuracy >85% by drones fitted with multispectral cameras ((Carrasco-Escobar *et al.*, 2019); see Case Study 1). In a challenging intertidal coastal saltmarsh landscape in South Australia, researchers were able to map areas of shallow inundation among hummocks of low vegetation, establishing a useful reflectance threshold with an overall classification accuracy of 80% to detect putative breeding sites with a minimum depth of 10 cm of local *Aedes* vectors of Ross River Virus (Sarira *et al.*, 2020). A brief description of the workflow for drone survey is presented in Fig. 5.2.

Spatial distribution of adult stages and abundance

Mosquitoes, like many organisms, are aggregated by available resources (e.g. resting and oviposition sites, hosts and sugar sources) within the landscape of their geographical range (Smith *et al.*, 2004; Ellis, 2008). Reisen (2010) recognized specific habitat characteristics of importance in predicting mosquito abundance and species composition. Vegetation indices, land use, canopy cover, elevation and hydrology can have significant effects on the availability of aquatic habitats suitable for immature stages or resting areas for adult mosquitoes. Heterogeneous micro-environmental factors associated with host availability around houses can also have a significant influence on the spatial distribution of adult malaria vectors (Kaindoa *et al.*, 2016). For instance, the number of people sleeping in the dwelling, housebuilding materials, availability of aquatic habitats and distance to them can all shape

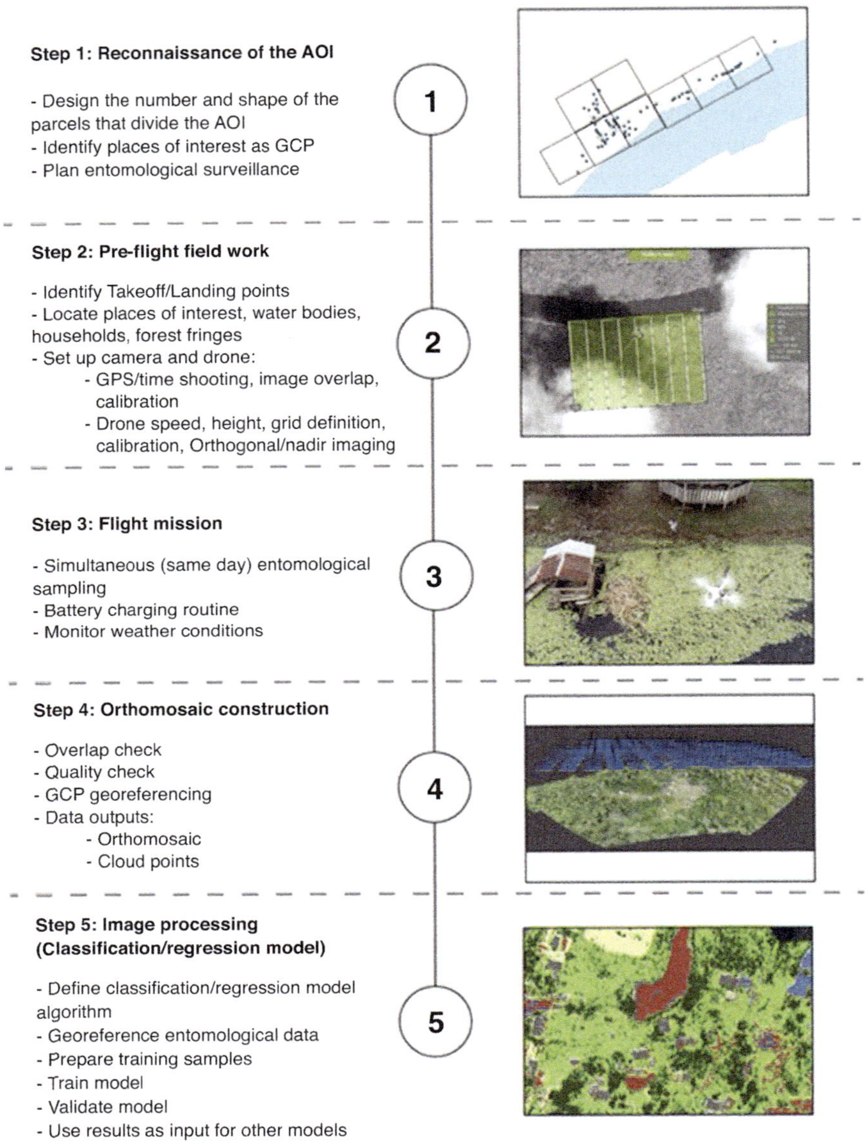

Fig. 5.2. To successfully map the environment where mosquitoes develop, it is important to do it in a systematic way, to guarantee repetition on subsequent sampling campaigns and to achieve the proposed goals of the survey. Steps 1–5 show an example of workflow that is used to map the aquatic habitats of anopheline mosquitoes.

anopheline distribution. In other mosquito species – *Culex territans* and *Culex peccator* – the distribution of host animals within the landscape is of paramount importance to explain mosquito spatial distribution (Burkett-Cadena *et al.*, 2013).

Characterization of land cover/use with remote-sensing data proved to be practical to estimate the diversity and abundance of adult *Anopheles* for the study of mosquito habitats at a local scale and subsequently to design appropriate sampling strategies that represent the heterogeneous environment in French Guiana (Roux *et al.*, 2013). More sophisticated analysis using light detection and ranging (LiDAR) technology has been able to detect and classify fine-scale and temporal distribution of malaria vectors by the difference in their wingbeat frequencies compared with those of other insects and between male and female mosquitoes (Brydegaard *et al.*, 2020).

Pre-Flight Considerations

Despite the utility of drones for vector surveillance and control, programmes need to consider several relevant aspects: local or regional regulations, safety, privacy and ethics, community acceptance and – as a final reality check – feasibility.

Drone regulations

Prior to flying a drone, the first step is to review and understand local regulations for the surveillance location proposed. Authorizations for equipment, pilot and activities that will be conducted with the drone, as well as type and characteristics of the drone (i.e. weight or number of rotors) vary between countries (Stöcker *et al.*, 2017). Generally, countries have been slow to establish national drone regulations, hindering widespread utilization of this technology (Stöcker *et al.*, 2017). Although a comprehensive analysis of drone regulations has been published (Stöcker *et al.*, 2017), both the technology and its regulation continue to change rapidly.

Safety

As popularity and volume of operations increase, concerns have been raised about drone safety and security. In contrast with other aerial vehicles, the only subjects at potential risk of a crash accident are people on the ground. Hence, new frameworks and strategies have been developed and are continuously being updated. Overall, the field is changing from crash-free to safe-to-crash operations addressed by software- and hardware-based solutions (Afman *et al.*, 2018). This is of particular importance for drones operating above populated areas. As will be discussed below, drones could be used to monitor water bodies as potential sites for mosquito oviposition (an example of an unpopulated scenario), as well as risk mapping of dwellings located along an environmental gradient (populated scenario). In this context, detect-and-avoid systems have received particular attention (Shakhatreh *et al.*, 2019). Pham *et al.* (2015) and Shakhatreh *et al.* (2019) reviewed major categories of collision avoidance methods that can be summarized as geometric, path planning, potential field approaches and vision-based approaches. Future scenarios that involve the simultaneous use of multiple drones require further detect-and-avoid strategies called swarm intelligence algorithms (Varela *et al.*, 2011) that are able to combine data from diverse sources including location and weather sensors, accelerometers, gyroscopes, RADARs and LiDAR (Shakhatreh *et al.*, 2019).

Privacy and ethical issues

Valid privacy concerns have emerged from the use of drones, particularly by the unprecedented spatial and temporal resolution at which data collection can be conducted. Recent sensors mounted on drones can photograph at sub-centimetre resolution in comparison to the 5- to 10 cm resolution from private or military satellites, or 5–10m from freely available satellite images (Case *et al.*, 2020). In many countries, there are legal requirements for data collection from private property using drones. However, in rural areas or marginalized populations (where most tropical diseases are transmitted), clear distinctions of private property are often lacking (Finn and Wright, 2012). Drones can (intentionally or unintentionally) collect data about private information such as location, behaviour, body characteristics, use of space or indirectly about income, habits or family composition, raising ethical concerns about restriction of freedom and privacy violation (Finn and Wright, 2012, Finn and Wright, 2016).

Community acceptance

In rural settings, besides authorization to fly drones, an important aspect to consider is the community acceptance of such activities. A difference with other remote-sensing instruments (such as satellites) is that drones fly in the aeronautic space (thus, national aeronautic authorization is required) and are perceivable by the population in the study area (Iskander, 2017). Even though the *area of interest (AOI)* of the flight plan cannot include dwellings, drone operations may be perceived as a violation of privacy, a concern that must be recognized by the research team (Peckham and Sinha, 2019). Prior to data collection, community meetings should be held to engage the local population and inform people about study objectives, benefits and the best practices that will be conducted (i.e. blurring personal information (like licence plates and faces), displacement, jittering and micro-spatial aggregation) to preserve privacy (Finn and Donovan, 2016; Case *et al.*, 2020). In urban areas, other aspects may include noise, design or patterns of movement of the device (Chang *et al.*, 2017; Eißfeldt *et al.*, 2020).

Feasibility and equipment

Considering the large variety of drone models (Aragão *et al.*, 2020) and specifications, it is important to consider the feasibility of the intended applications (i.e. mapping, surveillance and control). The two main drone types are multi-rotor and fixed wing. Multi-rotor drones are suitable for capturing stable images because of their great control over position and tilt; however, the energy consumption of their rotors impacts the flight time and speed and limits their use for long-distance inspection and monitoring over extended periods. On the other hand, the aerodynamic design (similar to airplanes) of fixed-wing drones provides efficient use of energy for longer flight time and greater speed. The main limitation of fixed-wing drones is the requirement of large, clear areas for take-off and landing (in contrast to multi-rotor drones, able to take-off and land vertically). In settings with dense forest cover, such as the Amazon or South-east Asia (Fornace *et al.*, 2014; Carrasco-Escobar *et al.*, 2019), multi-rotor drones might be suitable, and in areas such as open savannah, fixed wing might be most appropriate. Alongside the selection of drones is the cost of the equipment and sensors. Commercial multi-rotor drones range around US$1000–10,000 and fixed-wing $5000–20,000, commonly equipped with RGB cameras. It is important to consider the additional cost of supplementary sensors such as multispectral, thermal or LiDAR that range around $5000–20,000. Additional costs include spare parts for field surveys such as controllers, extra batteries and propellers, landing paths, data storage, charging supplements, repair kits, mounting kits and packaging, as well as costs for personnel, field survey costs, and analysis and computing requirements. While drones and sensors are becoming increasingly affordable, drone survey plans should include a budget that covers other associated costs.

Integrating Drones Into Vector Surveillance

Surveillance is defined as 'the continuous and systematic collection, analysis and interpretation of disease-specific data and the use of those data in the planning, implementation and evaluation of public health practice' and has been recognized by the World Health Organization as a core intervention for vector-borne disease control (World Health Organization, 2018).

Surveillance

A key surveillance priority is monitoring the distribution and abundance of local vector species to evaluate potential impacts on disease transmission. This requires keeping track of vector habitats and assessing changes, which may increase or decrease *vectorial capacity* directly (e.g. landscape change that creates suitable habitat) or indirectly (e.g. increasing availability of blood meals, resting sites or predation) or leads to ecological shifts in biting behaviour or predominant vector species (Ferguson *et al.*, 2010). Entomological surveillance is essential for monitoring and evaluating interventions, identifying priority areas for further surveillance, and assessing vector-borne transmission potential and whether migration of infected people or vectors into an area, or replacement of one vector species by another, is likely to lead to disease outbreaks. Generating accurate maps of mosquito breeding sites might also provide an opportunity to use them as information, education and communication materials to display specific risk conditions for a given locality (World Health Organization, 2018).

Earth Observation

Earth observation (EO) is the collection of information about the physical characteristics of Earth's surfaces using remote-sensing technology, such as satellites, manned aircraft or drones. *Raster data* from EO are used frequently to characterize vector habitats and monitor changing risks (Parselia *et al.*, 2019). EO uses sensors that measure reflectance or emitted radiation from a distance, with wavelength interval size and number of wavelengths measured on the electromagnetic spectrum defining the *spectral resolution* of the sensor (National Aeronautics and Space Administration, n.d.-a). These sensors are also characterized by *radiometric resolution*, the sensitivity of the sensor to measure differences in reflected or emitted energy. While RGB sensors measure only the visible range of the electromagnetic spectrum (similar to standard cameras), sensors with higher spectral resolution increase the capacity to detect specific habitat types and allow transformations, such as for vegetation indices. Additionally, EO data are characterized by *spatial resolution*, the size of a pixel within the image and *temporal resolution*, the frequency at which the data are collected. While satellite data often have much higher spectral resolution than data collected by drones, spatial resolution of satellite data is typically more constrained, because freely available data are limited to 20 m/pixel resolution or greater; higher resolution data are often prohibitively expensive. In contrast, typical aerial imagery collected by drones has less than 10 cm/pixel resolution. Additionally, drones offer distinct advantages for temporal resolution, as the user is able to define flight timing, whereas satellite data are only available when a satellite is passing over a specific location and may be further limited by cloud cover obscuring the target site (Fornace *et al.*, 2014). Despite these advantages, drones cover much smaller geographical areas than satellite imagery and are frequently best used to complement existing sources of EO data (Fig. 5.3).

Data sources and analysis methods to be used should be determined by the features to be identified and the resolution required for surveillance (e.g. direct observation of larval breeding sites at set time intervals for targeted larval source management, LSM). In other cases, the objective may be monitoring landscape changes or estimating numbers of people at risk or potential disease hosts. Based on these features and the technical capacity available,

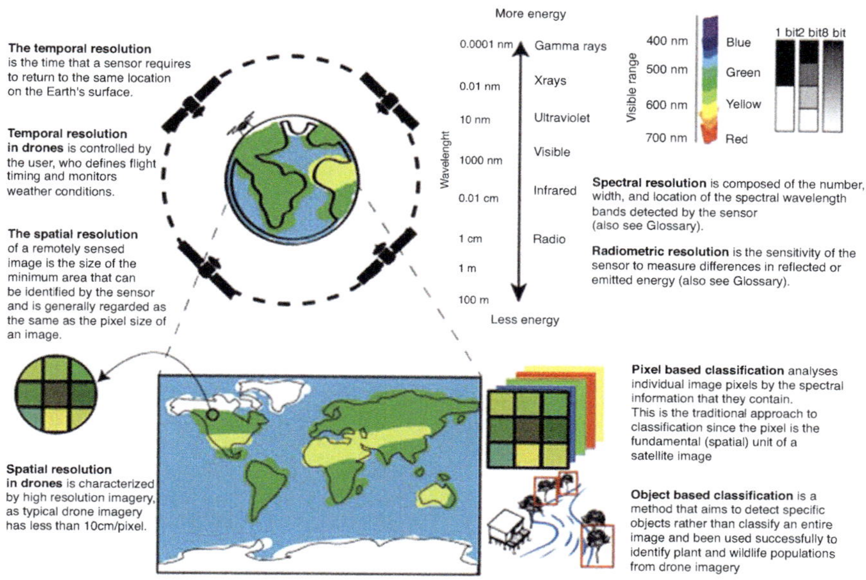

Fig. 5.3. Frequently used remote-sensing definitions.

a specific approach to analyse drone data should be determined. This can include:

- **Manual digitization:** The simplest approach that relies on visual identification of key features; this has been shown to be effective for pinpointing malaria vector breeding sites in Tanzania (Hardy *et al.*, 2017).
- **Unsupervised classification**: These methods group features within an image into various categories without the use of training data to define classes of interest and are used less frequently alone, as they require the user to determine subsequently what the categories represent.
- **Supervised classification:** Methods that use training data to fit a model classifying the imagery into predefined categories of interest. Training data may include specific categories or use ground-based data for imagery analysis. An example of pixel-based supervised classification (Case Study 1) is found in the following section.
- **Object detection:** These methods aim to detect specific objects such as plants and wildlife rather than to classify an entire image (Gray *et al.*, 2019; Kattenborn *et al.*, 2019).

While a wide range of different models can be used for all approaches, *machine-learning methods* are used increasingly for image analysis. These can be implemented in a range of open-source and commercially available software, with additional cloud-based computing power available through online platforms such as Google Earth Engine and Amazon Earth.

The features to be identified and the analytical approach should determine selection of the drone sensor. Commercially available drone sensors include *RGB*, *multispectral* and *thermal cameras* and increasingly a range of active sensors, such as 3D laser scanning (i.e. LiDAR) (Fig. 5.4). To collect data, the drone and sensor are flown over the AOI, typically in a grid-like configuration to allow collection of overlapping images. Using commercially available or open-source software, these images can then be processed to generate spatially referenced *orthomosaics*, images which have been stitched together and *ortho-rectified*. Additionally, photogrammetric methods (or Structure-from-Motion) can generate digital surface models and 3D representations of landscapes. These data can then be used as inputs into image analysis pathways or input directly into GIS software for viewing.

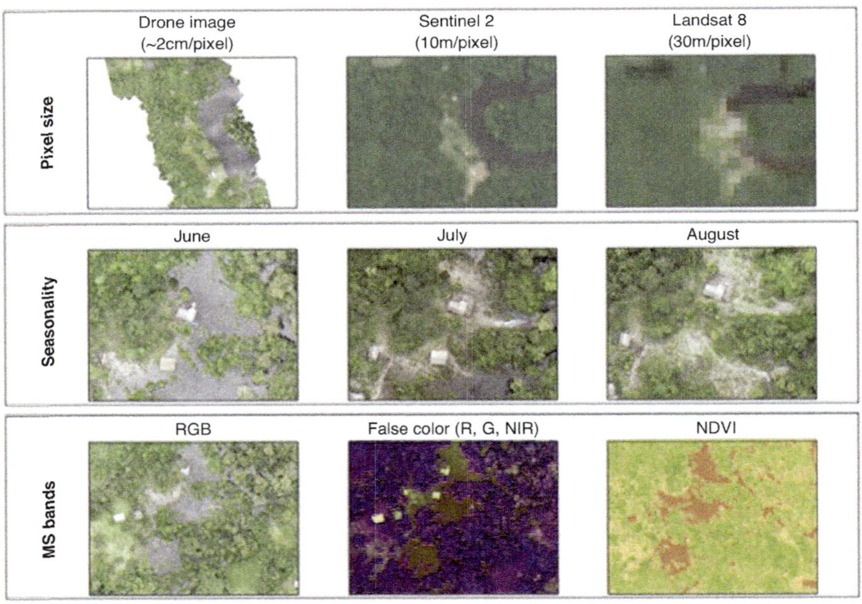

Fig. 5.4. Drones are advantageous in their ability to detect small-size features, as they produce high-resolution imagery at the sub-metre level. This is particularly important in vector-borne studies because water bodies/containers suitable for mosquito breeding are frequently small. The left panel shows a comparison of the pixel size produced by a drone in contrast to two commonly used freely available satellite imageries. The middle panel demonstrates changes in the landscape composition across a two-month window, captured by drones, that otherwise might be overlooked by satellite images. The right panel shows common bands that are available in the sensor often used in drones. RBG cameras (top), multispectral cameras that have near-infrared (NIR) and edge red bands (middle), and an NDVI composite using red and NIR bands (bottom).

Mapping juvenile aquatic habitats

Aquatic habitats used by malaria vectors have a spatially heterogeneous distribution controlled mostly by hydrological and geomorphological processes (Hardy *et al.*, 2015), together with regulated LU/LC (Minakawa *et al.*, 2005; Munga *et al.*, 2009). This results in a patchy distribution of water bodies – often spatially organized in clusters (Killeen *et al.*, 2002) – and is strongly correlated with rainfall that can dictate the temporal dynamics of anopheline larval habitats and adult abundance (Hii *et al.*, 1997; Minakawa *et al.*, 1999; Mutuku *et al.*, 2006). Mapping the spatial distribution of aquatic habitats preferred by mosquitoes is crucial to the successful deployment of different control and surveillance practices. Also, targeting the most productive habitats could increase the cost-effectiveness of mosquito larval control (Tusting *et al.*, 2013; World Health Organization, 2013).

Remote sensing and geospatial mapping are realistic approaches to identify larval habitats that cover large areas, overcoming some operational challenges of conventional ground-based surveys (Hayes *et al.*, 1985; Hay *et al.*, 1998; Minakawa *et al.*, 1999; Zou *et al.*, 2006). The identification of larval habitats using satellite high-spatial resolution may be challenging during the classification process, particularly when water bodies are small and turbidity is high (Mushinzimana *et al.*, 2006; Mutuku *et al.*, 2009). Other examples of the use of remote-sensing technology are malaria risk assessments through the study of anopheline larvae (Castro *et al.*, 2004; Sithiprasasna *et al.*, 2005). Recently, object-oriented machine-learning classification has been explored to overcome backscatter similarities when mapping open water or slightly vegetated areas using radar imagery between stretches of open water (Hardy *et al.*, 2019). Proof-of-concept studies have proven the feasibility of using drones to map aquatic larval habitats of African (Hardy *et al.*, 2017) and South American malaria vectors (Carrasco-Escobar *et al.*, 2019). The high-resolution imagery at relatively low cost compared with satellite images, the temporal flexibility to capture images by the drone user and the clear-sky conditions required for optical satellite imagery make this technology an effective operational tool for the identification of potential habitats (Fornace *et al.*, 2014) (Case Study 2; Fig. 5.5).

The proliferation of larval habitats of anopheline malaria vectors has been also associated with human activities, such as agriculture, settlement or other landscape alterations that incorporate deforestation and vegetation clearance (Coluzzi, 1994; Vittor *et al.*, 2006; Dos Reis *et al.*, 2015; Hawkes *et al.*, 2019). More frequent surveillance via drones will allow for the detection and subsequent development of quick responses to rapid changes and environmental alterations, such as urbanization, or changes in the use of rural-agricultural areas that might lead to new aquatic habitats favoured by certain mosquito species.

Aedes-borne disease surveillance programmes (e.g. dengue, chikungunya or Zika) commonly use periodic household inspections for

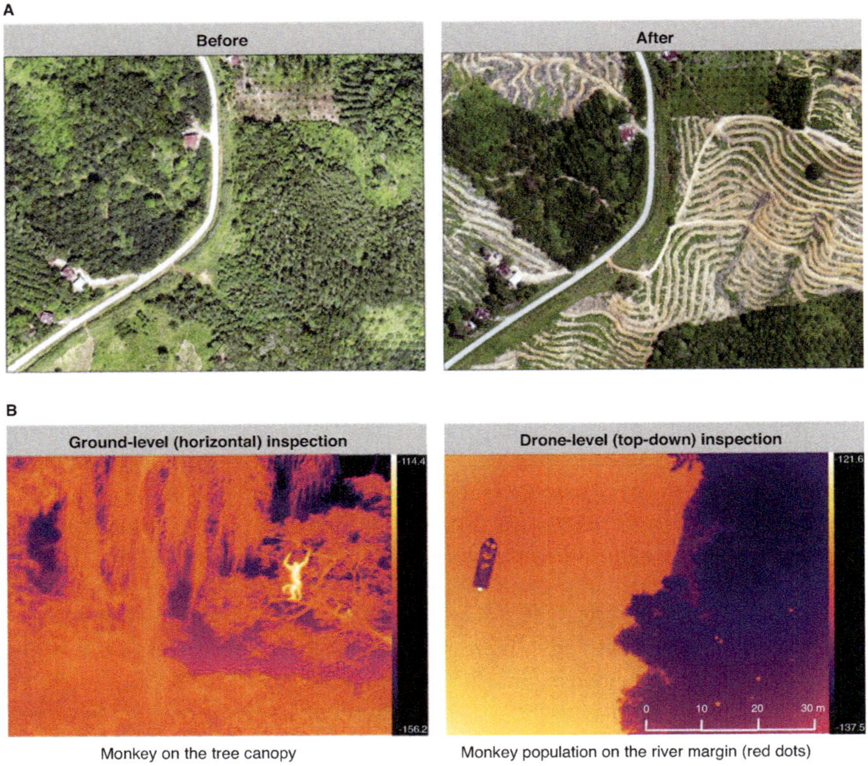

Fig. 5.5. A) Deforestation disrupts wildlife habitats and can bring human, mosquito and macaque populations in closer proximity, increasing the potential for disease transmission. Monitoring wildlife populations is essential for understanding disease dynamics in these changing landscapes. B) Inspection with thermal cameras.

detecting water-holding containers positive for *Aedes* larvae (World Health Organization, 2009; World Health Organization. Regional office for south-east, A, 2016). Estimation of *infestation indices*, used to define and inform control actions, is labour intensive for vector intervention and public health staff and can be inaccurate due to variation in searching efforts, householder availability or mosquito egg laying behaviour; they may also be intrusive for residents.

The centimetre scale of *Ae. aegypti* and *Ae. albopictus'* preferred oviposition habitat precludes the possibility of using high-resolution satellite imagery typically used for monitoring large water bodies for malaria vectors (Hardy *et al.*, 2017). Approaches that leverage the high-pixel resolution of drone imagery have been explored to identify and map potential breeding sites characteristic of *Aedes* such as cisterns, pots, tyres or flower pots. Ultimately, these approaches might support specific vector strategies, such as targeted application of larvicides (e.g. in rural areas of Central America against *Ae. aegypti*) (Rti International, n.d) or guide preventative actions and prioritize programmes that focus on controlling this species (e.g. in the city of Cuiabá, Brazil (Whittaker, 2018)).

In urban areas, drone imagery has also been used to locate preferred oviposition sites of *Ae. albopictus* (Case *et al.*, 2020). In a study combining aerial imagery and entomological surveys in New York State, up to 64% of the containers with water identified in ground surveys were detected with high-pixel resolution images and 14% were partially visible from the air. In this suburban setting, detection of containers depended on the location within the property, with difficulty viewing those located on porches and trees, underneath awnings or near walls. Visibility was also affected by common features, such as tree cover and the location of the drone when taking pictures (Case *et al.*, 2020).

Case study 1: Carrasco-Escobar et al., 2019

Carrasco-Escobar *et al.* (2019) used drones mounted with regular RGB and multispectral cameras to collect high-resolution images to determine the spectral signature (characteristics of the light back reflected from the land/water surface) of the most productive breeding sites of the malaria vector *Ny. darlingi* nearest to human habitation in Amazonian Peru. They detected distinctive spectral signatures (overall accuracy 86.73–96.98%) between *Ny. darlingi* positive and negative water bodies using machine-learning techniques. The methodology and analysis were a proof of concept for testing whether LSM can be combined successfully with LLINs and IRS to contribute to the elimination of transmission in malaria hotspots in the Amazon.

Case study 2: Stark et al., 2019, Jumail et al., 2021

In addition to collecting data on mosquito habitats, drones are valuable for monitoring wider environmental factors influencing disease dynamics. Vector-borne diseases such as *Plasmodium knowlesi* and Yellow Fever are maintained by wild non-human primate reservoirs. Drones enable mapping of fine-scale deforestation and other habitat modification that influence movement patterns of wildlife populations (Stark *et al.*, 2019). Specifically, Jumail *et al.* (2021) explored the use of thermal imagery to capture heat signatures of wild animals and estimate, in real time, wild primate populations within changing landscapes in the Lower Kinabatangan Wildlife Sanctuary in Malaysian Borneo (Zhang *et al.*, 2020; Jumail *et al.*, 2021). These methods were validated initially by ground-based surveys prior to deploying aerial thermal surveys, with the objective of developing non-invasive techniques to assess wildlife populations and study disease dynamics.

Vector Control

As discussed in the previous section, drone application in mosquito vector control has been related mainly to *LSM*. These strategies seek to reduce the immature stages of malaria vectors using habitat modification or manipulation, larviciding or biological control (Derua *et al.*, 2019). When implemented well, these strategies reduce the densities of indoor and outdoor biting mosquitoes and help decrease the dependence on insecticides, and therefore help prevent the emergence of insecticide-resistant mosquitoes. However, WHO recommends larviciding only for special cases 'when breeding sites are few, fixed and findable, and where the sites are easy to identify, map and treat' (World Health Organization, 2013). Such landscapes are uncommon not only in Africa but also in other malaria-endemic regions like South-east Asia and South America.

The application of drones in vector control can be divided into two: (i) use of the imagery to retrieve data that can serve as a source of visual inspection carried out before and after interventions, as a communication tool, to detect water bodies and positive breeding sites and to create detailed maps of the mosquito breeding habitats and landscape, and (ii) drones can be used to disperse the intervention (e.g. larvicide in liquid or granular form) (Johnson *et al.*, 2020) and lately as a potential means of deploying genetically modified mosquitoes (Bouyer *et al.*, 2020).

Some attempts have been made to use drones to deploy larvicides by leveraging the development of drones to apply pesticides for agriculture use (Qin *et al.*, 2016), such as the dispersal of a biological control product to control juvenile malaria vectors in irrigated rice fields in East Africa (Knisely, 2019) or the implementation in some mosquito control programmes for treating remote pools of standing water in brackish salt marsh with larvicides (Florida Keys Mosquito Control District). Using drones as deployment devices may overcome the challenges of treating and accessing large areas on foot and make aerial distribution affordable. Also, multispectral imagery collected with commercial drones has helped to predict the operational efficacy of some larvicide formulations against *Aedes vigilax* in complex environments with irregular canopy cover such as mangroves (Johnson *et al.*, 2020).

In drone spraying operations, flight parameters might influence the effectiveness of droplet deposition when using liquid products and might need adjustments according to the larval habitat characteristics (Guo *et al.*, 2019). Flight height and velocity have a remarkable influence on the deposition amount (Chen *et al.*, 2016); high-speed rotation of the rotor and the air (rotor wind field) (Xue *et al.*, 2014); and the downwash airflow may also affect intervention efficiency. Importantly, use of drones in targeted areas could reduce harm to the environment by limiting wildlife disturbance in sensitive habitats (U.S Fish and Wildlife Service, 2012; Ivošević *et al.*, 2015).

Release of adult mosquitoes

Vector control approaches built on sterile insect techniques and *Wolbachia*-based strategies rely on mass rearing and the dispersal of thousands of mosquitoes at a specific frequency, depending on the nature of the strategy (Dyck *et al.*, 2005; World Health Organization, 2012). Critical operational aspects need to be addressed before the deployment and scale up of these interventions in the areas to be treated, such as aerial coverage, operational costs associated with the number of release sites and mosquito transportation (Zheng *et al.*, 2019; Bouyer *et al.*, 2020). Operational programmes can benefit from using drones during the implementation of these interventions and to optimize some of the key technical aspects including remote preparation of the material and then expanding release and dispersion areas.

The World Mosquito Program (WMP) has employed drones to optimize the release of *Wolbachia*-carrying mosquitoes for control of dengue virus into the island of Fiji, South Pacific. There, the development of an aerial release mechanism that can store up to 160,000 mosquitoes and release 200 every 50m has proved to be much faster and more homogeneous than ground releases and also provided better coverage (We Robotics, 2018). In another epidemiological scenario, in the fight against dengue and Zika in Brazil, Bouyer *et al.* (2020) designed a mechanical compartment for dispersing sterile males by using drones. This study demonstrated minimum damage to mosquitoes and similar competitiveness between those released on the ground and by drone. Moreover, this drone-based dispersion modality was estimated to be a 20-fold reduction in SIT (sterile insect technique) cost estimates.

Challenges and Technological Developments

The drone industry is expanding and gaining increasing interest from a wide range of stakeholders for use in civil applications (Simic Milas *et al.*, 2018). As discussed above, important developments in the industry have been focusing on safety improvements. In this section, we detail additional challenges to drone operation, as well as recent technological developments for control and surveillance of vector-borne diseases.

Capacity building: Most countries require operator training for authorization of drone operations. However, there is still no consensus about minimum levels of training ensuring safe flight and ground operations (Szabolcsi, 2016). Despite algorithms that have been developed recently to automate common human tasks related to drone operations, over trust and automation bias in autonomous systems could be extremely dangerous in safety-critical settings (Cummings *et al.*, 2019). Recently, Kucherov *et al.* (2019) proposed and evaluated a training process for specialists in the maintenance and operation of unmanned aerial vehicles; however, this field is still under development.

Availability of training data: Accuracy of supervised image classification to detect breeding sites is largely dependent on the availability of training data (i.e. identified data of the features of interest used to fit predictive models) (Ma *et al.*, 2017; Elmes *et al.*, 2020). Recent initiatives have focused on developing large training data sets of such features including roads, housing or agricultural land types (MLHub, n.d.; Spacenet, n.d.). These data sets are typically labelled using standardized methodologies, such as the Spatiotemporal Asset Catalog (Spatiotemporal Asset Catalog, n.d), enabling utilization by different platforms for various purposes. However, there remains no standard image repository to share mosquito habitat training data, and most projects need to collect their own data.

Energy management: One important challenge of drone missions is the current trade-off between battery weight and time that limit the flight duration. To cover large areas, multiple returns to the charging station are needed, which is potentially problematic for drone operations in rural areas. Current developments on wireless charging (Dunbar *et al.*, 2015; Simic *et al.*, 2015) and lightweight solar powered battery components (Lee *et al.*, 2014) may improve drone operations.

Improved sensors: Multiple sensors such as visible RGB range, visible infrared region, thermal infrared region, visible near-infrared region, infrared laser region (3D camera), LiDAR and sound propagation (SONAR) are now used in drones. New nano- (Chen *et al.*, 2013) and ultra-high-resolution (Jiang and Song, 2015) sensors are in development.

Conclusions

In areas with endemic and residual transmission of vector-borne diseases, effective vector control measures need to be considered for moving towards elimination or, at the very least, dramatic reduction of pathogen transmission. A major consideration in this context for malaria control is targeting outdoor transmission by the identification, surveillance and treatment of aquatic habitats favourable for juvenile vector mosquito stages. Recent developments in drones (or UAVs) can be effective for accurately conducting surveillance, assessment of habitat suitability for larval and/or adult mosquitoes, and interventions. Examples that leverage this technology include high-resolution mapping of water bodies in the endemic areas in Latin America, South-east Asia and Africa. The implementation of this technology for current control activities requires specific considerations in terms of local regulations, safety, privacy and community acceptance. In this rapidly evolving field, emerging technological developments are addressing these concerns.

Glossary

Area of Interest (AOI) / Region of Interest (ROI): The geographic extent used to define a focus area for either a map or database production. The AOI/ROI could be based on an item in an index or a polygon feature.

Drone (also known as Unmanned Aerial Vehicle or uncrewed aerial vehicle – UAV): This is an aircraft without a human pilot on board that can fly autonomously or be controlled remotely. The two main drone types are *fixed-wing* and *multi-rotor drones*.

Earth Observation (EO): Earth observation is the gathering of information about planet Earth's physical, chemical and biological systems via remote-sensing technologies, usually involving satellites carrying imaging devices (National Aeronautics and Space Administration, n.d.-b). Examples include EOSDIS from NASA, European Space Agency, European Earth Observation Programme (Copernicus) and the Global Earth Observation System of Systems (GEOSS).

Ground truthing: The accuracy of remotely sensed or mathematically calculated data based on data measured in the field. In remote sensing, this is especially important to relate image data to real features and materials on the ground. The collection of ground-truthed data enables calibration of remote-sensing data and aids in the interpretation and analysis of what is being sensed.

Ground truthing is especially important in the initial supervised classification of an image. When the identity and location of land cover types are known through a combination of field work, maps and personal experience, these areas are known as training sites. The spectral characteristics of these areas are used to train the remote-sensing software using decision rules for classifying the rest of the image.

Image classification: The process of classifying image pixel data from raw remote sensing into various categories, labels or land cover types.

Indoor Residual Spraying (IRS): A malaria vector control intervention that involves coating the walls and other surfaces of a house with a residual insecticide (World Health Organization, 2015).

Larval Source Management (LSM): This is the management of aquatic habitats (water bodies) that are potential larval habitats for mosquitoes, in order to prevent the completion of development of the immature stages. LSM is only recommended as a supplementary malaria vector control measure and comprises temporal or permanent habitat modification and manipulation, larviciding and/or biological control (Tusting *et al.*, 2013; World Health Organization, 2013).

Long-Lasting Insecticidal Nets (LLINs): Nets treated with an insecticide incorporated into the net fabric, which retains the insecticide activity for up to 3 years, even after repeated washing. Mass distribution of LLINs has been the primary vector control tool for malaria-endemic countries for the last 20 years (World Health Organization, 2020).

Machine (Artificial) Intelligence (AI)/ Deep Learning: Machine learning is a branch of artificial intelligence focused on the study of algorithms which learn from data and automatically improve over time. Algorithms can be used to identify patterns from imagery and are widely used to classify EO data.

Deep learning is a type of algorithm which uses artificial neural networks and mimics the brain to process data and identify complex features.

Multispectral image Camera (MSC): A multispectral image camera captures image data within specific wavelength ranges in narrow bands across the electromagnetic spectrum.

Normalized Difference Vegetation Index (NDVI): NDVI quantifies vegetation and is an indicator of photosynthetically active biomass. It is calculated as a ratio between the absorbed red light (R) and the reflected near-infrared (NIR) values (NIR−R)/(NIR +R).

Orthomosaic: A georeferenced mosaic of overlapped images which includes correction for topographic distortions for each AOI or community.

Orthorectification: Adjustment for the topographical features of the Earth's surface and tilts of aerial or satellite sensors.

Parcels: Basic spatial units used for classification.

Spatial resolution: This is defined by the size of each pixel within a digital image and the area on the Earth's surface represented by that pixel (i.e. represents the area on the ground that each pixel covers). The finer the resolution, the more detail that can be seen.

Spectral resolution: This is the ability of a sensor to discriminate finer wavelengths (i.e. having more and narrower bands). Sensors with 3–10 bands are considered to be multispectral and with hundreds or even thousands of bands hyperspectral. The narrower the range of wavelengths for a given band, the finer the spectral resolution.

Radiometric resolution: This is the amount of information in each pixel or the number of bits representing the energy recorded and each bit records an exponent of power 2. The higher the radiometric resolution, the more values are available to store information and the greater detail and variation in light can be distinguished.

Raster data: Data stored in a grid of cells (or pixels) that contain information that represent a real-world phenomenon (i.e. climate or land cover).

Remote sensing: Process of scanning and detecting physical characteristics of an area from a distance. EO refers to remote sensing specifically used to characterize the Earth.

Remote sensors: Remote-sensing instruments (sensors) are of two primary types according to the source of signal they use to explore the object. *Active sensors* operate with their own source of energy to illuminate the objects they observe. *Passive sensors* detect natural energy (radiation) reflected from the target. Reflected sunlight is the most common source of radiation measured by passive sensors.

Passive sensors include several types of radiometers and spectrometers, and most of them operate in the visible, infrared, thermal infrared and microwave portions of the electromagnetic spectrum. Accelerometer, hyper-spectral radiometer and spectrometer are some examples of passive remote sensors.

The majority of the *active sensors* operate in the microwave portion of the electromagnetic spectrum, which makes them able to penetrate the atmosphere under most conditions. As they do not require sunlight, active sensors function at any time of the day, and they are relatively independent of atmospheric scatterings. Radar (a sensor used for detecting and tracking radio signals) and LiDAR (a light detection and ranging sensor which determines distance to the object by using the speed of light to calculate the distance travelled) are some examples of active sensors (National Aeronautics and Space Administration a,b).

RGB image: Also known as true colour image, these are regular images that use an additive colour model based on red (R), green (G) and blue (B).

Temporal resolution: The temporal resolution corresponds to the frequency of revisits over a site and is dependent on the orbitography of the platform or satellite on which the sensor is hosted, the sensor characteristics and the swath width.

Thermal imaging: Also called infrared thermography, this is imaging based on thermal radiation. The thermal radiation emitted by most objects is in the mid- to far-infrared spectral region.

Vectorial capacity: This is defined as the average number of inoculations with a specified parasite, originating from one case of malaria per unit time that the population would distribute if all the female vectors biting the initial case become infected (Cohuet *et al.*, 2010).

Unmanned Aerial/aircraft System (UAS): This includes a UAV, a ground-based controller and a system of communication between the two elements.

Acknowledgements

Funding for this project 'International collaborative network for the integration, standardization and assessment of the use of drones in malaria vector control strategies' was provided by the Global Challenges Research Fund (GCRF) Networking Grants (GCRFNGR4\1140). Photographs from Peru were taken under the auspices of Tropical Diseases Research (TDR) Contract 201,460,655. We would additionally like to acknowledge Nombre Apollinaire and Amaziasizamoria Jumail for images collected in Burkina Faso and Malaysia as well as the field teams and projects who supported this research. Additional funding support was provided by the CGIAR Research Programme on Agriculture for Nutrition and Health (A4NH; https://a4nh.cgiar.org/). The opinions expressed here belong to the authors and do not necessarily reflect those of A4NH or CGIAR.

References

Afman, J.-P., Ciarletta, L., Feron, E., Franklin, J., Gurriet, T. *et al.* (2018) *Towards a New Paradigm of UAV Safety*. Available at: ArXiv. arxiv.org/abs/1803.09026 (accessed 13 September 2023).

Aragão, F.V., Cavicchioli Zola, F., Nogueira Marinho, L.H., De Genaro Chiroli, D.M., Braghini Junior, A. *et al.* (2020) Choice of unmanned aerial vehicles for identification of mosquito breeding sites. *Geospatial Health* 15, 92–100.

Bhatt, S., Weiss, D.J., Cameron, E., Bisanzio, D., Mappin, B. *et al.* (2015) The effect of malaria control on *Plasmodium falciparum* in Africa between 2000 and 2015. *Nature* 526, 207–211.

Bouyer, J., Yamada, H., Pereira, R., Bourtzis, K. and Vreysen, M.J.B. (2020) Phased conditional approach for mosquito management using sterile insect technique. *Trends in Parasitology* 36, 325–336.

Brady, O.J. and Hay, S.I. (2020) The global expansion of dengue: How *Aedes aegypti* mosquitoes enabled the first pandemic arbovirus. *Annual Review of Entomology* 65, 191–208.

Brydegaard, M., Jansson, S., Malmqvist, E., Mlacha, Y.P., Gebru, A. *et al.* (2020) Lidar reveals activity anomaly of malaria vectors during pan-African eclipse. *Science Advances* 6, eaay5487.

Burkett-Cadena, N.D., Mcclure, C.J.W., Estep, L.K. and Eubanks, M.D. (2013) Hosts or habitats: What drives the spatial distribution of mosquitoes? *Ecosphere* 4, art30.

Carrasco-Escobar, G., Manrique, E., Ruiz-Cabrejos, J., Saavedra, M., Alava, F. *et al.* (2019) High-accuracy detection of malaria vector larval habitats using drone-based multispectral imagery. *PLoS Neglected Tropical Diseases* 13, e0007105.

Case, E., Shragai, T., Harrington, L., Ren, Y., Morreale, S. and Erickson, D. (2020) Evaluation of Unmanned Aerial Vehicles and neural networks for integrated mosquito management of *Aedes albopictus* (Diptera: Culicidae). *Journal of Medical Entomology* 57, 1588–1595.

Castro, M.C., Yamagata, Y., Mtasiwa, D., Tanner, M., Utzinger, J., Keiser, J. and Singer, B.H. (2004) Integrated urban malaria control: A case study in Dar es Salaam, Tanzania. *The American Journal of Tropical Medicine and Hygiene* 71, 103–117.

Chang, V., Chundury, P. and Chetty, M. (2017) Spiders in the sky: User perceptions of drones, privacy, and security. In: Mark, G., Fussell, S., Lampre, C., Schraefel, M.C., Houracade, J.P. *et al.* (eds) *Proceedings of the 2017 CHI Conference on Human Factors in Computing Systems*. Association for Computing Machinery, Denver, Colorado, pp. 6765–6776.

Chen, H., Xi, N., Song, B., Chen, L., Zhao, J., Lai, K.W.C. and Yang, R. (2013) Infrared camera using a single nano-photodetector. *IEEE Sensors Journal* 13, 949–958.

Chen, S.D., Lan, Y.B., Li, J.Y., Zhou, Z.Y., Jin, J. and Liu, A.M. (2016) Effect of spray parameters of small unmanned helicopter on distribution regularity of droplet deposition in hybrid rice canopy. *Transactions of the Chinese Society of Agricultural Engineering* 32, 40–46.

Cohuet, A., Harris, C., Robert, V. and Fontenille, D. (2010) Evolutionary forces on *Anopheles*: What makes a malaria vector? *Trends in Parasitology* 26, 130–136.

Coluzzi, M. (1994) Malaria and the Afrotropical ecosystems: Impact of man-made environmental changes. *Parassitologia* 36, 223–227.

Cummings, M., Huang, L., Zhu, H., Finkelstein, D. and Wei, R. (2019) The impact of increasing autonomy on training requirements in a UAV supervisory control task. *Journal of Cognitive Engineering and Decision Making* 13, 295–309.

Derua, Y.A., Kweka, E.J., Kisinza, W.N., Githeko, A.K. and Mosha, F.W. (2019) Bacterial larvicides used for malaria vector control in sub-Saharan Africa: Review of their effectiveness and operational feasibility. *Parasites and Vectors* 12:426.

Dos Reis, I.C., Codeco, C.T., Degener, C.M., Keppeler, E.C., Muniz, M.M. *et al.* (2015) Contribution of fish farming ponds to the production of immature *Anopheles* spp. in a malaria-endemic Amazonian town. *Malaria Journal* 14:452.

Dunbar, S., Wenzl, F., Hack, C., Hafeza, R., Esfeer, H. *et al.* (2015) Wireless far-field charging of a micro-UAV. In: *2015 IEEE Wireless Power Transfer Conference (WPTC), 13–15 May*, University of Colorado, pp. 1–4.

Durnez, L. and Coosemans, M. (2013) Residual transmission of malaria: an old issue for new approaches. In: Manguin, S. (ed.) *Anopheles Mosquitoes—New Insights into Malaria Vectors*. Intech Open, Rijeka, Croatia, pp. 671–704.

Dyck, V.A., Hendrichs, J. and Robinson, A.S. (eds) (2005) *Sterile Insect Technique: Principles and Practice in Area-wide Integrated Pest Management*. Springer, Dordrecht, The Netherlands.

Eißfeldt, H., Vogelpohl, V., Stolz, M., Papenfuß, A., Biella, M. *et al.* (2020) The acceptance of civil drones in Germany. *CEAS Aeronautical Journal* 11, 665–676.

Ellis, A.M. (2008) Linking movement and oviposition behaviour to spatial population distribution in the tree hole mosquito *Ochlerotatus triseriatus*. *Journal of Animal Ecology* 77, 156–166.

Elmes, A., Alemohammad, H., Avery, R., Caylor, K., Eastman, J.R. *et al.* (2020) Accounting for training data error in machine learning applied to earth observations. *Remote Sensing* 12, 1034.

Ferguson, H.M., Dornhaus, A., Beeche, A., Borgemeister, C., Gottlieb, M. *et al.* (2010) Ecology: A prerequisite for malaria elimination and eradication. *PLoS Medicine* 7, e1000303.

Finn, R. and Donovan, A. (2016) Big data, drone data: Privacy and ethical impacts of the intersection between big data and civil drone deployments. In: Custers, B. (ed.) *The Future of Drone Use: Opportunities and Threats from Ethical and Legal Perspectives*. T.M.C. Asser Press, The Hague, pp. 47–69.

Finn, R.L. and Wright, D. (2012) Unmanned aircraft systems: Surveillance, ethics and privacy in civil applications. *Computer Law & Security Review* 28, 184–194.

Finn, R.L. and Wright, D. (2016) Privacy, data protection and ethics for civil drone practice: A survey of industry, regulators and civil society organisations. *Computer Law & Security Review* 32, 577–586.

Fonseca, D.M., Unlu, I., Crepeau, T., Farajollahi, A., Healy, S.P. *et al.* (2013) Area-wide management of *Aedes albopictus*. part 2: Gauging the efficacy of traditional integrated pest control measures against urban container mosquitoes. *Pest Management Science* 69, 1351–1361.

Fornace, K.M., Drakeley, C.J., William, T., Espino, F. and Cox, J. (2014) Mapping infectious disease landscapes: unmanned aerial vehicles and epidemiology. *Trends in Parasitology* 30, 514–519.

Gray, P.C., Fleishman, A.B., Klein, D.J., Mckown, M.W., Bézy, V.S., Lohmann, K.J. and Johnston, D.W. (2019) A convolutional neural network for detecting sea turtles in drone imagery. *Methods in Ecology and Evolution* 10, 345–355.

Grisales, N., Poupardin, R., Gomez, S., Fonseca-Gonzalez, I., Ranson, H. and Lenhart, A. (2013) Temephos resistance in *Aedes aegypti* in Colombia compromises dengue vector control. *PLoS Neglected Tropical Diseases* 7:e2438.

Guo, S., Li, J., Yao, W., Zhan, Y., Li, Y. *et al.* (2019) Distribution characteristics on droplet deposition of wind field vortex formed by multi-rotor UAV. *PLoS ONE* 14, e0220024.

Haas-Stapleton, E.J., Barretto, M.C., Castillo, E.B., Clausnitzer, R.J. and Ferdan, R.L. (2019) Assessing mosquito breeding sites and abundance using an unmanned aircraft. *Journal of the American Mosquito Control Association* 35, 228–232.

Hardy, A., Mageni, Z., Dongus, S., Killeen, G., Macklin, M.G. *et al.* (2015) Mapping hotspots of malaria transmission from pre-existing hydrology, geology and geomorphology data in the pre-elimination context of Zanzibar, United Republic of Tanzania. *Parasites and Vectors* 8, 41.

Hardy, A., Makame, M., Cross, D., Majambere, S. and Msellem, M. (2017) Using low-cost drones to map malaria vector habitats. *Parasites and Vectors* 10, 29.

Hardy, A., Ettritch, G., Cross, D.E., Bunting, P., Liywalii, F. *et al.* (2019) Automatic detection of open and vegetated water bodies using Sentinel 1 to map African malaria vector mosquito breeding habitats. *Remote Sensing* 11, 593.

Hayes, R.O., Maxwell, E.L., Mitchell, C.J. and Woodzick, T.L. (1985) Detection, identification, and classification of mosquito larval habitats using remote sensing scanners in earth-orbiting satellites. *Bulletin of the World Health Organization* 63, 361–374.

Hawkes, F.M., Manin, B.O., Cooper, A., Daim, S.R.H., Jelip, J. *et al.* (2019) Vector compositions change across forested to deforested ecotones in emerging areas of zoonotic malaria transmission in Malaysia. *Scientific Reports* 9, 13312.

Hay, S.I., Snow, R.W. and Rogers, D.J. (1998) From predicting mosquito habitat to malaria seasons using remotely sensed data: practice, problems and perspectives. *Parasitology Today* 14, 306–313.

Hii, J.L., Smith, T., Mai, A., Mellor, S., Lewis, D., Alexander, N. and Alpers, M.P. (1997) Spatial and temporal variation in abundance of *Anopheles* (Diptera:Culicidae) in a malaria endemic area in Papua New Guinea. *Journal of Medical Entomology* 34, 193–205.

Iskander, D. (2017) Taking rumours seriously: Why organ thieves matter to malaria control. *Anthropology Today* 33, 9–12.

Ivošević, B., Han, Y.G., Cho, Y. and Kwon, O. (2015) The use of conservation drones in ecology and wildlife research. *Journal of Ecology and Environment* 38, 113–118.

Jiang, C. and Song, J. (2015) An ultrahigh-resolution digital image sensor with pixel size of 50 nm by vertical nanorod arrays. *Advanced Materials* 27, 4454–4460.

Johnson, B.J., Manby, R. and Devine, G.J. (2020) Performance of an aerially applied liquid *Bacillus thuringiensis* var. *Israelensis* formulation (strain AM65-52) against mosquitoes in mixed saltmarsh–mangrove systems and fine-scale mapping of mangrove canopy cover using affordable drone-based imagery. *Pest Management Science* 76, 3822–3831.

Jumail, A., Liew, T.S., Salgado-Lynn, M., Fornace, K.M. and Stark, D.J. (2021) A comparative evaluation of thermal camera and visual counting methods for primate census in a riparian forest at the lower Kinabatangan wildlife sanctuary (LKWS). Malaysian Borneo. *Primates* 62, 143–151.

Kaindoa, E.W., Mkandawile, G., Ligamba, G., Kelly-Hope, L.A. and Okumu, F.O. (2016) Correlations between household occupancy and malaria vector biting risk in rural Tanzanian villages: implications for high-resolution spatial targeting of control interventions. *Malaria Journal* 15:199.

Kattenborn, T., Eichel, J. and Fassnacht, F.E. (2019) Convolutional Neural Networks enable efficient, accurate and fine-grained segmentation of plant species and communities from high-resolution UAV imagery. *Scientific Reports* 9:17656.

Killeen, G., Fillinger, U. and Knols, B. (2002) Advantages of larval control for African malaria vectors: Low mobility and behavioural responsiveness of immature mosquito stages allow high effective coverage. *Malaria Journal* 1, 7.

Killeen, G.F., Marshall, J.M., Kiware, S.S., South, A.B., Tusting, L.S. *et al.* (2017) Measuring, manipulating and exploiting behaviours of adult mosquitoes to optimise malaria vector control impact. *BMJ Global Health* 2, e000212.

Knisely, T. (2019) DJI Joins the Fight Against Malaria: A New Use of Agricultural Drones for Public Health [Online]. Anti-Malaria Drones. Available at: https://www.antimalariadrones.com/blogs (accessed 9 March 2021).

Kraemer, M.U.G., Reiner, R.C., Brady, O.J., Messina, J.P., Gilbert, M. *et al.* (2019) Past and future spread of the arbovirus vectors *Aedes aegypti* and *Aedes albopictus*. *Nature Microbiology* 4, 854–863.

Kucherov, D., Sushchenko, O. and Kozub, A. (2019) Operator training for unmanned aerial vehicles control. In: *2019 IEEE 5th International Conference Actual Problems of Unmanned Aerial Vehicles Developments (APUAVD), 22–24 October 2019*, Kiev, Ukraine, pp. 1–4. Available at: (accessed 9 March 2021).

Lee, B., Park, P., Kim, K. and Kwon, S. (2014) The flight test and power simulations of an UAV powered by solar cells, a fuel cell and batteries. *Journal of Mechanical Science and Technology* 28, 399–405.

Lopes, R.P., Lima, J.B.P. and Martins, A.J. (2019) Insecticide resistance in *Culex quinquefasciatus* Say, 1823 in Brazil: a review. *Parasites and Vectors* 12:591.

Ma, L., Li, M., Ma, X., Cheng, L., Du, P. and Liu, Y. (2017) A review of supervised object-based land-cover image classification. *ISPRS Journal of Photogrammetry and Remote Sensing* 130, 277–293.

Messina, J.P., Brady, O.J., Pigott, D.M., Golding, N., Kraemer, M.U. *et al.* (2015) The many projected futures of dengue. *Nature Reviews Microbiology* 13, 230–239.

Minakawa, N., Mutero, C.M., Githure, J.I., Beier, J.C. and Yan, G. (1999) Spatial distribution and habitat characterization of anopheline mosquito larvae in Western Kenya. *The American Journal of Tropical Medicine and Hygiene* 61, 1010–1016.

Minakawa, N., Munga, S., Atieli, F., Mushinzimana, E., Zhou, G., Githeko, A.K. and Yan, G. (2005) Spatial distribution of anopheline larval habitats in Western Kenyan highlands: Effects of land cover types and topography. *The American Journal of Tropical Medicine and Hygiene* 73, 157–165.

MLHub (n.d) *Radiant MLHub: Open Library for Earth Observations Machine Learning. Radiant Earth Foundation*. Available at: https://mlhub.earth/ (accessed 9 March 2021).

Moyes, C.L., Vontas, J., Martins, A.J., Ng, L.C., Koou, S.Y. *et al.* (2017) Contemporary status of insecticide resistance in the major *Aedes* vectors of arboviruses infecting humans. *PLOS Neglected Tropical Diseases* 11, e0005625.

Munga, S., Yakob, L., Mushinzimana, E., Zhou, G., Ouna, T. *et al.* (2009) Land use and land cover changes and spatiotemporal dynamics of anopheline larval habitats during a four-year period in a highland community of Africa. *The American Journal of Tropical Medicine and Hygiene* 81, 1079–1084.

Mushinzimana, E., Munga, S., Minakawa, N., Li, L., Feng, C.-C. *et al.* (2006) Landscape determinants and remote sensing of anopheline mosquito larval habitats in the western Kenya highlands. *Malaria Journal* 5, 13.

Mutuku, F.M., Alaii, J.A., Bayoh, M.N., Gimnig, J.E., Vulule, J.M. *et al.* (2006) Distribution, description, and local knowledge of larval habitats of *Anopheles gambiae* s.l. in a village in western Kenya. *The American Journal of Tropical Medicine and Hygiene* 74, 44–53.

Mutuku, F.M., Bayoh, M.N., Hightower, A.W., Vulule, J.M., Gimnig, J.E. *et al.* (2009) A supervised land cover classification of a western Kenya lowland endemic for human malaria: associations of land cover with larval *Anopheles* habitats. *International Journal of Health Geographics* 8:13.

National Aeronautics and Space Administration (n.d.-a) EARTHDATA: Open Access for Open Science NASA. Available at: https://earthdata.nasa.gov/learn/backgrounders/remote-sensing#:~:text=Remote%20sensing%20is%20the%20acquiring,record%20reflected%20or%20emitted%20ener (accessed 9 March 2021).

National Aeronautics and Space Administration (n.d.-b) EARTHDATA: Open Access for Open Science [Online]. NASA. Available at: https://earthdata.nasa.gov/learn/remote-sensors (accessed 9 March 2021).

Palaniyandi, M. (2012) The role of remote sensing and GIS for spatial prediction of vector-borne diseases transmission: A systematic review. *Journal of Vector Borne Diseases* 49, 197–204.

Parselia, E., Kontoes, C., Tsouni, A., Hadjichristodoulou, C., Kioutsioukis, I. *et al.* (2019) Satellite earth observation data in epidemiological modeling of malaria, dengue and West Nile virus: A scoping review. *Remote Sensing* 11, 1862.

Peckham, R. and Sinha, R. (2019) Anarchitectures of health: Futures for the biomedical drone. *Global Public Health* 14, 1204–1219.

Pham, H., Smolka, S.A., Stoller, S.D., Phan, D. and Yang, J. (2015) A survey on unmanned aerial vehicle collision avoidance systems. Available at: https://ui.adsabs.harvard.edu/abs/2015arXiv150807723P (accessed 1 August 2015).

Qin, W.-C., Qiu, B.-J., Xue, X.-Y., Chen, C., Xu, Z.-F. and Zhou, Q.-Q. (2016) Droplet deposition and control effect of insecticides sprayed with an unmanned aerial vehicle against plant hoppers. *Crop Protection* 85, 79–88.

Ranson, H. and Lissenden, N. (2016) Insecticide resistance in African *Anopheles* mosquitoes: a worsening situation that needs urgent action to maintain malaria control. *Trends in Parasitology* 32, 187–196.

Reisen, W.K. (2010) Landscape epidemiology of vector-borne diseases. *Annual Review of Entomology* 55, 461–483.

Roux, E., Gaborit, P., Romaña, C.A., Girod, R., Dessay, N. and Dusfour, I. (2013) Objective sampling design in a highly heterogeneous landscape – characterizing environmental determinants of malaria vector distribution in French Guiana, in the Amazonian region. *BMC Ecology* 13:45.

Rti International (n.d) Using Drones for Vector Control and Surveillance of *Aedes* Mosquitoes in Guatemala. Research Triangle Institute. Available at: https://www.rti.org/impact/using-drones-vector-control-and-surveillance-aedes-mosquitoes-guatemala (accessed 9 March 2021).

Russell, B.M., Kay, B.H. and Shipton, W. (2001) Survival of *Aedes aegypti* (Diptera: Culicidae) eggs in surface and subterranean breeding sites during the northern Queensland dry season. *Journal of Medical Entomology* 38, 441–445.

Sarira, T.V., Clarke, K., Weinstein, P., Koh, L.P. and Lewis, M. (2020) Rapid identification of shallow inundation for mosquito disease mitigation using drone-derived multispectral imagery. *Geospatial Health* 15, 101–108.

Shakhatreh, H., Sawalmeh, A.H., Al-Fuqaha, A., Dou, Z., Almaita, E. *et al.* (2019) Unmanned aerial vehicles (UAVS): A survey on civil applications and key research challenges. *IEEE Access* 7, 48572–48634.

Sherrard-Smith, E., Skarp, J.E., Beale, A.D., Fornadel, C., Norris, L.C. *et al.* (2019) Mosquito feeding behavior and how it influences residual malaria transmission across Africa. *Proceedings of the National Academy of Sciences of the United States of America* 116, 15086–15095.

Simic, M., Bil, C. and Vojisavljevic, V. (2015) Investigation in wireless power transmission for UAV charging. *Procedia Computer Science* 60, 1846–1855.

Simic Milas, A., Cracknell, A.P. and Warner, T.A. (2018) Drones – the third generation source of remote sensing data. *International Journal of Remote Sensing* 39, 7125–7137.

Sithiprasasna, R., Lee, W.J., Ugsang, D.M. and Linthicum, K.J. (2005) Identification and characterization of larval and adult anopheline mosquito habitats in the Republic of Korea: Potential use of remotely sensed data to estimate mosquito distributions. *International Journal of Health Geographics* 4:17.

Smith, D.L., Dushoff, J. and Mckenzie, F.E. (2004) The risk of a mosquito-borne infection in a heterogeneous environment. *PLoS Biology* 2, e368.

Spacenet (n.d) SpaceNet: Accelerating Geospatial Machine Learning. SpaceNet. Available at: https://spacenet.ai/ (accessed 9 March 2021).

Spatiotemporal Asset Catalog (n.d) SpatioTemporal AssetCatalogs. Available at: https://stacspec.org/ (accessed 9 March 2021).

Stark, D.J., Fornace, K.M., Brock, P.M., Abidin, T.R., Gilhooly, L. *et al.* (2019) Long-tailed macaque response to deforestation in a *Plasmodium knowlesi*-endemic area. *EcoHealth* 16, 638–646.

Stöcker, C., Bennett, R., Nex, F., Gerke, M. and Zevenbergen, J. (2017) Review of the current state of UAV regulations. *Remote Sensing* 9, 459.

Stresman, G., Bousema, T. and Cook, J. (2019) Malaria hotspots: is there epidemiological evidence for fine-scale spatial targeting of interventions? *Trends in Parasitology* 35, 822–834.

Szabolcsi, R. (2016) Beyond training minimums – a new concept of the UAV operator training program. *International Conference Knowledge Based Organization XXII*, 560–566.

Tusting, L.S., Thwing, J., Sinclair, D., Fillinger, U., Gimnig, J. *et al.* (2013) Mosquito larval source management for controlling malaria. *Cochrane Database of Systematic Reviews* 8, CD008923.

U.S Fish and Wildlife Service (2012) *Don Edwards San Francisco Bay National Wildlife Refuge Comprehensive Conservation Plan: Final Mosquito Management Plan*. Appendix K. US Fish and Wildlife Service, Washington, DC.

Valle, D., Bellinato, D.F., Viana-Medeiros, P.F., Lima, J.B.P. and Martins Junior, A.J. (2019) Resistance to temephos and deltamethrin in *Aedes aegypti* from Brazil between 1985 and 2017. *Memórias do Instituto Oswaldo Cruz* 114:e180544.

Varela, G., Caamaño, P., Orjales, F., Deibe, Á., López-Peña, F. *et al.* (2011) Swarm intelligence based approach for real time UAV team coordination in search operations. In: *2011 Third World Congress on Nature and Biologically Inspired Computing, 19–21 October 2011.*, pp. 365–370.

Viana, J., Santos, J.V., Neiva, R.M., Souza, J., Duarte, L. *et al.* (2017) Remote sensing in human health: A 10-year bibliometric analysis. *Remote Sensing* 9, 1225.

Vittor, A.Y., Gilman, R.H., Tielsch, J., Glass, G., Shields, T. *et al.* (2006) The effect of deforestation on the human-biting rate of *Anopheles darlingi*, the primary vector of *Falciparum malaria* in the Peruvian Amazon. *The American Journal of Tropical Medicine and Hygiene* 74, 3–11.

We Robotics (2018) How Local Drone Pilots Are Helping to Reduce Dengue in Fiji. Wilmington, Delaware. We Robotics. Available at: https://blog.werobotics.org/2018/12/13/how-local-drone-pilots-are-helping-to-reduce-dengue-in-fiji/ (accessed 9 March 2021).

Whittaker, S. (2018) *Drones Combat Dengue and Zika Outbreaks in Brazil*. Available at: https://dronebelow.com/2018/03/15/drones-combat-dengue-and-zika-outbreaks-in-brazil/ (accessed 13 September 2023).

World Health Organization. (2009) *Dengue guidelines for diagnosis, treatment, prevention and control* (new edn). World Health Organization, Geneva, Switzerland.

World Health Organization. (2012) *Handbook for Integrated Vector Management*. World Health Organization, Geneva, Switzerland.

World Health Organization. (2013) *Larval Source Management: A Supplementary Malaria Vector Control Measure: An Operational Manual*. World Health Organization, Geneva, Switzerland.

World Health Organization. (2015) *Indoor Residual Spraying: An Operational Manual for Indoor Residual Spraying (IRS) for Malaria Transmission Control and Elimination*. 2nd edn. World Health Organization, Geneva, Switzerland.

World Health Organization. (2018) *Malaria Surveillance, Monitoring and Evaluation: A Reference Manual*. World Health Organization, Geneva, Switzerland.

World Health Organization. (2020) *World Malaria Report 2020: 20 Years of Global Progress and Challenges*. World Health Organization, Geneva, Switzerland.

World Health Organization. Regional office for south-east, A (2016) *Dengue Bulletin*. Vol. 39. World Health Organization, Regional Office for South-East Asia, New Delhi, India.

Xue, X.Y., Tu, K., Qin, W.C., Lan, Y.B. and Zhang, H.H. (2014) Drift and deposition of ultra-low altitude and low volume application in paddy field. *International Journal of Agricultural & Biological Engineering* 7, 23–28.

Yadouléton, A., Badirou, K., Agbanrin, R., Jöst, H., Attolou, R. *et al.* (2015) Insecticide resistance status in *Culex quinquefasciatus* in Benin. *Parasites and Vectors*, 8, 17.

Zhang, H., Wang, C., Turvey, S.T., Sun, Z., Tan, Z. *et al.* (2020) Thermal infrared imaging from drones can detect individuals and nocturnal behavior of the world's rarest primate. *Global Ecology and Conservation* 23:e01101.

Zheng, X., Zhang, D., Li, Y., Yang, C., Wu, Y. *et al.* (2019) Incompatible and sterile insect techniques combined eliminate mosquitoes. *Nature* 572, 56–61.

Zou, L., Miller, S.N. and Schmidtmann, E.T. (2006) Mosquito larval habitat mapping using remote sensing and GIS: implications of coalbed methane development and West Nile virus. *Journal of Medical Entomology* 43, 1034–1041.

6

The Use of *Wolbachia* for Mosquito Control

Perran A. Ross

Pest and Environmental Adaptation Research Group, Bio21 Institute and the University of Melbourne, Parkville, Victoria, Australia

Abstract

Wolbachia are endosymbiotic bacteria that occur naturally in many insects. Mosquitoes transinfected with *Wolbachia* are being released in mosquito and arbovirus control programmes around the world. *Wolbachia* are maternally inherited and cause cytoplasmic incompatibility, where females that lack *Wolbachia* do not produce viable offspring when they mate with *Wolbachia*-infected males. Releases of male-only *Wolbachia*-infected mosquitoes can be used to suppress mosquito populations, but the reproductive advantage provided by cytoplasmic incompatibility also allows *Wolbachia* to spread through mosquito populations. *Wolbachia*-infected female mosquitoes have a reduced ability to transmit arboviruses including dengue and Zika, making them an effective tool for disease control.

A variety of *Wolbachia* infections in *Aedes* mosquitoes have been developed and deployed in open-field trials. Releases of *Wolbachia*-infected male mosquitoes have achieved over 95% population suppression, while the replacement of populations with *Wolbachia*-infected females is self-sustaining and can greatly reduce local dengue transmission. Despite these successes, significant questions and challenges lie ahead. *Wolbachia*, viruses and their mosquito hosts can evolve, leading to uncertainty around the long-term effectiveness of a given *Wolbachia* strain. Further, few ecological impacts of *Wolbachia* releases have been explored. *Wolbachia* strains are diverse, and the choice of strain to release should be made carefully, taking environmental conditions and the release objective into account. Mosquito quality control, thoughtful community awareness programmes and long-term monitoring of populations are essential for all types of *Wolbachia* intervention. Releases of *Wolbachia*-infected mosquitoes show great promise, but existing control measures remain an important way to reduce the burden of mosquito-borne disease.

*Corresponding author: perran.ross@unimelb.edu.au

 Practical Control of Mosquito Disease Vectors (eds J.D. Charlwood)
DOI: 10.1079/9781789248845.0006

A Brief Introduction to *Wolbachia*

Wolbachia are a group of Gram-negative bacteria that live inside the cells of insects and other arthropods. *Wolbachia* are normally maternally inherited, where females that carry *Wolbachia* transmit the infection to their offspring (Newton *et al.*, 2015). On rare occasions, *Wolbachia* can spread between species through routes such as predation and parasitism (Sanaei *et al.*, 2020). *Wolbachia* are best known for their effects on host reproduction. *Wolbachia* may increase the proportion of females produced by their host by killing or feminizing male offspring (Hurst and Jiggins, 2000). *Wolbachia* can also trigger asexual reproduction in females, removing the need for males entirely (Huigens and Stouthamer, 2003; Ma and Schwander, 2017). More commonly, *Wolbachia* induce cytoplasmic incompatibility, where males infected with *Wolbachia* do not produce viable offspring with females that are not infected (Shropshire *et al.*, 2020). These effects on reproduction aid the spread of *Wolbachia* through populations by making the females that carry and transmit *Wolbachia* relatively more fit.

Wolbachia can also provide direct host fitness benefits (Zug and Hammerstein, 2015). For instance, *Wolbachia* can increase host lifespan or fecundity through nutritional provisioning (Brownlie *et al.*, 2009; Ju *et al.*, 2020) or altering programmed cell death (Fast *et al.*, 2011; Guo *et al.*, 2018). Furthermore, *Wolbachia* can protect their hosts against natural pathogens, including viruses (Hedges *et al.*, 2008; Teixeira *et al.*, 2008). The dramatic effects of *Wolbachia* on their hosts are being used to control mosquito populations through releases of *Wolbachia*-infected mosquitoes into the environment (Fig. 6.1B).

Effects of Natural and Artificial *Wolbachia* Infections in Mosquitoes

Wolbachia are found naturally in about 30% of mosquito species (Silva *et al.*, 2021). Although *Wolbachia* are absent from many important vectors, including *Aedes aegypti* (Gloria-Soria *et al.*, 2018; Ross *et al.*, 2020c) and most *Anopheles* species (Chrostek and Gerth, 2019; Walker *et al.*, 2020), *Wolbachia* can be introduced artificially through a process called transinfection (Hughes and Rasgon, 2014). To transfer *Wolbachia* to a new host, cytoplasm or tissue homogenate from a *Wolbachia*-infected insect is injected into mosquito embryos using a fine needle. Stably infected lines can be generated by selecting *Wolbachia*-infected females each generation. Both natural and artificial *Wolbachia* infections can have effects that are useful for mosquito control programmes, including cytoplasmic incompatibility, pathogen blocking and fitness costs (Fig. 6.1B).

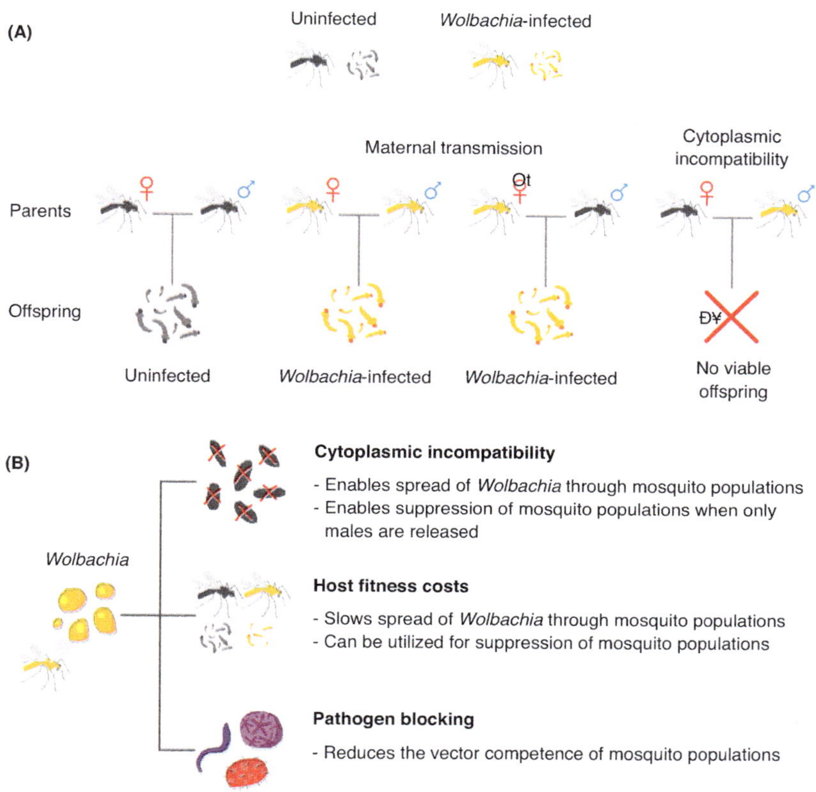

Fig. 6.1. *Wolbachia* transmission and effects in mosquitoes. (A) *Wolbachia* are maternally inherited and cause cytoplasmic incompatibility. *Wolbachia*-infected females transmit *Wolbachia* to their offspring, regardless of whether males are *Wolbachia*-infected or uninfected. Uninfected females that mate with *Wolbachia*-infected males have fewer or no viable offspring due to cytoplasmic incompatibility. (B) *Wolbachia* have three key effects in mosquitoes that can be utilized to reduce pathogen transmission or mosquito population sizes.

Cytoplasmic incompatibility

Most *Wolbachia* infections in mosquitoes cause cytoplasmic incompatibility, where uninfected females that mate with *Wolbachia*-infected males produce fewer (or no) viable offspring (Fig. 6.1A). Female mosquitoes carrying *Wolbachia* are compatible with both *Wolbachia*-infected and uninfected males. Cytoplasmic incompatibility provides *Wolbachia*-infected females with a reproductive advantage, allowing *Wolbachia* to spread through mosquito populations. This advantage is frequency-dependent (Hoffmann and Turell, 1997);

when *Wolbachia* is at a high frequency in the population, uninfected females have a high likelihood of mating with *Wolbachia*-infected males and becoming infertile. When *Wolbachia* is at a low frequency, cytoplasmic incompatibility provides little advantage, because most uninfected females will mate with uninfected males.

Cytoplasmic incompatibility is key to all *Wolbachia*-based mosquito control programmes (Fig. 6.1B). It can drive *Wolbachia* through mosquito populations, replacing the native population with mosquitoes that are less capable of transmitting viruses. When only *Wolbachia*-infected male mosquitoes are released, cytoplasmic incompatibility can reduce the fertility of wild female mosquitoes and their overall population size. Cytoplasmic incompatibility can occur between mosquitoes carrying different *Wolbachia* strains; when the infertility occurs in both directions, this is called bidirectional incompatibility (Atyame *et al.*, 2014; Joubert *et al.*, 2016). By using *Wolbachia* infections with different patterns of cytoplasmic incompatibility, it is possible to replace or crash mosquito populations, even if they already carry *Wolbachia* (Fu *et al.*, 2010; Zheng *et al.*, 2019).

The strength of cytoplasmic incompatibility (in other words, the proportion of eggs hatching from a cross between a *Wolbachia*-infected male and an uninfected female) depends on the *Wolbachia* strain and mosquito species. Most characterized *Wolbachia* infections in mosquitoes cause strong or complete cytoplasmic incompatibility (Dobson *et al.*, 2001; Atyame *et al.*, 2011; Fraser *et al.*, 2017), but some do not cause any (Ant *et al.*, 2018; Wu, 2020). The strength of cytoplasmic incompatibility is also related to the density of *Wolbachia* inside the mosquito, which can be influenced by environmental conditions. The *w*Mel strain in *Ae. aegypti* shows complete cytoplasmic incompatibility under standard laboratory conditions but weakens when mosquitoes experience high temperatures (Ross *et al.*, 2017) or hatch from quiescent eggs (Lau *et al.*, 2020a). For some natural *Wolbachia* infections, cytoplasmic incompatibility weakens as males age, likely corresponding to a decline in *Wolbachia* density (Kittayapong *et al.*, 2002).

There are no known cases of *Wolbachia* causing feminization, male killing or asexual reproduction in mosquitoes. *Wolbachia* strains with these effects could be useful for driving *Wolbachia* into mosquito populations, and strains causing these effects could plausibly be introduced from other insects, though such effects may not persist in the mosquito.

Pathogen blocking

Wolbachia are renowned for their ability to protect their hosts against and interfere with the transmission of pathogens. *Wolbachia*-induced pathogen protection was first discovered in *Drosophila*, where flies carrying *Wolbachia* had higher survival compared to uninfected flies when challenged with RNA viruses (Hedges *et al.*, 2008; Teixeira *et al.*, 2008) or the fungus *Beauveria*

Table 6.1. *Wolbachia* infections can block human pathogens transmitted by mosquitoes. Blocking is defined by a reduction in pathogen titre, infection rate or transmission rate in *Wolbachia*-infected mosquitoes compared to uninfected controls. Note that cases where *Wolbachia* infections have no effect or enhance pathogen infection are not included here, but see Pan *et al.* (2017) and Caragata and Moreira (2017).

Pathogen	*Wolbachia* strains (mosquito host species)	Selected references
Dengue virus	*w*AlbB (*Aedes aegypti*)	Bian *et al.* (2010)
	*w*MelPop (*Aedes aegypti*)	(Moreira *et al.*, 2009a; Ferguson *et al.*, 2015)
	*w*Mel (*Aedes aegypti*)	(Walker *et al.*, 2011; Carrington *et al.*, 2017)
	*w*MelCS (*Aedes aegypti*)	(Fraser *et al.*, 2017)
	*w*Ri (*Aedes aegypti*)	(Fraser *et al.*, 2017)
	*w*Au (*Aedes aegypti*)	(Ant *et al.*, 2018)
	*w*Mel/*w*AlbB (*Aedes aegypti*)	(Joubert *et al.*, 2016)
	*w*Mel (*Aedes albopictus*)	(Blagrove *et al.*, 2012)
	*w*Mel/*w*Pip (*Aedes albopictus*)	(Moretti *et al.*, 2018b)
	*w*AlbA/*w*AlbB (*Aedes albopictus*)	(Mousson *et al.*, 2012)
	*w*Au/*w*AlbA/*w*AlbB (*Aedes albopictus*)	(Mancini *et al.*, 2020)
	*w*AlbB (*Aedes polynesiensis*)	(Bian *et al.*, 2013b)
Zika virus	*w*AlbB (*Aedes aegypti*)	(Ant *et al.*, 2018)
	*w*Mel (*Aedes aegypti*)	(Aliota *et al.*, 2016a; Dutra *et al.*, 2016)
	*w*AlbA (*Aedes aegypti*)	(Chouin-Carneiro *et al.*, 2020)
	*w*Au (*Aedes aegypti*)	(Ant *et al.*, 2018)
	*w*AlbB (*Aedes albopictus*[a])	(Schultz *et al.*, 2017)
	*w*Stri (*Aedes albopictus*[a])	(Schultz *et al.*, 2017)
	*w*Au/*w*AlbA/*w*AlbB (*Aedes albopictus*)	(Mancini *et al.*, 2020)
Chikungunya virus	*w*MelPop (*Aedes aegypti*)	(Moreira *et al.*, 2009a)
	*w*Mel (*Aedes aegypti*)	(Hurk *et al.*, 2012; Aliota *et al.*, 2016b)
	*w*Mel (*Aedes albopictus*)	(Blagrove *et al.*, 2013)
	*w*AlbB (*Aedes albopictus*[a])	(Raquin *et al.*, 2015)
	*w*Mel/*w*Pip (*Aedes albopictus*)	(Moretti *et al.*, 2018b)
West Nile virus	*w*AlbB (*Aedes aegypti*)	(Joubert and O'Neill, 2017)
	*w*MelPop (*Aedes aegypti*)	(Hussain *et al.*, 2013)
	*w*Pip (*Culex quinquefasciatus*)	(Glaser and Meola, 2010)
Yellow Fever virus	*w*MelPop (*Aedes aegypti*)	(Hurk *et al.*, 2012)
	*w*Mel (*Aedes aegypti*)	(Rocha *et al.*, 2019)

Continued

Table 6.1. Continued

Pathogen	*Wolbachia* strains (mosquito host species)	Selected references
Sindbis virus	*w*AlbB (*Aedes aegypti*)	(Bhattacharya *et al.*, 2020)
	*w*AlbB (*Aedes albopictus*[a])	(Ekwudu *et al.*, 2020)
	*w*Stri (*Aedes albopictus*[a])	(Bhattacharya *et al.*, 2020)
	*w*Mel (*Drosophila melanogaster*[b])	(Bhattacharya *et al.*, 2017; Nainu *et al.*, 2019)
Semliki forest virus	*w*AlbB (*Aedes aegypti*)	(Ant *et al.*, 2018)
	*w*Mel (*Aedes aegypti*)	(Ant *et al.*, 2018)
	*w*Au (*Aedes aegypti*)	(Ant *et al.*, 2018)
	*w*Mel (*Drosophila melanogaster*[b])	(Rainey *et al.*, 2016)
Ross River virus	*w*AlbB (*Aedes albopictus*[a])	(Ekwudu *et al.*, 2020)
Barmah Forest virus	*w*AlbB (*Aedes albopictus*[a])	(Ekwudu *et al.*, 2020)
Mayaro virus	*w*Mel (*Aedes aegypti*)	(Pereira *et al.*, 2018; Sucupira *et al.*, 2020)
Plasmodium falciparum (human malaria parasite)	*w*AlbB[c] (*Anopheles gambiae*)	(Hughes *et al.*, 2011)
	*w*MelPop[c] (*Anopheles gambiae*)	(Hughes *et al.*, 2011)
	*w*AlbB (*Anopheles stephensi*)	(Bian *et al.*, 2013a)

[a]mosquito cell line
[b]non-mosquito cell line model
[c]somatic (transient) *Wolbachia* infection

bassiana (Panteleev *et al.*, 2007). In mosquitoes, *Wolbachia* infections can interfere with the replication and transmission of human pathogens (Table 6.1), insect-specific viruses (Schnettler *et al.*, 2016) and non-human animal pathogens, including the filarial nematode *Brugia pahangi* (Kambris *et al.*, 2009; Andrews *et al.*, 2012) and rodent malaria parasites (Joshi *et al.*, 2017). How *Wolbachia* can block pathogen transmission is not fully understood, but there appear to be multiple mechanisms involved (Terradas and McGraw, 2017; Lindsey *et al.*, 2018), including immune upregulation (Rances *et al.*, 2012) and interactions with host lipids (Geoghegan *et al.*, 2017; Koh *et al.*, 2020; Manokaran *et al.*, 2020).

The pathogen-blocking effects of some *Wolbachia* strains form the basis of mosquito release programmes to reduce dengue transmission (Fig. 6.1B). When natural *Ae. aegypti* populations carry the *w*Mel or *w*AlbB *Wolbachia* strains at a high frequency, this reduces the vector competence of the population, resulting in fewer dengue cases (Nazni *et al.*, 2019; Ryan *et al.*, 2019;

Indriani *et al.*, 2020). *Wolbachia* infections do not always provide protection against pathogens, and the strength of blocking depends on the pathogen (Teixeira *et al.*, 2008), insect host (Lu *et al.*, 2012), *Wolbachia* strain (Osborne *et al.*, 2009; Ferguson *et al.*, 2015) and environmental conditions (Caragata *et al.*, 2013; Murdock *et al.*, 2014; Chrostek *et al.*, 2020; Mancini *et al.*, 2021). The strength of virus blocking may also depend on the novelty of the *Wolbachia*-mosquito association (Terradas and McGraw, 2017); *Wolbachia* strains that show weak blockage in their native host can block much more effectively when transferred to a new host (Lu *et al.*, 2012). However, some *Wolbachia* strains like *w*Pip (originating from *Culex quinquefasciatus*) do not block, even in novel hosts (Fraser *et al.*, 2020).

Some *Wolbachia* infections can increase the probability of pathogen infection or transmission by mosquitoes (Hughes *et al.*, 2012; Dodson *et al.*, 2014), and there is a risk that releases of *Wolbachia*-infected mosquitoes could increase rather than prevent disease. At low virus doses, the *w*Mel strain in *Ae. aegypti* can increase the susceptibility of mosquitoes to dengue infection (King *et al.*, 2018). Importantly, the *Wolbachia* strains in *Ae. aegypti* that are being released into the field, including *w*Mel, have an overall suppressive effect on dengue transmission, both in laboratory (Ferguson *et al.*, 2015; Carrington *et al.*, 2017; Flores *et al.*, 2020) and field trials (O'Neill *et al.*, 2018; Nazni *et al.*, 2019; Ryan *et al.*, 2019; Indriani *et al.*, 2020). However, given that pathogen blocking by *Wolbachia* is rarely complete, it is important to consider this variability when planning releases of *Wolbachia*-infected mosquitoes.

Host fitness costs

Wolbachia infections may alter the fitness of their hosts, with effects depending on the *Wolbachia* strain and insect host. Natural *Wolbachia* infections in mosquitoes tend to be benign (Rasgon and Scott, 2003; Baton *et al.*, 2013) or even beneficial, with *Wolbachia* infections conferring increased fertility or lifespans (Dobson *et al.*, 2002; Brelsfoard and Dobson, 2011). In contrast, *Wolbachia* infections typically have harmful effects when transferred to novel mosquito hosts (Ross *et al.*, 2019c). Most drastic are their effects on female fertility, with almost all *Wolbachia* transinfections reducing fecundity or egg hatch (Axford *et al.*, 2016; Fraser *et al.*, 2017; Ant *et al.*, 2018). These negative effects become pronounced with age, such as when eggs are quiescent (Allman *et al.*, 2020), or appear under poor nutritional conditions (Caragata *et al.*, 2014). Consequently, the negative effects of *Wolbachia* strains are likely to be underestimated when evaluated under standard laboratory conditions.

Host fitness costs of *Wolbachia* can be utilized during mosquito release programmes (Fig. 6.1B). Before the discovery that *Wolbachia* could block pathogen transmission, a deleterious strain of *Wolbachia* called *w*MelPop (Min and Benzer, 1997) was introduced to *Ae. aegypti* mosquitoes. *w*MelPop shortens mosquito lifespan (McMeniman *et al.*, 2009; Ross *et al.*, 2020a) and reduces the blood-feeding success of older females (Moreira *et al.*, 2009b; Turley *et al.*,

2009). These effects could indirectly reduce dengue transmission, because *w*MelPop-infected mosquitoes are less likely to survive long enough to feed on an infected person, incubate the virus and then transmit it to a new host (Brownstein *et al.*, 2003; Rasgon *et al.*, 2003; Cook *et al.*, 2008). Host fitness costs can also be used to suppress mosquito populations once a *Wolbachia* strain is established (Rašić *et al.*, 2014; Ritchie *et al.*, 2015). However, these same effects could prevent the establishment of *Wolbachia* in the population if they are too severe (Nguyen *et al.*, 2015).

The Use of *Wolbachia* for Mosquito Control

Wolbachia-infected mosquitoes can be used in three main ways to control mosquito populations, with each approach having distinct objectives that rely on different *Wolbachia* traits (Fig. 6.2). All types of *Wolbachia* intervention involve the release of *Wolbachia*-infected mosquitoes (either naturally occurring or transinfected) from the laboratory into the field. Unlike conventional vector control, *Wolbachia* releases target a single mosquito species. The released mosquitoes are intended to mate with wild mosquitoes and/or reproduce in the natural environment. Releases can either introduce *Wolbachia* infections into the target species, suppress populations of the target species or both (Fig. 6.2).

Population replacement

Population replacement programmes (Fig. 6.2A) use cytoplasmic incompatibility to spread *Wolbachia* through a mosquito population to reach a stable high frequency, with the overall goal of reducing the vector competence of mosquitoes. This approach requires a *Wolbachia* strain that directly blocks human pathogen transmission (Table 6.1) or indirectly reduces mosquito vector competence, such as by shortening adult lifespan (McMeniman *et al.*, 2009). Alternatively, the *Wolbachia* strain could be coupled with an anti-pathogen transgene (Curtis and Sinkins, 1998; Buchman *et al.*, 2020).

Population replacement programmes involve the release of both male and female *Wolbachia*-infected mosquitoes. Both sexes contribute to *Wolbachia* proliferation; *Wolbachia*-infected females transmit *Wolbachia* to their offspring, while *Wolbachia*-infected males mate with wild females and induce cytoplasmic incompatibility, increasing the relative fitness of *Wolbachia*-infected females. Once *Wolbachia*-infected mosquitoes reach a threshold frequency in the population, *Wolbachia* can spread by itself without further releases of mosquitoes. The threshold frequency depends on the fidelity of *Wolbachia* maternal transmission and cytoplasmic incompatibility, as well as the host fitness costs of *Wolbachia* infection (Caspari and Watson, 1959; Hoffmann and Turell, 1997). For the *w*Mel strain in *Ae. aegypti*, which has relatively low host fitness costs, the threshold frequency is estimated to be 20–30% (Hoffmann *et al.*, 2011). If the frequency of *Wolbachia* in the population is below this point, the infection will likely be lost without further releases of *Wolbachia*-infected mosquitoes.

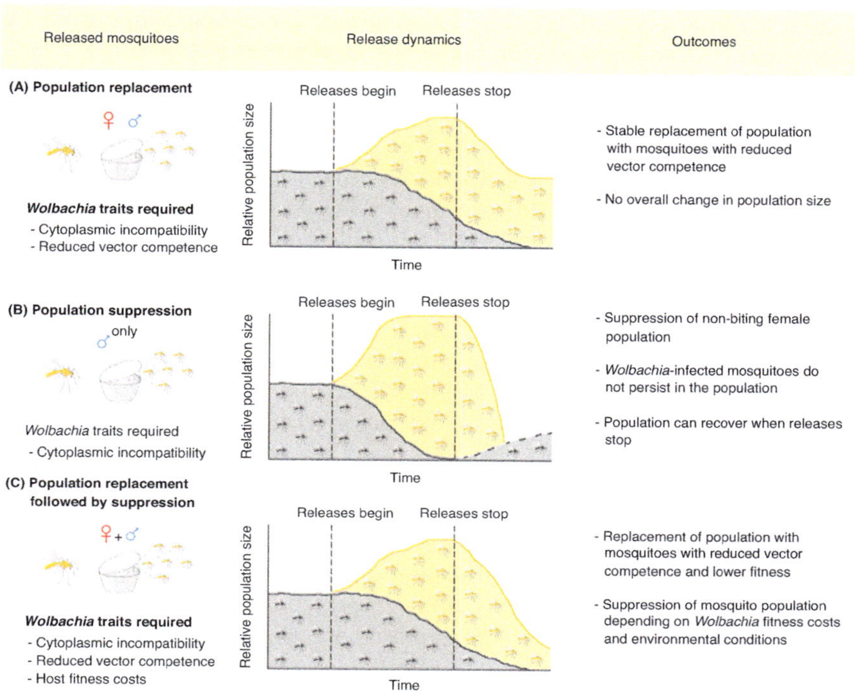

Fig. 6.2. Three ways to use *Wolbachia* for mosquito and mosquito-borne pathogen control. (A) 'Population replacement' involves releasing *Wolbachia*-infected mosquitoes with the aim of spreading *Wolbachia* through the population to reduce mosquito vector competence. (B) 'Population suppression' involves releasing only male mosquitoes that cause cytoplasmic incompatibility to reduce the fertility of wild mosquitoes and temporarily crash the population. (C) 'Population replacement followed by suppression' involves releasing *Wolbachia*-infected mosquitoes that cause substantial host fitness costs with the aim of directly reducing the vector competence of mosquitoes as well as reducing the population size. In the 'Release dynamics' column, yellow- and grey-shaded regions show the proportion of *Wolbachia*-infected and wild mosquitoes respectively in the population.

The number of *Wolbachia*-infected mosquitoes required for successful population replacement depends on the native mosquito population density and the size of the release area. Releases can be timed so that the mosquito population is at its lowest point; alternative mosquito control efforts prior to releases can aid population replacement (Hoffmann *et al*., 2011). Population replacement releases can involve hundreds of thousands to millions of mosquitoes. The first *Wolbachia* population replacement programme in Cairns, Australia, involved the weekly release of ~30,000 *Wolbachia*-infected *Ae. aegypti* for 10 weeks in two suburbs with an estimated population size of ~20,000 *Ae. aegypti* (Hoffmann *et al*., 2011; Ritchie *et al*., 2013). Releases and monitoring (i.e.

regular *Wolbachia* screening of wild mosquitoes) should continue until the frequency of *Wolbachia* is well above the threshold for several weeks.

Population replacement is irreversible; once a *Wolbachia* infection has established, it may persist indefinitely. The *w*Mel strain of *Wolbachia* has remained at high frequencies in Cairns for a decade (Ryan *et al.*, 2019), with other programmes also showing persistence of *Wolbachia* across multiple years (Tantowijoyo *et al.*, 2020; Ahmad *et al.*, 2021). In continuous populations, where an area with *Wolbachia*-infected mosquitoes is adjacent to an area with uninfected mosquitoes, *Wolbachia* can continue to spread outside the release area (Schmidt *et al.*, 2017). The rate of spread depends on *Wolbachia* traits such as host fitness costs (Turelli and Barton, 2017) but also the environment; heterogeneity in mosquito population densities (Hancock *et al.*, 2019) and landscape features that inhibit mosquito dispersal (Schmidt *et al.*, 2018) can both slow spread. In *Ae. aegypti*, *w*Mel spreads at a rate of 100–200m per year (Schmidt *et al.*, 2017) or less (Tantowijoyo *et al.*, 2020), making it possible to contain *Wolbachia* within a local area. Consequently, releases across most of the target location may be required for complete *Wolbachia* coverage. Once established, *Wolbachia* infections can only drop out of the population if incomplete maternal transmission and/or immigration of uninfected mosquitoes push the *Wolbachia* frequency below the threshold required for spread. This irreversibility means that population replacement can provide ongoing protection from dengue without the need for further intervention.

Population suppression

Population suppression programmes (Fig. 6.2B) rely on the release of *Wolbachia*-infected males to induce cytoplasmic incompatibility with wild females, making them infertile. *Wolbachia*-based population suppression (also known as the incompatible insect technique) is like the sterile insect technique but uses cytoplasmic incompatibility rather than irradiation as the method of reducing female fertility. Population suppression programmes require accurate sex sorting so that only male mosquitoes are released. Sex sorting is typically achieved through mechanical separation at the pupal stage, followed by visual inspection of the adults to remove any remaining females (O'Connor *et al.*, 2012; Mains *et al.*, 2016). Recent releases have also used automated processes, including machine learning, to allow for larger-scale and more efficient sex sorting (Crawford *et al.*, 2020).

For population suppression to be effective, large numbers of *Wolbachia*-infected males need to be released over a sustained period. Suppression is easier to achieve when the mosquito population is low, such as during the dry season, because higher ratios of *Wolbachia*-infected males to wild males will increase the likelihood of female infertility (Chambers *et al.*, 2011). Ratios of at least 5:1 (*Wolbachia*-infected to wild males) are thought to be required for suppression (Pagendam *et al.*, 2020), but a recent programme that achieved greater than 95% suppression involved release ratios that exceeded 50:1

(Crawford *et al.*, 2020). The released males must be able to survive under field conditions, inseminate wild females and induce cytoplasmic incompatibility, so unfit males will reduce the efficiency of population suppression. Although *Wolbachia* infections do not typically affect male mating success (Segoli *et al.*, 2014), the fitness of released males could decrease as a consequence of mass rearing, inbreeding or adaptation to laboratory conditions (Qureshi *et al.*, 2019; Ross *et al.*, 2019a).

Population suppression is substantially more labour intensive than population replacement; it requires higher numbers of mosquitoes as well as accurate sex sorting. Population suppression is also temporary, requiring ongoing releases to maintain effectiveness. Elimination of a population is unlikely; *Wolbachia*-infected males may not reach all wild females in the population if coverage is incomplete or if females exhibit mating preferences. Since *Aedes* eggs may undergo diapause or quiescence, fertile adults could appear for months after releases begin. Even if elimination is successful, the population can re-establish from immigrant mosquitoes if suitable habitat remains. Consequently, suppression is likely to be more effective in locations that are geographically isolated from other mosquito populations (O'Connor *et al.*, 2012; Crawford *et al.*, 2020).

Population suppression, even if temporary, is still an effective way to control mosquito populations and brings several advantages over population replacement. Population suppression may be perceived more favourably by the community, because it can reduce nuisance biting by mosquitoes as well as the threat of disease. Because releases do not involve biting females, obtaining regulatory approval is also more straightforward (Murray *et al.*, 2016). Furthermore, the *Wolbachia*-infected mosquitoes are not intended to persist in the population, which may ease potential concerns about irreversible changes to the ecosystem. Because direct pathogen blocking by females is not required, any *Wolbachia* strain that induces cytoplasmic incompatibility (even natural *Wolbachia* infections) can potentially be used for population suppression. This means that population suppression can target a much broader range of mosquito species, including mosquitoes where *Wolbachia* has no effect on pathogen transmission or mosquitoes that are more important as nuisance pests than vectors.

Population replacement followed by suppression

'Population replacement followed by suppression' aims to reduce the mosquito population size after the *Wolbachia* infection has already been established in the population (Fig. 6.2C). This strategy requires a *Wolbachia* strain that causes host fitness costs and involves the release of both males and females. Once the *Wolbachia* infection is at a high frequency, host fitness costs can reduce the size of the population by decreasing mosquito survival or fertility. If the *Wolbachia* infection is at an intermediate frequency in the population, cytoplasmic incompatibility can also contribute to population suppression by making some females infertile, particularly when incompatibility is bidirectional (Moretti *et al.*, 2018a). Furthermore, *Wolbachia* strains that have pathogen-blocking effects can also reduce the vector competence of the remaining mosquitoes.

The extent of population suppression will depend on the nature and severity of the host fitness costs. When costs are triggered by environmental conditions, such as high temperatures or low rainfall, releases of *Wolbachia*-infected mosquitoes can be timed to facilitate the spread of *Wolbachia* when costs are low, followed by seasonal suppression when costs are high (Rašić *et al.*, 2014; Ritchie *et al.*, 2015). Most *Wolbachia* transinfections in mosquitoes reduce the viability of quiescent eggs (Allman *et al.*, 2020), making it possible to use these strains for population suppression during long dry seasons when rainfall is infrequent. *Ae. aegypti* carrying the *w*Mel strain of *Wolbachia* become partially self-incompatible at high temperatures (Ross *et al.*, 2019b), which may decrease population sizes during the hottest times of the year. If host fitness costs occur regardless of conditions, ongoing suppression could be achieved without further mosquito releases if the *Wolbachia* infection remains in the population. If host fitness costs are too severe, stable population replacement may not be possible (Nguyen *et al.*, 2015), but releases may still achieve temporary suppression.

Replacement followed by suppression has not yet been undertaken deliberately in the field, but semifield trials demonstrate its potential (Ritchie *et al.*, 2015). Some population replacement releases may have already achieved unintended suppression. Releases of *w*AlbB-infected *Ae. aegypti* in Malaysia decreased the size of the *Ae. aegypti* population in some locations (Nazni *et al.*, 2019), possibly due to host fitness costs in mosquitoes hatching from quiescent eggs (Lau *et al.*, 2020a). Because all *Wolbachia* strains being used for population replacement have host fitness costs, observed decreases in dengue transmission could in part be explained by decreased vector population sizes. Population replacement releases should therefore include monitoring of mosquito abundance in addition to *Wolbachia* frequencies.

Field Releases Around the World

Population replacement

Mosquitoes carrying *Wolbachia* infections are now being released in open-field trials (Fig. 6.3). The first population replacement release was conducted in 2011, leading to establishment of the *w*Mel *Wolbachia* strain in two *Ae. aegypti* populations in Cairns, Australia (Hoffmann *et al.*, 2011). Following this success, releases expanded to cover a total area of 150 km^2 throughout Australia's current *Ae. aegypti* distribution (O'Neill *et al.*, 2018; Ryan *et al.*, 2019). Although dengue is not endemic in Australia, imported dengue cases frequently result in local outbreaks. Today, local dengue transmission in Australia has almost completely been eliminated in locations where *Wolbachia* infections have been established (O'Neill *et al.*, 2018; Ryan *et al.*, 2019).

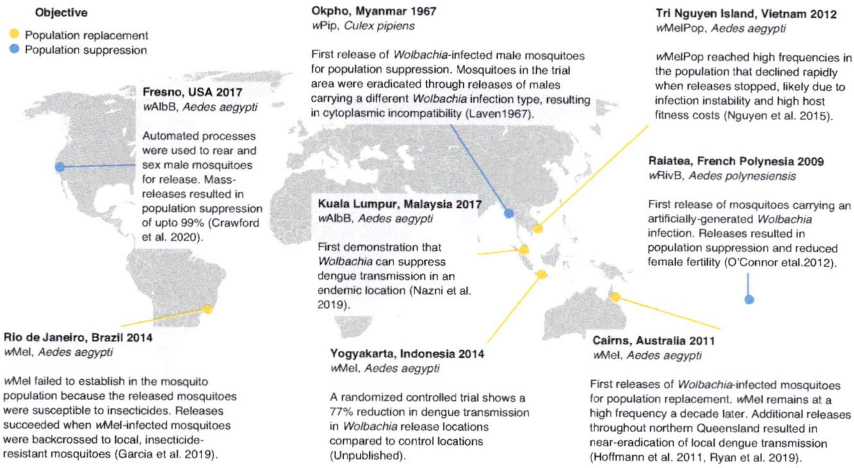

Fig. 6.3. Milestone releases of *Wolbachia*-infected mosquitoes for population replacement or population suppression.

Population replacement programmes are now operating in over 10 countries (http://worldmosquitoprogram.org, Nazni *et al.*, 2019). *Wolbachia* infections can invade and persist in natural populations (Garcia *et al.*, 2019; Tantowijoyo *et al.*, 2020) and reduce dengue transmission (Nazni *et al.*, 2019; Indriani *et al.*, 2020) in locations where dengue is endemic. Randomized controlled trials are also under way (Anders *et al.*, 2018), with (unpublished) results from Yogyakarta, Indonesia, indicating a 77% reduction in dengue transmission in locations where *Wolbachia*-infected mosquitoes have been released (https://www.worldmosquitoprogram.org/en/news-stories/media-releases/world-mosquito-programs-wolbachia-method-dramatically-reduces-dengue). Because *Wolbachia* infections remain in the population, population replacement is likely to provide long-term protection against dengue.

Not all population replacement releases have been successful. The *w*MelPop *Wolbachia* infection dropped out of natural populations after reaching high frequencies in two locations, likely because the host fitness costs of this strain are severe (Nguyen *et al.*, 2015). Releases with *w*Mel in Rio de Janeiro, Brazil, initially failed because insecticide-susceptible mosquitoes were being released into locations with heavy insecticide use, greatly reducing their fitness compared to the insecticide-resistant wild population (Garcia *et al.*, 2019). *Wolbachia* may also drop out from some locations if immigration of uninfected mosquitoes is too high, requiring additional releases (Nazni *et al.*, 2019).

To date, population replacement programmes have only been undertaken in *Ae. aegypti*, with almost all involving the *w*Mel *Wolbachia* strain. Although most research has focused on dengue, *Wolbachia* can block a variety of human pathogens in several vector species (Table 6.1), raising opportunities to expand release programmes to target alternative mosquito hosts and human pathogens. Given the diversity of *Wolbachia* strains in nature, strains may exist that could more effectively block pathogen transmission and persist in populations.

Population suppression

Wolbachia-based population suppression has a long history. The first releases took place during the 1960s in *Culex pipiens* mosquitoes (Laven, 1967), before the discovery that cytoplasmic incompatibility was caused by infection with *Wolbachia* (Yen and Barr, 1973). *Cx. pipiens* carry several related *Wolbachia* strains with diverse patterns of incompatibility (Atyame *et al.*, 2014). Population suppression was achieved by releasing male mosquitoes from a population that carried a different *Wolbachia* strain to the target population, resulting in cytoplasmic incompatibility (Laven, 1967). A later study achieved suppression by releasing *Aedes polynesiensis* carrying a *Wolbachia* strain from *Aedes riversi*, a closely related species (O'Connor *et al.*, 2012). These mosquitoes were generated by introducing the *Wolbachia* into an *Ae. polynesiensis* background through repeated backcrossing (Brelsfoard *et al.*, 2008).

Population suppression programmes experienced a resurgence in the 2010s after the first successful *Wolbachia* transinfections in mosquitoes (Xi *et al.*, 2005, 2006; McMeniman *et al.*, 2009), making population suppression with *Wolbachia* now possible for *Ae. aegypti* and *Aedes albopictus*. Recent studies show that releases of *Wolbachia*-infected males can directly reduce female fertility as well as population sizes (Mains *et al.*, 2016, 2019; Caputo *et al.*, 2020; Crawford *et al.*, 2020). Although population suppression can reduce the incidence of biting by female mosquitoes (Zheng *et al.*, 2019), no studies have linked reductions in mosquito populations to decreased disease incidence. Unfortunately, studies rarely include data from population monitoring after the release period, raising the question of how long suppression will last.

A common criticism of *Wolbachia*-based population suppression is that the accidental release of *Wolbachia*-infected females (which are fertile) could result in unintended population replacement. Although low levels of contamination during sex sorting are common, unintended population replacement has not yet been observed. Population replacement will only occur if the frequency of *Wolbachia*-infected females exceeds the threshold required for *Wolbachia* to spread, meaning that a low level of contamination is unlikely to pose a threat (Pagendam *et al.*, 2020). Furthermore, if the target mosquito population carries a different *Wolbachia* strain, this will further reduce the probability of unintended replacement, because the presence of bidirectional cytoplasmic incompatibility will increase the threshold frequency (Brelsfoard *et al.*, 2008; Moretti *et al.*, 2018a). To mitigate this risk, two recent population suppression programmes irradiated all released mosquitoes to ensure that any released females are sterile (Kittayapong *et al.*, 2019; Zheng *et al.*, 2019). However, these studies are more akin to the sterile insect technique, because the doses of irradiation used cause complete sterility in both sexes (Bond *et al.*, 2019), making the use of *Wolbachia* superfluous.

Choosing the right *Wolbachia* strain for the job

Over 25 artificial *Wolbachia* infections have been generated in mosquitoes through transinfection, mainly in *Ae. aegypti* and *Ae. albopictus* (Ross *et al.*, 2019c). These infections are diverse in their effects on pathogen blocking and host fitness costs, as well as in their stability in different environments. *Wolbachia* superinfections, where more than one *Wolbachia* strain occupies a single mosquito, have also been generated, which can have cumulative effects (Joubert *et al.*, 2016). For population suppression programmes, the choice of *Wolbachia* strain is relatively simple: the best strain will induce complete cytoplasmic incompatibility with the target population and have as few host fitness costs as possible to increase the potential competitiveness of males. However, ideal *Wolbachia* strains for population replacement do not exist, requiring a trade-off between *Wolbachia* infection stability, host fitness costs and pathogen blocking. The choice of *Wolbachia* strain must be made carefully.

A successful population replacement programme relies on *Wolbachia* infections invading and persisting in natural mosquito populations and blocking pathogen transmission. Three different *Wolbachia* strains in *Ae. aegypti* have been released in population replacement programmes, to varying levels of success (Table 6.2).

*w*Mel is the strain of choice for the World Mosquito Programme, who have been involved in most population replacement releases to date (https://www.worldmosquitoprogram.org). *w*Mel has relatively few host fitness costs, at least when tested under standard laboratory conditions (Ross *et al.*, 2019c). *w*Mel reduces the transmission of several arboviruses (Table 6.1), although blocking is incomplete and can be quite variable when tested under realistic conditions (Carrington *et al.*, 2017). *w*Mel's greatest strength is perhaps its reputation, with demonstrated success in multiple field trials. *w*Mel is used widely in basic research, including as a model for studying the mechanisms of pathogen blocking (Ford *et al.*, 2019; Koh *et al.*, 2020; Manokaran *et al.*, 2020) and cytoplasmic incompatibility (LePage *et al.*, 2017; Shropshire *et al.*, 2018). A key limitation of *w*Mel is its instability at high temperatures (Ulrich *et al.*, 2016; Ross *et al.*, 2017). Although *w*Mel has successfully invaded multiple *Ae. aegypti* populations in the tropics (O'Neill *et al.*, 2018; Ryan *et al.*, 2019; Indriani *et al.*, 2020), cytoplasmic incompatibility (Ross *et al.*, 2019b) and dengue blocking (Mancini *et al.*, 2021) could weaken after *w*Mel is established. Instability at high temperatures could help to explain the slow spread of *w*Mel and fluctuating infection frequencies in some trial locations (Ross *et al.*, 2020b; Tantowijoyo *et al.*, 2020), though direct comparisons between strains under field conditions have not yet been performed.

*w*AlbB was the first *Wolbachia* strain to be introduced to *Ae. aegypti* (Xi *et al.*, 2005) and is similar to *w*Mel in many ways; both strains induce complete

Table 6.2. *Wolbachia* strains are diverse in their effects. Comparison of the three *Wolbachia* strains in *Aedes aegypti* mosquitoes that have been released in open-field trials. For a comprehensive list of *Wolbachia* infections generated in mosquitoes, see Table S1 of Ross *et al.* (2019c).

	***Wolbachia* strain**		
	*w*AlbB	*w*MelPop	*w*Mel
Original host	*Aedes albopictus*	*Drosophila melanogaster*	*Drosophila melanogaster*
Reference for first transinfection	Xi *et al.* (2005)	McMeniman *et al.* (2009)	Walker *et al.* (2011)
Features	Cytoplasmic incompatibility, maternal transmission, pathogen blocking.	Cytoplasmic incompatibility, maternal transmission, pathogen blocking.	Cytoplasmic incompatibility, maternal transmission, pathogen blocking.
Key advantages	Heat and antibiotic resistant, with *Wolbachia* phenotypes likely to be stable in different environments (Ross *et al.*, 2017; Endersby-Harshman *et al.*, 2019). Host fitness costs when eggs are quiescent could be utilized for seasonal population suppression (Lau *et al.*, 2020a).	Strong pathogen blocking compared to other strains (Walker *et al.*, 2011; Ferguson *et al.*, 2015). Substantial host fitness costs could be effective for temporary population suppression (Ritchie *et al.*, 2015).	Limited host fitness costs allow *w*Mel to invade natural populations with relative ease (Walker *et al.*, 2011). Demonstrated success in multiple field trials around the world (worldmosquitoprogram.org).

Key disadvantages	*w*AlbB-infected females hatching from quiescent eggs become infertile. This fitness cost could reduce *Wolbachia* invasion and persistence in locations with long dry seasons (Lau *et al.*, 2020a).	Severe costs to host fitness, particularly when mosquitoes age, make stable population replacement unlikely (McMeniman and O'Neill, 2010; Yeap *et al.*, 2011). Occasional maternal transmission instability, even under benign conditions (Ross *et al.*, 2020a).	Instability at high temperatures could result in weakened cytoplasmic incompatibility, maternal transmission and pathogen blocking in tropical environments (Ross *et al.*, 2017, 2019b, ; Mancini *et al.*, 2021).
Field release outcomes	*w*AlbB reached stable, high frequencies in some release locations in a population replacement programme in Malaysia (Nazni *et al.*, 2019). *w*AlbB suppressed populations in Singapore, Australia and the USA through male-only releases (Mains *et al.*, 2019; Crawford *et al.*, 2020, unpublished).	*w*MelPop reached high frequencies in population replacement programmes in Australia and Vietnam but failed to persist (Nguyen *et al.*, 2015).	*w*Mel reached stable, high frequencies in population replacement programmes in several countries including Australia, Indonesia and Brazil (Hoffmann *et al.*, 2011; Garcia *et al.*, 2019; Tantowijoyo *et al.*, 2020), worldmosquitoprogram.org).

cytoplasmic incompatibility and have similar effects on pathogen blocking (Joubert *et al.*, 2016; Flores *et al.*, 2020). However, *w*AlbB is relatively stable at high temperatures (Ross *et al.*, 2017; Ant *et al.*, 2018), and its effects on host fitness can differ to *w*Mel under some environmental conditions (Lau *et al.*, 2020a; b). When *w*AlbB-infected *Ae. aegypti* hatch from eggs that are more than a few weeks old, a substantial proportion of females that hatch are infertile (Lau *et al.*, 2020a). These traits could make *w*AlbB more effective *Wolbachia* than *w*Mel for reducing dengue transmission in hot environments, but population replacement with *w*AlbB could be more challenging in locations with a long dry season.

The *w*MelPop strain of *Wolbachia* is a variant of *w*Mel that imposes a suite of costly effects on mosquitoes, including dramatically reduced lifespan and fertility (McMeniman *et al.*, 2009; Turley *et al.*, 2009; Yeap *et al.*, 2011). *w*MelPop is no longer being considered for field releases due to previous failures (Nguyen *et al.*, 2015), but it remains a popular choice for basic research due to its strong and far-reaching effects on insect hosts.

Planning an Effective *Wolbachia* Release Programme

Releases of mosquitoes carrying *Wolbachia* aim to reduce the incidence of disease (or nuisance biting) in a target population, but many factors will influence these objectives (Fig. 6.4). Anyone undertaking a *Wolbachia* release programme should first decide on the approach, whether it be replacement, suppression or a combination. The choice will depend on aspects of the target population including the incidence of disease, local expertise and facilities, as well as public perception. Population replacement is more cost-effective (Brady *et al.*, 2020) but may have greater perceived risks (Murphy *et al.*, 2010). Suppression could be preferred when nuisance biting is a key concern, when the release area is small and isolated or when the costs associated with ongoing releases can be met. Population suppression is only possible in locations with the resources to mass-rear and sex mosquitoes, but population replacement can be done almost anywhere. Since sex sorting is not required, mosquitoes can be mass-reared remotely, shipped as eggs to the release site and hatched locally. This approach was recently used for *Wolbachia* replacement releases in the Pacific Islands (http://www.eliminatedengue.com/pi/Progress/view/news/1063/pg/1). However, because storage and transport can reduce *Wolbachia*-infected egg viability (Allman *et al.*, 2020), remote mass rearing may increase production requirements and produce lower quality mosquitoes.

Controlling mosquito populations by releasing even more mosquitoes may seem counter-intuitive, especially when biting female mosquitoes are released. There are also potential risks unique to *Wolbachia* releases, such as the possibility of *Wolbachia* spreading to non-target insects or enhancing rather than suppressing virus transmission. Community engagement programmes are essential to ensure that the general public are informed and approve of releases (O'Neill *et al.*, 2018; Costa *et al.*, 2020). Engagement activities should

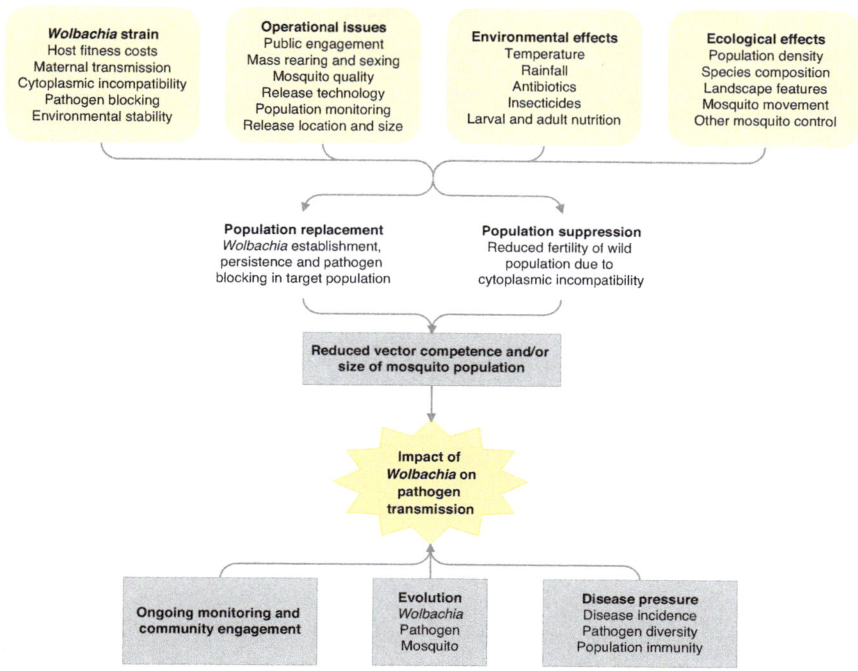

Fig. 6.4. Factors affecting the success of *Wolbachia* releases. This figure has been modified and expanded from Ross *et al.* (2019c).

clearly explain the *Wolbachia* technology and adequately address concerns from the community. Obtaining regulatory approval for *Wolbachia* releases is also necessary, but it is not enough. If the public oppose the technology or are unaware of the releases taking place, this will build mistrust and jeopardize future releases (Nading, 2014).

Successful population replacement or suppression requires more than just releasing as many mosquitoes as possible. Mosquitoes must be of high quality and released with adequate coverage of the target area. *Wolbachia* releases must consider the host fitness costs of the *Wolbachia* strain but also the fitness of the mosquitoes they inhabit. Mosquitoes that are inbred or that lack local genetic traits (such as insecticide resistance) are unlikely to perform well in the field (Garcia *et al.*, 2019; Ross *et al.*, 2019a). Backcrossing mosquitoes to the target population can improve fitness (Yeap *et al.*, 2011), but the *Wolbachia* infection must also be maintained at a high frequency in release stocks. Mosquito genetic background can influence pathogen blocking by *Wolbachia* (Terradas *et al.*, 2017; Ford *et al.*, 2019), as well as host fitness costs (Ritchie *et al.*, 2015; Carvalho *et al.*, 2020), making it important to evaluate strains in local mosquitoes before releases take place. Handling conditions are also crucial, since rearing, storage, transport and release procedures can affect mosquito quality (Chung *et al.*, 2018; Culbert *et al.*, 2018).

Release planning should involve careful consideration of the local environment, including landscape features and climate. Gaps in *Wolbachia* coverage may occur in areas where mosquito dispersal is limited (Jasper *et al.*, 2019) or where population densities are heterogeneous (Hancock *et al.*, 2019), requiring additional targeted releases. *Wolbachia* spread may depend on the distribution and productivity of larval habitats, with slower spread expected when larval competition is high (Hancock *et al.*, 2016) and when habitats are unsheltered and reach high temperatures (Ross *et al.*, 2019b). Verticality is also an important consideration; population suppression programmes in dense urban environments with high-rise apartments may be ineffective if mosquitoes are only released at ground level (https://www.straitstimes.com/singapore/male-wolbachia-carrying-mozzies-to-be-released-in-tampines-west-and-nee-soon-east-in-april). The nature of the local mosquito population can also affect release outcomes. In locations with multiple vector species, there is a risk that suppression of one species could increase the prevalence of others that occupy a similar niche, which could have unexpected consequences for pathogen transmission (Ritchie *et al.*, 2018). Natural *Wolbachia* infections in the target mosquito population could also be an issue; if the resident *Wolbachia* strain is compatible with the release strain, this can prevent replacement or suppression (Ross *et al.*, 2020c).

Releases of *Wolbachia*-infected mosquitoes for population replacement began only a decade ago, so how long will *Wolbachia* remain effective for? Over time, the phenotypic effects of *Wolbachia* are expected to weaken due to evolutionary changes in *Wolbachia*, mosquitoes or pathogens (Bull and Turelli, 2013). In *Ae. aegypti*, *Wolbachia* have so far remained stable, with no observed changes in cytoplasmic incompatibility, fitness costs or virus blocking (Frentiu *et al.*, 2014; Hoffmann *et al.*, 2014; Ross and Hoffmann, 2018; Ahmad *et al.*, 2021) and a few changes in the *Wolbachia* genome (Huang *et al.*, 2020) after field releases. No studies have tested whether *Wolbachia* releases cause genetic changes in mosquito populations, but mosquitoes could plausibly adapt to *Wolbachia* infection (White, 2011). Because *Wolbachia* and mitochondria are both maternally inherited, mitochondrial variants associated with the release strain will spread alongside *Wolbachia*, reducing the native diversity (Yeap *et al.*, 2016; Huang *et al.*, 2020). Since pathogen blocking by *Wolbachia* is usually incomplete, there is a risk that pathogens could evolve to bypass *Wolbachia*-mediated blocking. Laboratory selection experiments have failed to generate *Wolbachia*-resistant virus strains (Koh *et al.*, 2019; Martinez *et al.*, 2019), but virus diversity is much greater in natural populations. There is also a risk that wild mosquitoes could evolve mating preferences against the released mosquitoes (Cator *et al.*, 2020), decreasing the effectiveness of *Wolbachia* releases over time. With the potential for evolutionary changes, ongoing monitoring of *Wolbachia*, pathogen and mosquito populations is crucial to ensure that releases continue to achieve their objective (Ritchie *et al.*, 2018). If a *Wolbachia* strain loses its effectiveness, releases of different *Wolbachia* strains might extend the longevity of disease suppression (Joubert *et al.*, 2016), but this will depend on the mechanisms involved.

Conclusions

Releases of *Wolbachia*-infected mosquitoes can be an effective tool for disease control, with field trials showing tangible impacts on mosquito populations and dengue transmission. *Wolbachia* releases are best used in conjunction with existing vector control measures and are not intended to replace them (Ritchie and Johnson, 2017). Source reduction, surveillance and community awareness programmes will continue to be effective ways to reduce mosquito-borne disease (Fonseca *et al.*, 2013; Healy *et al.*, 2014; Trewin *et al.*, 2017; Forsyth *et al.*, 2020), and public awareness programmes for *Wolbachia* releases should continue to emphasize the importance of these measures. *Wolbachia* releases currently focus on a handful of mosquito species, but there is great potential for *Wolbachia* to control other vectors, pests and diseases (Kamtchum-Tatuene *et al.*, 2016; Gong *et al.*, 2020; Mateos *et al.*, 2020). Expansions of *Wolbachia* releases around the world will likely have great benefits, but releases must be supported by ongoing research, monitoring and community engagement.

References

Ahmad, N.A., Mancini, M.V., Ant, T.H., Martinez, J., Kamarul, G.M.R. *et al.* (2021) *Wolbachia* strain walbb maintains high density and dengue inhibition following introduction into a field population of *Aedes aegypti*. *Philosophical Transactions of the Royal Society of London. Series B, Biological Sciences* 376, 20190809.

Aliota, M.T., Peinado, S.A., Velez, I.D. and Osorio, J.E. (2016a) The wmel strain of *Wolbachia* reduces transmission of zika virus by *Aedes aegypti*. *Scientific Reports* 6, 28792.

Aliota, M.T., Walker, E.C., Uribe Yepes, A., Dario Velez, I., Christensen, B.M. *et al.* (2016b) The wmel strain of *Wolbachia* reduces transmission of chikungunya virus in *Aedes aegypti*. *PLoS Neglected Tropical Diseases* 10, e0004677.

Allman, M.J., Fraser, J.E., Ritchie, S.A., Joubert, D.A., Simmons, C.P. *et al.* (2020) *Wolbachia*'s deleterious impact on *Aedes aegypti* egg development: the potential role of nutritional parasitism. *Insects* 11, 735.

Anders, K.L., Indriani, C., Ahmad, R.A., Tantowijoyo, W., Arguni, E. *et al.* (2018) The AWED trial (applying *Wolbachia* to eliminate dengue) to assess the efficacy of *Wolbachia*-infected mosquito deployments to reduce dengue incidence in Yogyakarta, Indonesia: study protocol for a cluster randomised controlled trial. *Trials* 19, 302.

Andrews, E.S., Crain, P.R., Fu, Y., Howe, D.K. and Dobson, S.L. (2012) Reactive oxygen species production and *Brugia pahangi* survivorship in *Aedes polynesiensis* with artificial *Wolbachia* infection types. *PLoS Pathogens* 8, e1003075.

Ant, T.H., Herd, C.S., Geoghegan, V., Hoffmann, A.A. and Sinkins, S.P. (2018) The *Wolbachia* strain wau provides highly efficient virus transmission blocking in *Aedes aegypti*. *PLoS Pathogens* 14, e1006815.

Atyame, C.M., Pasteur, N., Dumas, E., Tortosa, P., Tantely, M.L. *et al.* (2011) Cytoplasmic incompatibility as a means of controlling *Culex pipiens quinquefasciatus* mosquito in the islands of the South-Western Indian ocean. *PLoS Neglected Tropical Diseases* 5, e1440.

Atyame, C.M., Labbe, P., Dumas, E., Milesi, P., Charlat, S. *et al.* (2014) *Wolbachia* divergence and the evolution of cytoplasmic incompatibility in *Culex pipiens*. *PLoS One* 9, e87336.

Axford, J.K., Ross, P.A., Yeap, H.L., Callahan, A.G. and Hoffmann, A.A. (2016) Fitness of walbb *Wolbachia* infection in *Aedes aegypti*: Parameter estimates in an outcrossed background and potential for population invasion. *American Journal of Tropical Medicine and Hygiene* 94, 507–516.

Baton, L.A., Pacidônio, E.C., Silva Gonçalves, D. and Moreira, L.A. (2013) WFlu: characterization and evaluation of a native *Wolbachia* from the mosquito *Aedes fluviatilis* as a potential vector control agent. *PLoS One* 8, e59619.

Bhattacharya, T., Newton, I.L.G. and Hardy, R.W. (2017) *Wolbachia* elevates host methyltransferase expression to block an RNA virus early during infection. *PLoS Pathogens* 13, e1006427.

Bhattacharya, T., Newton, I.L. and Hardy, R.W. (2020) Viral RNA is a target for *Wolbachia*-mediated pathogen blocking. *PLoS Pathogens* 16, e1008513.

Bian, G., Xu, Y., Lu, P., Xie, Y. and Xi, Z. (2010) The endosymbiotic bacterium *Wolbachia* induces resistance to dengue virus in *Aedes aegypti*. *PLoS Pathogens* 6, e1000833.

Bian, G., Joshi, D., Dong, Y., Lu, P., Zhou, G. *et al.* (2013a) *Wolbachia* invades *Anopheles stephensi* populations and induces refractoriness to *Plasmodium* infection. *Science* 340, 748–751.

Bian, G., Zhou, G., Lu, P. and Xi, Z. (2013b) Replacing a native *Wolbachia* with a novel strain results in an increase in endosymbiont load and resistance to dengue virus in a mosquito vector. *PLoS Neglected Tropical Diseases* 7, e2250.

Blagrove, M.S., Arias-Goeta, C., Failloux, A.B. and Sinkins, S.P. (2012) *Wolbachia* strain wmel induces cytoplasmic incompatibility and blocks dengue transmission in *Aedes albopictus*. *Proceedings of the National Academy of Sciences of the United States of America* 109, 255–260.

Blagrove, M.S., Arias-Goeta, C., Genua, C., Failloux, A.B. and Sinkins, S.P. (2013) A *Wolbachia* wmel transinfection in *Aedes albopictus* is not detrimental to host fitness and inhibits chikungunya virus. *PLoS Neglected Tropical Diseases* 7, e2152.

Bond, J.G., Osorio, A.R., Avila, N., Gomez-Simuta, Y., Marina, C.F. *et al.* (2019) Optimization of irradiation dose to *Aedes aegypti* and *Ae. albopictus* in a sterile insect technique program. *PLoS One* 14, e0212520.

Brady, O.J., Kharisma, D.D., Wilastonegoro, N.N., O'Reilly, K.M., Hendrickx, E. *et al.* (2020) The cost-effectiveness of controlling dengue in Indonesia using *w*Mel *Wolbachia* released at scale: a modelling study. *BMC Medicine* 18, 1–12.

Brelsfoard, C.L. and Dobson, S.L. (2011) *Wolbachia* effects on host fitness and the influence of male aging on cytoplasmic incompatibility in *Aedes polynesiensis* (Diptera: Culicidae). *Journal of Medical Entomology* 48, 1008–1015.

Brelsfoard, C.L., Séchan, Y. and Dobson, S.L. (2008) Interspecific hybridization yields strategy for South Pacific filariasis vector elimination. *PLoS Neglected Tropical Diseases* 2, e129.

Brownlie, J.C., Cass, B.N., Riegler, M., Witsenburg, J.J., Iturbe-Ormaetxe, I. *et al.* (2009) Evidence for metabolic provisioning by a common invertebrate endosymbiont,

Wolbachia pipientis, during periods of nutritional stress. *PLoS Pathogens* 5, e1000368.

Brownstein, J.S., Hett, E. and O'Neill, S.L. (2003) The potential of virulent *Wolbachia* to modulate disease transmission by insects. *Journal of Invertebrate Pathology* 84, 24–29.

Buchman, A., Gamez, S., Li, M., Antoshechkin, I., Li, H.H. *et al.* (2020) Broad dengue neutralization in mosquitoes expressing an engineered antibody. *PLoS Pathogens* 16, e1008103.

Bull, J.J. and Turelli, M. (2013) *Wolbachia* versus dengue: evolutionary forecasts. *Evolution, Medicine, and Public Health* 197–207.

Caputo, B., Moretti, R., Manica, M., Serini, P., Lampazzi, E. *et al.* (2020) A bacterium against the tiger: preliminary evidence of fertility reduction after release of *Aedes albopictus* males with manipulated *Wolbachia* infection in an italian urban area. *Pest Management Science* 76, 1324–1332.

Caragata, E.P., Rances, E., Hedges, L.M., Gofton, A.W., Johnson, K.N. *et al.* (2013) Dietary cholesterol modulates pathogen blocking by *Wolbachia*. *PLoS Pathogens* 9, e1003459.

Caragata, E.P., Rances, E., O'Neill, S.L. and McGraw, E.A. (2014) Competition for amino acids between wolbachia and the mosquito host, *Aedes aegypti*. *Microbial Ecology* 67, 205–218.

Caragata, E.P. and Moreira, L.A. (2017) Using an endosymbiont to control mosquito-transmitted disease. In: Wikel, S.K., Aksoy, S. and Dimopoulos, G. (eds) *Arthropod Vector: Controller of Disease Transmission*, Vol. 1. Academic Press, Cambridge, Massachusetts, pp. 123–142.

Carrington, L.B., Tran, B.C.N., Le, N.T.H., Luong, T.T.H., Nguyen, T.T. *et al.* (2017) Field- and clinically derived estimates of *Wolbachia*-mediated blocking of dengue virus transmission potential in *Aedes aegypti* mosquitoes. *Proceedings of the National Academy of Sciences of the United States of America* 115, 361–366.

Carvalho, D.O., Torres-Monzon, J.A., Koskinioti, P., Dilrukshi Wijegunawardana, N.D.A., Liang, X. *et al.* (2020) *Aedes aegypti* lines for combined sterile insect technique and incompatible insect technique applications: the importance of host genomic background. *Entomologia Experimentalis et Applicata* 168, 560–572.

Caspari, E. and Watson, G. (1959) On the evolutionary importance of cytoplasmic sterility in mosquitoes. *Evolution* 13, 568–570.

Cator, L.J., Wyer, C.A. and Harrington, L.C. (2020) Mosquito sexual selection and reproductive control programs. *Trends in Parasitology* 37, 330–339.

Chambers, E.W., Hapairai, L., Peel, B.A., Bossin, H. and Dobson, S.L. (2011) Male mating competitiveness of a *Wolbachia*-introgressed aedes polynesiensis strain under semi-field conditions. *PLoS Neglected Tropical Diseases* 5, e1271.

Chouin-Carneiro, T., Ant, T.H., Herd, C., Louis, F., Failloux, A.B. *et al.* (2020) *Wolbachia* strain walba blocks zika virus transmission in *Aedes aegypti*. *Medical and Veterinary Entomology* 34, 116–119.

Chrostek, E. and Gerth, M. (2019) Is anopheles gambiae a natural host of *Wolbachia*. *mBio* 10, e00784-19.

Chrostek, E., Martins, N.E., Marialva, M.S. and Teixeira, L. (2020) *Wolbachia*-conferred antiviral protection is determined by developmental temperature. *bioRxiv* 169169.

Chung, H.N., Rodriguez, S.D., Gonzales, K.K., Vulcan, J., Cordova, J.J. *et al.* (2018) Toward implementation of mosquito sterile insect technique: the effect of storage conditions on survival of male aedes *Aegypti* mosquitoes (Diptera: Culicidae) during transport. *Journal of Insect Science* 18, 2.

Cook, P.E., McMeniman, C.J. and O'Neill, S.L. (2008) Modifying insect population age structure to control vector-borne disease. In: Aksoy, S. (ed.) *Transgenesis and the Management of Vector-Borne Disease.* Springer, New York, pp. 126–140.

Costa, G.B., Smithyman, R., O'Neill, S.L. and Moreira, L.A. (2020) How to engage communities on a large scale? lessons from world mosquito program in rio de janeiro. *Brazil. Gates Open Research* 4, 109.

Crawford, J.E., Clarke, D.W., Criswell, V., Desnoyer, M., Cornel, D. *et al.* (2020) Efficient production of male *Wolbachia*-infected *Aedes aegypti* mosquitoes enables large-scale suppression of wild populations. *Nature Biotechnology* 38, 482–492.

Culbert, N.J., Maiga, H., Somda, N.S.B., Gilles, J.R.L., Bouyer, J. *et al.* (2018) Longevity of mass-reared, irradiated and packed male *Anopheles arabiensis* and *Aedes aegypti* under simulated environmental field conditions. *Parasites and Vectors* 11, 603.

Curtis, C. and Sinkins, S. (1998) *Wolbachia* as a possible means of driving genes into populations. *Parasitology* 116, S111–S115.

Dobson, S.L., Marsland, E.J. and Rattanadechakul, W. (2001) *Wolbachia*-induced cytoplasmic incompatibility in single- and superinfected *Aedes albopictus* (Diptera: Culicidae). *Journal of Medical Entomology* 38, 382–387.

Dobson, S.L., Marsland, E.J. and Rattanadechakul, W. (2002) Mutualistic *Wolbachia* infection in *Aedes albopictus*: accelerating cytoplasmic drive. *Genetics* 160, 1087–1094.

Dodson, B.L., Hughes, G.L., Paul, O., Matacchiero, A.C., Kramer, L.D. *et al.* (2014) *Wolbachia* enhances west nile virus (WNV) infection in the mosquito *Culex tarsalis*. *PLoS Neglected Tropical Diseases* 8, e2965.

Dutra, H.L., Rocha, M.N., Dias, F.B., Mansur, S.B., Caragata, E.P. *et al.* (2016) *Wolbachia* blocks currently circulating zika virus isolates in Brazilian *Aedes aegypti* mosquitoes. *Cell Host and Microbe* 19, 771–774.

Ekwudu, O., Devine, G.J., Aaskov, J.G. and Frentiu, F.D. (2020) *Wolbachia* strain *w*AlBB blocks replication of flaviviruses and alphaviruses in mosquito cell culture. *Parasites and Vectors* 13, 54.

Endersby-Harshman, N.M., Axford, J.K. and Hoffmann, A.A. (2019) Environmental concentrations of antibiotics may diminish *Wolbachia* infections in *Aedes aegypti* (Diptera: Culicidae). *Journal of Medical Entomology* 56, 1078–1086.

Fast, E.M., Toomey, M.E., Panaram, K., Desjardins, D., Kolaczyk, E.D. *et al.* (2011) *Wolbachia* enhance *Drosophila* stem cell proliferation and target the germline stem cell niche. *Science* 334, 990–992.

Ferguson, N.M., Hue Kien, D.T., Clapham, H., Aguas, R., Trung, V.T. *et al.* (2015) Modeling the impact on virus transmission of *Wolbachia*-mediated blocking of dengue virus infection of *Aedes aegypti*. *Science Translational Medicine* 7.

Flores, H.A., Bruyne, J., O'Donnell, T.B., Tuyet Nhu, V., Thi Giang, N. *et al.* (2020) Multiple *Wolbachia* strains provide comparative levels of protection against dengue virus infection in *Aedes aegypti*. *PLoS Pathogens* 16, e1008433.

Fonseca, D.M., Unlu, I., Crepeau, T., Farajollahi, A., Healy, S.P. *et al.* (2013) Area-wide management of *Aedes albopictus*. Part 2: gauging the efficacy of traditional integrated pest control measures against urban container mosquitoes. *Pest Management Science* 69, 1351–1361.

Ford, S.A., Allen, S.L., Ohm, J.R., Sigle, L.T., Sebastian, A. *et al.* (2019) Selection on *Aedes aegypti* alters *Wolbachia*-mediated dengue virus blocking and fitness. *Nature Microbiology* 4, 1832–1839.

Forsyth, J.E., Mutuku, F.M., Kibe, L., Mwashee, L., Bongo, J. *et al.* (2020) Source reduction with a purpose: mosquito ecology and community perspectives offer insights for improving household mosquito management in Coastal Kenya. *PLoS Neglected Tropical Diseases* 14, e0008239.

Fraser, J.E., Bruyne, J.T., Iturbe-Ormaetxe, I., Stepnell, J., Burns, R.L. *et al.* (2017) Novel *Wolbachia*-transinfected *Aedes aegypti* mosquitoes possess diverse fitness and vector competence phenotypes. *PLoS Pathogens* 13, e1006751.

Fraser, J.E., O'Donnell, T.B., Duyvestyn, J.M., O'Neill, S.L., Simmons, C.P. *et al.* (2020) Novel phenotype of *Wolbachia* strain wpip in *Aedes aegypti* challenges assumptions on mechanisms of *Wolbachia*-mediated dengue virus inhibition. *PLoS Pathogens* 16, e1008410.

Frentiu, F.D., Zakir, T., Walker, T., Popovici, J., Pyke, A.T. *et al.* (2014) Limited dengue virus replication in field-collected *Aedes aegypti* mosquitoes infected with *Wolbachia. PLoS Neglected Tropical Diseases* 8, e2688.

Fu, Y., Gavotte, L., Mercer, D.R. and Dobson, S.L. (2010) Artificial triple *Wolbachia* infection in *Aedes albopictus* yields a new pattern of unidirectional cytoplasmic incompatibility. *Applied and Environmental Microbiology* 76, 5887–5891.

Garcia, G.D.A., Sylvestre, G., Aguiar, R., Costa, G.B., Martins, A.J. *et al.* (2019) Matching the genetics of released and local *Aedes aegypti* populations is critical to assure *Wolbachia* invasion. *PLoS Neglected Tropical Diseases* 13, e0007023.

Geoghegan, V., Stainton, K., Rainey, S.M., Ant, T.H., Dowle, A.A. *et al.* (2017) Perturbed cholesterol and vesicular trafficking associated with dengue blocking in *Wolbachia*-infected *Aedes aegypti* cells. *Nature Communications* 8, 526.

Glaser, R.L. and Meola, M.A. (2010) The native *Wolbachia* endosymbionts of *Drosophila melanogaster* and *Culex quinquefasciatus* increase host resistance to West Nile virus infection. *PLoS One* 5, e11977.

Gloria-Soria, A., Chiodo, T.G. and Powell, J.R. (2018) Lack of evidence for natural *Wolbachia* infections in *Aedes aegypti* (Diptera: Culicidae). *Journal of Medical Entomology* 55, 1354–1356.

Gong, J.T., Li, Y., Li, T.P., Liang, Y., Hu, L. *et al.* (2020) Stable introduction of plant-virus-inhibiting wolbachia into planthoppers for rice protection. *Current Biology* 30, 4837–4845.

Guo, Y., Hoffmann, A.A., Xu, X.Q., Zhang, X., Huang, H.J. *et al.* (2018) *Wolbachia*-induced apoptosis associated with increased fecundity in *Laodelphax striatellus* (Hemiptera: Delphacidae). *Insect Molecular Biology* 27, 796–807.

Hancock, P.A., Linley-White, V., Callahan, A.G., Godfray, H.C.J., Hoffmann, A.A. *et al.* (2016) Density-dependent population dynamics in *Aedes aegypti* slow the spread of *w*Mel *Wolbachia. Journal of Applied Ecology* 53, 785–793.

Hancock, P.A., Ritchie, S.A., Koenraadt, C.J.M., Scott, T.W., Hoffmann, A.A. *et al.* (2019) Predicting the spatial dynamics of wolbachia infections in aedes aegypti arbovirus vector populations in heterogeneous landscapes. *Journal of Applied Ecology* 56, 1674–1686.

Healy, K., Hamilton, G., Crepeau, T., Healy, S., Unlu, I. *et al.* (2014) Integrating the public in mosquito management: active education by community peers can lead to significant reduction in peridomestic container mosquito habitats. *PLoS One* 9, e108504.

Hedges, L.M., Brownlie, J.C., O'Neill, S.L. and Johnson, K.N. (2008) *Wolbachia* and virus protection in insects. *Science* 322, 702.

Hoffmann, A.A. and Turell, M.J. (1997) Cytoplasmic incompatibility in insects. In: O'Neill, S.L., Hoffmann, A.A. and Werren, J.H. (eds) *Influential Passengers: Inherited Microorganisms and Arthropod Reproduction.* Oxford University Press, Oxford, UK, pp. 42–80.

Hoffmann, A.A., Montgomery, B.L., Popovici, J., Iturbe-Ormaetxe, I., Johnson, P.H. *et al.* (2011) Successful establishment of *Wolbachia* in *Aedes* populations to suppress dengue transmission. *Nature* 476, 454–457.

Hoffmann, A.A., Iturbe-Ormaetxe, I., Callahan, A.G., Phillips, B.L., Billington, K. *et al.* (2014) Stability of the wmel *Wolbachia* infection following invasion into *Aedes aegypti* populations. *PLoS Neglected Tropical Diseases* 8, e3115.

Huang, B., Yang, Q., Hoffmann, A.A., Ritchie, S.A., Hurk, A.F. *et al.* (2020) *Wolbachia* genome stability and mtdna variants in *Aedes aegypti* field populations eight years after release. *iScience* 23, 101572.

Hughes, G.L. and Rasgon, J.L. (2014) Transinfection: a method to investigate *Wolbachia*-host interactions and control arthropod-borne disease. *Insect Molecular Biology* 23, 141–151.

Hughes, G.L., Koga, R., Xue, P., Fukatsu, T. and Rasgon, J.L. (2011) *Wolbachia* infections are virulent and inhibit the human malaria parasite *Plasmodium falciparum* in *Anopheles* gambiae. *PLoS Pathogens* 7, e1002043.

Hughes, G.L., Vega-Rodriguez, J., Xue, P. and Rasgon, J.L. (2012) *Wolbachia* strain *w*AlBB enhances infection by the rodent malaria parasite *Plasmodium berghei* in *Anopheles* gambiae mosquitoes. *Applied and Environmental Microbiology* 78, 1491–1495.

Huigens, M.E. and Stouthamer, R. (2003) Parthenogenesis associated with *Wolbachia*. *Insect Symbiosis* 1, 247–266.

Hurk, A.F., Hall-Mendelin, S., Pyke, A.T., Frentiu, F.D., Mcelroy, K. *et al.* (2012) Impact of *Wolbachia* on infection with Chikungunya and yellow fever viruses in the mosquito vector *Aedes aegypti*. *PLoS Neglected Tropical Diseases* 6, e1892.

Hurst, G. and Jiggins, F.M. (2000) Male-killing bacteria in insects: mechanisms, incidence, and implications. *Emerging Infectious Diseases* 6, 329.

Hussain, M., Lu, G., Torres, S., Edmonds, J.H., Kay, B.H. *et al.* (2013) Effect of *Wolbachia* on replication of West Nile virus in a mosquito cell line and adult mosquitoes. *Journal of Virology* 87, 851–858.

Indriani, C., Tantowijoyo, W., Rancès, E., Andari, B., Prabowo, E. *et al.* (2020) Reduced dengue incidence following deployments of *Wolbachia*-infected *Aedes aegypti* in Yogyakarta, Indonesia: a quasi-experimental trial using controlled interrupted time series analysis. *Gates Open Research* 4, 50.

Jasper, M., Schmidt, T.L., Ahmad, N.W., Sinkins, S.P. and Hoffmann, A.A. (2019) A genomic approach to inferring kinship reveals limited intergenerational dispersal in the yellow fever mosquito. *Molecular Ecology Resources* 19, 1254–1264.

Joshi, D., Pan, X., McFadden, M.J., Bevins, D., Liang, X. *et al.* (2017) The maternally inheritable *Wolbachia w*AlBB induces refractoriness to *Plasmodium berghei* in *Anopheles stephensi*. *Frontiers in Microbiology* 8, 366.

Joubert, D.A. and O'Neill, S.L. (2017) Comparison of stable and transient *Wolbachia* infection models in *Aedes aegypti* to block dengue and West Nile viruses. *PLoS Neglected Tropical Diseases* 11, e0005275.

Joubert, D.A., Walker, T., Carrington, L.B., Bruyne, J.T., Kien, D.H. *et al.* (2016) Establishment of a *Wolbachia* superinfection in *Aedes aegypti* mosquitoes as a potential approach for future resistance management. *PLoS Pathogens* 12, e1005434.

Ju, J.F., Bing, X.L., Zhao, D.S., Guo, Y., Xi, Z. *et al.* (2020) *Wolbachia* supplement biotin and riboflavin to enhance reproduction in planthoppers. *The ISME Journal* 14, 676–687.

Kambris, Z., Cook, P.E., Phuc, H.K. and Sinkins, S.P. (2009) Immune activation by life-shortening *Wolbachia* and reduced filarial competence in mosquitoes. *Science* 326, 134–136.

Kamtchum-Tatuene, J., Makepeace, B.L., Benjamin, L., Baylis, M. and Solomon, T. (2016) The potential role of *Wolbachia* in controlling the transmission of emerging human arboviral infections. *Current Opinion in Infectious Diseases* 30, 108–116.

King, J.G., Souto-Maior, C., Sartori, L.M., Maciel-De-Freitas, R. and Gomes, M.G.M. (2018) Variation in *Wolbachia* effects on *Aedes* mosquitoes as a determinant of invasiveness and vectorial capacity. *Nature Communications* 9:1483.

Kittayapong, P., Mongkalangoon, P., Baimai, V. and O'Neill, S. (2002) Host age effect and expression of cytoplasmic incompatibility in field populations of *Wolbachia*-superinfected *Aedes albopictus*. Heredity 88, 270–274.

Kittayapong, P., Ninphanomchai, S., Limohpasmanee, W., Chansang, C., Chansang, U. *et al.* (2019) Combined sterile insect technique and incompatible insect technique: the first proof-of-concept to suppress *Aedes aegypti* vector populations in semi-rural settings in Thailand. *PLoS Neglected Tropical Diseases* 13, e0007771.

Koh, C., Audsley, M.D., Di Giallonardo, F., Kerton, E.J., Young, P.R. *et al.* (2019) Sustained *Wolbachia*-mediated blocking of dengue virus isolates following serial passage in *Aedes aegypti* cell culture. *Virus Evolution* 5, vez012.

Koh, C., Islam, M.N., Ye, Y.H., Chotiwan, N., Graham, B. *et al.* (2020) Dengue virus dominates lipid metabolism modulations in *Wolbachia*-coinfected *Aedes aegypti*. Communications Biology 3:518.

Lau, M.-J., Ross, P.A. and Hoffmann, A.A. (2020a) Infertility and fecundity loss of *Wolbachia*-infected *Aedes aegypti* hatched from quiescent eggs is expected to alter invasion dynamics. *bioRxiv* 397240.

Lau, M.J., Ross, P.A., Endersby-Harshman, N.M. and Hoffmann, A.A. (2020b) Impacts of low temperatures on *Wolbachia* (Rickettsiales: Rickettsiaceae)-infected *Aedes aegypti* (Diptera: Culicidae). *Journal of Medical Entomology* 57, 1567–1574.

Laven, H. (1967) Eradication of *Culex pipiens fatigans* through cytoplasmic incompatibility. *Nature* 216, 383–384.

LePage, D.P., Metcalf, J.A., Bordenstein, S.R., On, J., Perlmutter, J.I. *et al.* (2017) Prophage WO genes recapitulate and enhance *Wolbachia*-induced cytoplasmic incompatibility. *Nature* 543, 243–247.

Lindsey, A.R., Bhattacharya, T., Newton, I.L. and Hardy, R.W. (2018) Conflict in the intracellular lives of endosymbionts and viruses: a mechanistic look at *Wolbachia*-mediated pathogen-blocking. *Viruses* 10, 141.

Lu, P., Bian, G., Pan, X. and Xi, Z. (2012) *Wolbachia* induces density-dependent inhibition to dengue virus in mosquito cells. *PLoS Neglected Tropical Diseases* 6, e1754.

Ma, W.J. and Schwander, T. (2017) Patterns and mechanisms in instances of endosymbiont-induced parthenogenesis. *Journal of Evolutionary Biology* 30, 868–888.

Mains, J.W., Brelsfoard, C.L., Rose, R.I. and Dobson, S.L. (2016) Female adult *Aedes albopictus* suppression by *Wolbachia*-infected male mosquitoes. *Scientific Reports* 6, 33846.

Mains, J.W., Kelly, P.H., Dobson, K.L., Petrie, W.D. and Dobson, S.L. (2019) Localized control of *Aedes aegypti* (Diptera: Culicidae) in Miami, FL, via inundative releases of *Wolbachia*-infected male mosquitoes. *Journal of Medical Entomology* 56, 1296–1303.

Mancini, M.V., Herd, C.S., Ant, T.H., Murdochy, S.M. and Sinkins, S.P. (2020) *Wolbachia* strain wau efficiently blocks arbovirus transmission in *Aedes albopictus. PLoS Neglected Tropical Diseases* 14, e0007926.

Mancini, M.V., Ant, T.H., Herd, C.S., Martinez, J., Murdochy, S.M. *et al.* (2021) High temperature cycles result in maternal transmission and dengue infection differences between *Wolbachia* strains in *Aedes aegypti. mBio* 12, e0025021.

Manokaran, G., Flores, H.A., Dickson, C.T., Narayana, V.K., Kanojia, K. *et al.* (2020) Modulation of acyl-carnitines, the broad mechanism behind *Wolbachia*-mediated inhibition of medically important flaviviruses in *Aedes aegypti. Proceedings of the National Academy of Sciences of the United States of America* 117, 24475–24483.

Martinez, J., Bruner-Montero, G., Arunkumar, R., Smith, S.C.L., Day, J.P. *et al.* (2019) Virus evolution in *Wolbachia*-infected *Drosophila. Proceedings of the Royal Society B: Biological Sciences*286, 20192117.

Mateos, M., Martinez Montoya, H., Lanzavecchia, S.B., Conte, C., Guillén, K. *et al.* (2020) *Wolbachia pipientis* associated with tephritid fruit fly pests: from basic research to applications. *Frontiers in Microbiology* 11, 1080.

McMeniman, C.J. and O'Neill, S.L. (2010) A virulent *Wolbachia* infection decreases the viability of the dengue vector *Aedes aegypti* during periods of embryonic quiescence. *PLoS Neglected Tropical Diseases* 4, e748.

McMeniman, C.J., Lane, R.V., Cass, B.N., Fong, A.W., Sidhu, M. *et al.* (2009) Stable introduction of a life-shortening *Wolbachia* infection into the mosquito *Aedes aegypti. Science* 323, 141–144.

Min, K.-T. and Benzer, S. (1997) *Wolbachia*, normally a symbiont of drosophila, can be virulent, causing degeneration and early death. *Proceedings of the National Academy of Sciences of the United States of America* 94, 10792–10796.

Moreira, L.A., Iturbe-Ormaetxe, I., Jeffery, J.A., Lu, G., Pyke, A.T. *et al.* (2009a) A *Wolbachia* symbiont in *Aedes aegypti* limits infection with dengue, Chikungunya, and Plasmodium. *Cell* 139, 1268–1278.

Moreira, L.A., Saig, E., Turley, A.P., Ribeiro, J.M., O'Neill, S.L. *et al.* (2009b) Human probing behavior of *Aedes aegypti* when infected with a life-shortening strain of *Wolbachia. PLoS Neglected Tropical Diseases* 3, e568.

Moretti, R., Marzo, G.A., Lampazzi, E. and Calvitti, M. (2018a) Cytoplasmic incompatibility management to support incompatible insect technique against *Aedes albopictus. Parasites and Vectors* 11, 649.

Moretti, R., Yen, P.S., Houe, V., Lampazzi, E., Desiderio, A. *et al.* (2018b) Combining *Wolbachia*-induced sterility and virus protection to fight *Aedes albopictus*-borne viruses. *PLoS Neglected Tropical Diseases* 12, e0006626.

Mousson, L., Zouache, K., Arias-Goeta, C., Raquin, V., Mavingui, P. *et al.* (2012) The native *Wolbachia* symbionts limit transmission of dengue virus in *Aedes albopictus. PLoS Neglected Tropical Diseases* 6, e1989.

Murdock, C.C., Blanford, S., Hughes, G.L., Rasgon, J.L. and Thomas, M.B. (2014) Temperature alters *Plasmodium* blocking by *Wolbachia. Scientific Reports* 4, 3932.

Murphy, B., Jansen, C., Murray, J. and Barro, P. (2010) *Risk Analysis on the Australian Release of Aedes aegypti (L.)(Diptera: Culicidae) Containing Wolbachia.* CSIRO Entomology, Indooroopilly, Australia.

Murray, J.V., Jansen, C.C. and Barro, P. (2016) Risk associated with the release of *Wolbachia*-infected *Aedes aegypti* mosquitoes into the environment in an effort to control dengue. *Frontiers in Public Health* 4, 43.

Nading, A.M. (2014) The lively ethics of global health GMOS: the case of the Oxitec mosquito. *BioSocieties* 10, 24–47.

Nainu, F., Trenerry, A. and Johnson, K.N. (2019) *Wolbachia*-mediated antiviral protection is cell-autonomous. *Journal of General Virology* 100, 1587–1592.

Nazni, W.A., Hoffmann, A.A., Noorafizah, A., Cheong, Y.L., Mancini, M.V. *et al.* (2019) Establishment of *Wolbachia* strain walbb in Malaysian populations of *Aedes aegypti* for dengue control. *Current Biology* 29, 4241–4248.

Newton, I.L., Savytskyy, O. and Sheehan, K.B. (2015) *Wolbachia* utilize host actin for efficient maternal transmission in *Drosophila melanogaster*. *PLoS Pathogens* 11, e1004798.

Nguyen, T.H., Nguyen, H.L., Nguyen, T.Y., Vu, S.N., Tran, N.D. *et al.* (2015) Field evaluation of the establishment potential of *w*MelPop *Wolbachia* in Australia and Vietnam for dengue control. *Parasites and Vectors* 8, 563.

O'Connor, L., Plichart, C., Sang, A.C., Brelsfoard, C.L., Bossin, H.C. *et al.* (2012) Open release of male mosquitoes infected with a *Wolbachia* biopesticide: field performance and infection containment. *PLoS Neglected Tropical Diseases* 6, e1797.

O'Neill, S.L., Ryan, P.A., Turley, A.P., Wilson, G., Retzki, K. *et al.* (2018) Scaled deployment of *Wolbachia* to protect the community from dengue and other *Aedes* transmitted arboviruses. *Gates Open Research* 2, 36.

Osborne, S.E., San Leong, Y., O'Neill, S.L. and Johnson, K.N. (2009) Variation in antiviral protection mediated by different *Wolbachia* strains in *Drosophila* simulans. *PLoS Pathogens* 5, e1000656.

Pagendam, D.E., Trewin, B.J., Snoad, N., Ritchie, S.A., Hoffmann, A.A. *et al.* (2020) Modelling the *Wolbachia* incompatible insect technique: strategies for effective mosquito population elimination. *BMC Biology* 18, 161.

Pan, X., Thiem, S. and Xi, Z. (2017) *Wolbachia*-mediated immunity induction in mosquito vectors. In: Wikel, S.K., Aksoy, S. and Dimopoulos, G. (eds) *Arthropod Vector: Controller of Disease Transmission*, Vol. 1. Academic Press, Cambridge Massachusetts, pp. 35–58.

Panteleev, D.Y., Goryacheva, I.I., Andrianov, B.V., Reznik, N.L., Lazebny, O.E. *et al.* (2007) The endosymbiotic bacterium *Wolbachia* enhances the nonspecific resistance to insect pathogens and alters behavior of *Drosophila melanogaster*. *Russian Journal of Genetics* 43, 1066–1069.

Pereira, T.N., Rocha, M.N., Sucupira, P.H.F., Carvalho, F.D. and Moreira, L.A. (2018) *Wolbachia* significantly impacts the vector competence of *Aedes aegypti* for mayaro virus. *Scientific Reports* 8, 1–9.

Qureshi, A., Aldersley, A., Hollis, B., Ponlawat, A. and Cator, L.J. (2019) Male competition and the evolution of mating and life-history traits in experimental populations of *Aedes aegypti*. *Proceedings of the Royal Society B: Biological Sciences 286*, 20190591.

Rainey, S.M., Martinez, J., McFarlane, M., Juneja, P., Sarkies, P. *et al.* (2016) *Wolbachia* blocks viral genome replication early in infection without a transcriptional response by the endosymbiont or host small RNA pathways. *PLoS Pathogens* 12, e1005536.

Rances, E., Ye, Y.H., Woolfit, M., McGraw, E.A. and O'Neill, S.L. (2012) The relative importance of innate immune priming in *Wolbachia*-mediated dengue interference. *PLoS Pathogens* 8, e1002548.

Raquin, V., Valiente Moro, C., Saucereau, Y., Tran, F.H., Potier, P. *et al.* (2015) Native *Wolbachia* from *Aedes albopictus* blocks Chikungunya virus infection in cellulo. *PLoS One* 10, e0125066.

Rasgon, J.L. and Scott, T.W. (2003) *Wolbachia* and cytoplasmic incompatibility in the California *Culex pipiens* mosquito species complex: parameter estimates and infection dynamics in natural populations. *Genetics* 165, 2029–2038.

Rasgon, J.L., Styer, L.M. and Scott, T.W. (2003) *Wolbachia*-induced mortality as a mechanism to modulate pathogen transmission by vector arthropods. *Journal of Medical Entomology* 40, 125–132.

Rašić, G., Endersby, N.M., Williams, C. and Hoffmann, A.A. (2014) Using *Wolbachia*-based release for suppression of *Aedes* mosquitoes: insights from genetic data and population simulations. *Ecological Applications* 24, 1226–1234.

Ritchie, S.A. and Johnson, B.J. (2017) Advances in vector control science: rear-and-release strategies show promise... but don't forget the basics. *The Journal of Infectious Diseases* 215, S103–S108.

Ritchie, S.A., Montgomery, B.L. and Hoffmann, A.A. (2013) Novel estimates of *Aedes aegypti* (Diptera: Culicidae) population size and adult survival based on *Wolbachia* releases. *Journal of Medical Entomology* 50, 624–631.

Ritchie, S.A., Townsend, M., Paton, C.J., Callahan, A.G. and Hoffmann, A.A. (2015) Application of wmelpop *Wolbachia* strain to crash local populations of *Aedes aegypti*. *PLoS Neglected Tropical Diseases* 9, e0003930.

Ritchie, S.A., Hurk, A.F., Smout, M.J., Staunton, K.M. and Hoffmann, A.A. (2018) Mission accomplished? We need a guide to the 'post release' world of *Wolbachia* for *Aedes*-borne disease control. *Trends in Parasitology* 34, 217–226.

Rocha, M.N., Duarte, M.M., Mansur, S.B., Silva, B.D.M.E., Pereira, T.N. *et al.* (2019) Pluripotency of *Wolbachia* against arboviruses: the case of yellow fever. *Gates Open Research* 3, 161.

Ross, P.A. and Hoffmann, A.A. (2018) Continued susceptibility of the wmel *Wolbachia* infection in *Aedes aegypti* to heat stress following field deployment and selection. *Insects* 9, 78.

Ross, P.A., Wiwatanaratanabutr, I., Axford, J.K., White, V.L., Endersby-Harshman, N.M. *et al.* (2017) *Wolbachia* infections in *Aedes aegypti* differ markedly in their response to cyclical heat stress. *PLoS Pathogens* 13, e1006006.

Ross, P.A., Endersby-Harshman, N.M. and Hoffmann, A.A. (2019a) A comprehensive assessment of inbreeding and laboratory adaptation in *Aedes aegypti* mosquitoes. *Evolutionary Applications* 12, 572–586.

Ross, P.A., Ritchie, S.A., Axford, J.K. and Hoffmann, A.A. (2019b) Loss of cytoplasmic incompatibility in *Wolbachia*-infected *Aedes aegypti* under field conditions. *PLoS Neglected Tropical Diseases* 13, e0007357.

Ross, P.A., Turelli, M. and Hoffmann, A.A. (2019c) Evolutionary ecology of *Wolbachia* releases for disease control. *Annual Review of Genetics* 53, 93–116.

Ross, P.A., Axford, J.K., Callahan, A.G., Richardson, K.M. and Hoffmann, A.A. (2020a) Persistent deleterious effects of a deleterious *Wolbachia* infection. *PLoS Neglected Tropical Diseases* 14, e0008204.

Ross, P.A., Axford, J.K., Yang, Q., Staunton, K.M., Ritchie, S.A. *et al.* (2020b) Heatwaves cause fluctuations in *w*Mel *Wolbachia* densities and frequencies in *Aedes aegypti*. *PLoS Neglected Tropical Diseases* 14, e0007958.

Ross, P.A., Callahan, A.G., Yang, Q., Jasper, M., Arif, M.A.K. *et al.* (2020c) An elusive endosymbiont: does *Wolbachia* occur naturally in *Aedes aegypti*. *Ecology and Evolution* 10, 1581–1591.

Ryan, P.A., Turley, A.P., Wilson, G., Hurst, T.P., Retzki, K. *et al.* (2019) Establishment of *w*Mel *Wolbachia* in *Aedes aegypti* mosquitoes and reduction of local dengue transmission in Cairns and surrounding locations in Northern Queensland, Australia. *Gates Open Research* 3, 1547.

Sanaei, E., Charlat, S. and Engelstadter, J. (2020) *Wolbachia* host shifts: routes, mechanisms, constraints and evolutionary consequences. *Biological Reviews of the Cambridge Philosophical Society* 96, 433–453.

Schmidt, T.L., Barton, N.H., Rasic, G., Turley, A.P., Montgomery, B.L. *et al.* (2017) Local introduction and heterogeneous spatial spread of dengue-suppressing *Wolbachia* through an urban population of *Aedes aegypti*. *PLoS Biology* 15, e2001894.

Schmidt, T.L., Filipovic, I., Hoffmann, A.A. and Rasic, G. (2018) Fine-scale landscape genomics helps explain the slow spatial spread of *Wolbachia* through the *Aedes aegypti* population in Cairns, Australia. *Heredity* 120, 386–395.

Schnettler, E., Sreenu, V.B., Mottram, T. and Mcfarlane, M. (2016) *Wolbachia* restricts insect-specific flavivirus infection in *Aedes aegypti* cells. *Journal of General Virology* 97, 3024.

Schultz, M.J., Isern, S., Michael, S.F., Corley, R.B., Connor, J. *et al.* (2017) Variable inhibition of zika virus replication by different *Wolbachia* strains in mosquito cell cultures. *Journal of Virology* 91, e00339–17.

Segoli, M., Hoffmann, A.A., Lloyd, J., Omodei, G.J. and Ritchie, S.A. (2014) The effect of virus-blocking *Wolbachia* on male competitiveness of the dengue vector mosquito, *Aedes aegypti*. *PLoS Neglected Tropical Diseases* 8, e3294.

Shropshire, J.D., On, J., Layton, E.M., Zhou, H. and Bordenstein, S.R. (2018) One prophage WO gene rescues cytoplasmic incompatibility in *Drosophila melanogaster*. *Proceedings of the National Academy of Sciences of the United States of America* 115, 4987–4991.

Shropshire, J.D., Leigh, B. and Bordenstein, S.R. (2020) Symbiont-mediated cytoplasmic incompatibility: what have we learned in 50 years? *eLife* 9, e61989.

Silva, L.M., Dezordi, F.Z., Paiva, M.H.S. and Wallau, G.L. (2021) Systematic review of *Wolbachia* symbiont detection in mosquitoes: an entangled topic about methodological power and true symbiosis. *Pathogens* 10, 39.

Sucupira, P.H.F., Ferreira, A.G.A., Leite, T., Mendonca, S.F., Ferreira, F.V. *et al.* (2020) The RNAi pathway is important to control mayaro virus infection in *Aedes aegypti* but not for *Wolbachia*-mediated protection. *Viruses* 12, 871.

Tantowijoyo, W., Andari, B., Arguni, E., Budiwati, N., Nurhayati, I. *et al.* (2020) Stable establishment of *w*Mel *Wolbachia* in *Aedes aegypti* populations in Yogyakarta. *PLoS Neglected Tropical Diseases* 14, e0008157.

Teixeira, L., Ferreira, A. and Ashburner, M. (2008) The bacterial symbiont *Wolbachia* induces resistance to RNA viral infections in *Drosophila melanogaster*. *PLoS Biology* 6, e1000002.

Terradas, G. and McGraw, E.A. (2017) *Wolbachia*-mediated virus blocking in the mosquito vector *Aedes aegypti*. *Current Opinion in Insect Science* 22, 37–44.

Terradas, G., Allen, S.L., Chenoweth, S.F. and McGraw, E.A. (2017) Family level variation in *Wolbachia*-mediated dengue virus blocking in *Aedes aegypti*. *Parasites and Vectors* 10, 622.

Trewin, B.J., Darbro, J.M., Jansen, C.C., Schellhorn, N.A., Zalucki, M.P. *et al.* (2017) The elimination of the dengue vector, *Aedes aegypti*, from Brisbane, Australia: the role of surveillance, larval habitat removal and policy. *PLoS Neglected Tropical Diseases* 11, e0005848.

Turelli, M. and Barton, N.H. (2017) Deploying dengue-suppressing *Wolbachia*: Robust models predict slow but effective spatial spread in *Aedes aegypti*. *Theoretical Population Biology* 115, 45–60.

Turley, A.P., Moreira, L.A., O'Neill, S.L. and Mcgraw, E.A. (2009) *Wolbachia* infection reduces blood-feeding success in the dengue fever mosquito, *Aedes aegypti*. *PLoS Neglected Tropical Diseases* 3, e516.

Ulrich, J.N., Beier, J.C., Devine, G.J. and Hugo, L.E. (2016) Heat sensitivity of *w*Mel *Wolbachia* during *Aedes aegypti* development. *PLoS Neglected Tropical Diseases* 10, e0004873.

Walker, T., Johnson, P.H., Moreira, L.A., Iturbe-Ormaetxe, I., Frentiu, F.D. *et al.* (2011) The *w*Mel *Wolbachia* strain blocks dengue and invades caged *Aedes aegypti* populations. *Nature* 476, 450–453.

Walker, T., Quek, S., Jeffries, C.L., Bandibabone, J., Dhokiya, V. *et al.* (2020) Genomic and microscopic evidence of stable high density and maternally inherited *Wolbachia* infections in *Anopheles* mosquitoes. *bioRxiv* 357400.

White, J.A. (2011) Caught in the act: rapid, symbiont-driven evolution: Endosymbiont infection is a mechanism generating rapid evolution in some arthropods – but how widespread is the phenomenon. *Bioessays* 33, 823–829.

Wu, C. (2020) *Characterization of Wolbachia infections from native Australian mosquitoes*. PhD, The University of Queensland.

Xi, Z., Khoo, C.C. and Dobson, S.L. (2005) *Wolbachia* establishment and invasion in an *Aedes aegypti* laboratory population. *Science* 310, 326–328.

Xi, Z., Khoo, C.C. and Dobson, S.L. (2006) Interspecific transfer of *Wolbachia* into the mosquito disease vector *Aedes albopictus*. *Proceedings of the Royal Society B: Biological Sciences* 273, 1317–1322.

Yeap, H.L., Mee, P., Walker, T., Weeks, A.R., O'Neill, S.L. *et al.* (2011) Dynamics of the "popcorn" *Wolbachia* infection in outbred *Aedes aegypti* informs prospects for mosquito vector control. *Genetics* 187, 583–595.

Yeap, H.L., Rasic, G., Endersby-Harshman, N.M., Lee, S.F., Arguni, E. *et al.* (2016) Mitochondrial DNA variants help monitor the dynamics of *Wolbachia* invasion into host populations. *Heredity (Edinb)* 116, 265–276.

Yen, J.H. and Barr, A.R. (1973) The etiological agent of cytoplasmic incompatibility in *Culex pipiens*. *Journal of Invertebrate Pathology* 22, 242–250.

Zheng, X., Zhang, D., Li, Y., Yang, C., Wu, Y. *et al.* (2019) Incompatible and sterile insect techniques combined eliminate mosquitoes. *Nature* 572, 56–61.

Zug, R. and Hammerstein, P. (2015) Bad guys turned nice? A critical assessment of *Wolbachia* mutualisms in arthropod hosts. *Biological Reviews of the Cambridge Philosophical Society* 90, 89–111.

Ivermectin for the Control and Elimination of Malaria

7

MARTA MAIA[1,2*] AND KEVIN KOBYLINSKI[3]

[1]*Kenya Medical Research Institute (KEMRI) Wellcome Trust Research Programme, Kilifi, Kenya*
[2]*University of Oxford, Centre for Tropical Medicine and Global Health, Oxford, United Kingdom*
[3]*Mahidol Oxford Research Unit, Bangkok, Thailand*

Abstract

Ivermectin, a widely used endectocide, holds promise for malaria control by rendering blood-feeding Anopheles mosquitoes lethal, suggesting potential for mass drug administration (MDA) and mass ivermectin treatment of livestock (ITL). Discovered in 1973 from Streptomyces avermitilis , ivermectin targets glutamate-gated chloride ion channels, offering a novel mechanism against malaria vectors. Initial studies demonstrated significant Anopheles mortality when feeding on ivermectin-treated hosts, with field trials confirming reduced mosquito survival. Regional research highlights varied impacts, with significant reductions in key malaria vectors across Africa, Latin America and South-east Asia. Beyond lethality, ivermectin's sublethal effects include delayed feeding, inhibited parasite development and reduced fecundity, which collectively diminish malaria transmission. Combining ivermectin with antimalarials like dihydroartemisinin-piperaquine (DP) enhances mosquito mortality, while ITL provides a viable strategy for outdoor-feeding vectors. This chapter explores ivermectin's multifaceted role in accelerating global malaria elimination efforts.

Introduction

Ivermectin is a safe endectocidal drug used widely in human and animal medicine to treat a wide range of internal and external parasitic diseases.

*Corresponding author: MMaia@kemri-wellcome.org

 Practical Control of Mosquito Disease Vectors (ed J.D. Charlwood)
DOI: 10.1079/9781789248845.0007

Ivermectin-treated hosts are lethal to blood-feeding *Anopheles* mosquitoes. Thus, ivermectin mass drug administration (MDA) and mass ivermectin treatment of livestock (ITL) hold promise as novel vector control tools to aid malaria control and elimination efforts. Large-scale MDA programmes utilize ivermectin for the elimination of onchocerciasis and lymphatic filariasis. Ivermectin affects the glutamate-gated chloride ion channels of nematodes and insects, which is a novel mode of action from the insecticides currently used for mosquito control. Delivering ivermectin at the point of mosquito blood feeding represents a novel route of delivery that can directly target residual malaria transmission. This chapter reviews how ivermectin MDA and mass ITL could aid and accelerate global malaria elimination efforts.

Ivermectin Discovery

The bacteria, *Streptomyces avermitilis* was first isolated from a soil sample on a Japanese golf course by the Kitasato Institute in 1973 (Burg *et al.*, 1979). A partnership between Merck, Sharpe and Dohme, Inc. (MSD) and the Kitasato Institute was formed in 1973 specifically to test novel bacterial samples for anti-nematocidal and anti-protozoal properties (Omura and Crump, 2004). A tandem assay was developed whereby bacterial culture and fermentation broth was fed to mice infected with either *Eimeria muris* (Protozoan) or *Nematospiroides dubius* (Helminth), and subsequent monitoring was used to determine if the broth reduced parasite loads in the mice (Stapley and Woodruff, 1982). Of the 40,000 bacterial broths tested, only one (OS-3153) reduced *N. dubius* parasite loads, which was from the Japanese golf course (Burg *et al.*, 1979). In 1975, the active anthelmintic compound produced was isolated and named avermectin (Burg *et al.*, 1979; Egerton *et al.*, 1979; Miller *et al.*, 1979). Avermectins are macrocyclic lactones comprised of a series of pentacyclic macrolactones with a disaccharide of methylated deoxysugar L-oleandrose attached at C-13. Fermentation of *S. avermitilis* produces four pairs of homologous compounds (i.e. A_{1a}, A_{1b}, A_{2a}, A_{2b}, B_{1a}, B_{1b}, B_{2a} and B_{2b}), and various combinations of these compounds produce several commercialized avermectins, including ivermectin, selamectin, doramectin, eprinomectin, abamectin and emamectin. Ivermectin is composed of the B_1 fractions, a 22,23-di-hydro B_1 complex ($B_{1a} > 80\%$, $B_{1b} < 20\%$) (Egerton *et al.*, 1979; Yoon *et al.*, 2004; Wei *et al.*, 2005), which is the only avermectin licensed for human use (Pitterna *et al.*, 2009).

Initial human trials in the early 1980s demonstrated that ivermectin was safe, well tolerated and effective at controlling *Onchocerca volvulus*, the causative agent of onchocerciasis. Ivermectin is lethal to the *O. volvulus* microfilariae and limits release of microfilariae for up to 9 months but is not lethal to the adult female (Duke *et al.*, 1992). In 1987, ivermectin was licensed for human use, and shortly thereafter, MSD announced that they would donate ivermectin for the control of onchocerciasis for the duration of the elimination efforts (Collins, 2004). In 1998, ivermectin was registered for the human treatment of lymphatic filariasis (LF) (Dreyer *et al.*, 1995) and MSD pledged to donate ivermectin for the treatment of LF in all areas where it was co-endemic with onchocerciasis in Africa (Crump and Ōmura, 2011). Both onchocerciasis and LF elimination programmes conduct once- or twice-yearly ivermectin MDAs for elimination efforts. It was recently shown that triple therapy with ivermectin, diethylcarbamazine and albendazole was more effective for LF control than dual combinations, as it is lethal to the adult nematode (Kanamitie *et al.*, 2017; Weil *et al.*, 2020). Therefore, MSD has pledged an additional 100 million ivermectin treatments per year for LF MDAs outside of Africa (MDP, 2017, Gustavsen *et al.*, 2018). In 2019, over 400 million ivermectin treatments were administered by MDA, with four billion treatments given to date (MDP, 2020). There are numerous secondary health benefits from these MDA programmes as ivermectin can affect a wide range of neglected tropical diseases, including onchocerciasis, LF, scabies, lice, strongyloidiasis, cutaneous larval migrans, myiasis, ascariasis, trichuriasis and gnathostomiasis.

Early *Anopheles* Ivermectin Research

Ivermectin has been documented to effect the survivorship or fecundity of numerous arthropod vectors of medical and veterinary importance, including mites (Mounsey *et al.*, 2008), Ixodid ticks (Drummond *et al.*, 1981), lice (Glaziou *et al.*, 1994), triatomine bugs (de Azambuja *et al.*, 1985), botflies, calliphorids (Drummond, 1984), stable flies, horn flies (Miller *et al.*, 1986), tsetse flies (Distelmans *et al.*, 1983), ceratopogonids (Standfast *et al.*, 1984), sand flies (Mascari *et al.*, 2008; Mascari and Foil, 2010) and mosquitoes (Ali and Nayar, 1985; Pampiglione *et al.*, 1985; Iakubovich *et al.*, 1989; Tesh and Guzman, 1990; Cartel *et al.*, 1991; Focks *et al.*, 1991; Mahmood *et al.*, 1991; Jones *et al.*, 1992; Gardner *et al.*, 1993; Bockarie *et al.*, 1999; Chandre and Hougard, 1999; Foley *et al.*, 2000; Fritz *et al.*, 2009; Chaccour *et al.*, 2010). Interestingly in mosquitoes, with the exception of *Aedes polynesiensis* (Cartel *et al.*, 1991), the *Aedes* and *Culex* genera are not nearly as susceptible to ivermectin as are the *Anopheles*, a point that remains unresolved to this day.

The first report of the effect of ivermectin on mosquitoes demonstrated that the survival of colony-raised *Anopheles stephensi*, *Aedes aegypti* and *Culex quinquefasciatus* was reduced after feeding on ivermectin-injected mice or on ivermectin soaked cotton, with *An. stephensi* being the most susceptible species (Pampiglione *et al.*, 1985). Iakubovich *et al.* (1989)

fed colonized *Anopheles sacharovi*, *Anopheles atroparvus* and *An. stephensi* on rabbits injected with 340 μg/kg of ivermectin. Six days after blood meal ingestion, 93% of the *An. stephensi*, 70% of the *An. atroparvus* and 79% of the *An. sacharovi* had died. Jones *et al.* (1992) found that wild, field-caught *Anopheles quadrimaculatus* blood fed on ivermectin-injected dogs (10–2500 μg/kg) had 90.5% mortality within 24 hour and 98.6% mortality within 48 hour. This demonstrates for the first time that field-caught *Anopheles* are susceptible to ivermectin at concentrations found in a treated vertebrate (Jones *et al.*, 1992). A follow-up study demonstrated dogs orally treated with lower doses of ivermectin at 6, 12 and 24 μg/kg, corresponding to 6 ± 1, 11 ± 2 and 16 ± 7.5 ng/ml in blood at 4 hour post ingestion, to be lethal to field-caught *An. quadrimaculatus* held for 48 hour (mortality 33.2, 66.9, 65.3% respectively) (Gardner *et al.*, 1993). One issue with the above studies is that although ivermectin susceptibility of the mosquitoes was demonstrated, the lethal concentration that killed 50% (LC_{50}) of the mosquitoes was not calculated, making comparisons of ivermectin susceptibility across species difficult. Bockarie *et al.* (1999) performed the first field study on the effects of ivermectin MDA on wild *Anopheles* survival. The study investigated *Anopheles punctulatus* survival at the time of MDA (single-dose 400 μg/kg ivermectin plus 6 mg/kg diethylcarbamazine) for LF control in Papua New Guinea. Blood-fed mosquitoes were aspirated from nine houses in one village for three consecutive days before and after MDA and again at 28 days after MDA. Within 1 day post collection, 70% of the *An. punctulatus* collected up to 3 days after the MDA died, while none of the *An. punctulatus* pre-MDA died within 1 day. By the 9th day post collection, none of the *An. punctulatus* collected within 3 days post-MDA were alive, while 67% of *An. punctulatus* caught pre-MDA survived. This study, albeit underpowered, demonstrates that field-caught *Anopheles* are susceptible to ivermectin-treated humans during MDA. Foley *et al.* (2000) fed colonized *Anopheles farauti* (Sarong strain) on a single human administered ivermectin (250 μg/kg) and demonstrated their blood was lethal up to 14 days post ingestion. The reduction of the daily probability of mosquito survival (p) is the most influential variable in the vectorial capacity equation, and demonstration that ivermectin MDAs can reduce *Anopheles* survival suggests a potential role for ivermectin MDA or mass ITL in malaria control.

Ivermectin Impacts on *Anopheles* Survival by Region

It would be nearly a decade before additional evidence was published on ivermectin impact on *Anopheles*, with three different research teams separately converging on the mosquito-lethal effect of ivermectin on the primary African malaria vector, *Anopheles gambiae*. Fritz *et al.* (2009) documented, for the first time, the effects of ivermectin on colonized *An. gambiae* s.s. and *Anopheles arabiensis*, which are members of the *An. gambiae* complex.

Ivermectin compound was fed to *An. gambiae* s.s. (Kisumu strain) and *An. arabiensis* (Dongola strain), and both were stated to have the same 9-day-LC_{50} of 19.8 ng/ml. *An. gambiae* s.s. were also fed on Zebu bulls injected with 600 µg/kg ivermectin, which is three times the recommended standard concentration for cattle. The median survival of *An. gambiae* s.s. was reduced for mosquitoes that fed on ivermectin-injected bulls compared to saline-injected control bulls up to 20 days post treatment. Based on these results, the authors conclude that mass ivermectin treatment of cattle in Africa could reduce the transmission of malaria parasites (Fritz *et al.*, 2009). A laboratory study demonstrated that *An. gambiae* s.s. (G3 strain) had an ivermectin 5-day-LC_{50} of 22.4 [95% CI 18.0–26.9] ng/ml and that two consecutive ivermectin blood meals with the second at sublethal concentrations can compound mortality (Kobylinski *et al.*, 2010). Chaccour *et al.* (2010) fed *An. gambiae* s.s. (G3 strain) directly on people that ingested 200 µg/kg ivermectin 1 and 14 days post ingestion, with significant reduction in mosquito survival when fed on people 1 day but not 14 days post treatment. Field evidence from south-eastern Senegal examined the impact of ivermectin (150 µg/kg) MDAs for onchocerciasis control on wild *An. gambiae* s.l. survival. Five villages were investigated across two field seasons, with paired mosquito collections occurring in a control and treatment village. Blood-fed mosquitoes were aspirated from homes and held in an insectary for 5 days, demonstrating an *An. gambiae* s.s. and *An. arabiensis* reduction in survival of 50% and 38% respectively, up to 6 days post administration (Sylla *et al.*, 2010). This same field investigation also demonstrated that single ivermectin MDAs reduced the proportion of *An. gambiae* s.s. that were *Plasmodium falciparum* infectious by 79% for at least 2 weeks (Kobylinski *et al.*, 2011), demonstrating that ivermectin MDAs can reduce *Plasmodium* transmission. Follow-on trials with similar study design in Senegal, Liberia and Burkina Faso provided further evidence that single ivermectin (150–200 µg/kg) with or without albendazole (400 mg) MDAs reduced *An. gambiae* s.l. survival by 34% for 1 week post-MDA and the sporozoite rate by 77% for 2 weeks. This reduction in *An. gambiae* survival led to a shift in the mosquito population age structure as measured by the parity rate (i.e. the proportion of older parous mosquitoes that have laid eggs compared to younger nulliparous mosquitoes that have not laid eggs), which was reduced by 30% for 2 weeks post-MDA (Alout *et al.*, 2014). This parity analysis is instrumental, as the sporozoite rate can fluctuate dramatically from day to day and can be very low, especially in areas nearing malaria elimination. Thus, a proxy such as the parity rate can be used to demonstrate impact on vector populations, and by extension on transmission (Slater *et al.*, 2020). In Nigeria, following ivermectin (150–200 µg/kg) MDAs for onchocerciasis in two treatment villages compared to two control villages, there was a significant reduction in *Anopheles* parity for up to 2 weeks post-MDA (Omitola *et al.*, 2021). In Burkina Faso, a repeated ivermectin (150–200 µg/kg) MDA trial for malaria (RIMDMAL) control administered ivermectin MDAs six times every 3 weeks, with four treated and four control villages while monitoring malaria incidence in children 1–5 years old. Results from this trial showed that in villages treated with ivermectin, there was a 20% decrease in cumulative clinical *Plasmodium falciparum* cases. Additionally, the proportion of children who remained malaria-free doubled, and the median time to first infection was significantly extended from 38 days to 47 days (Foy *et al.*, 2019). This is the first field evidence to demonstrate that ivermectin MDAs can reduce clinical malaria burden in human populations. Survival of colonized *Anopheles coluzzii*, an important vector and member of the *An. gambiae* complex, was reduced when fed upon cattle up to 3 weeks after injection with ivermectin at 200 µg/kg (Pooda *et al.*, 2015). Another important African malaria vector *Anopheles funestus* was shown to be susceptible to ivermectin when fed blood from treated malaria patients that ingested ivermectin (200 µg/kg) and was roughly as susceptible as *An. gambiae* s.s. (Ouédraogo *et al.*, 2015). Taken together, this information suggests that either ivermectin MDA or mass ITL holds promise for malaria control efforts in Africa.

Residual malaria transmission, which is the *Plasmodium* transmission that occurs following full implementation of vector control tools (e.g. long-lasting insecticide treated nets (LLINs) and/or indoor residual spraying (IRS) with insecticides), is driven primarily by early-biting and outdoor-feeding *Anopheles*. While IRS and LLINs have caused impressive reductions in malaria burden in Africa, these reductions have stalled as of 2017, with some countries now facing increased transmission (WHO, 2018). In Africa, there is evidence that mass LLIN rollout has caused species composition shifts within the *An. gambiae* complex (Mwangangi *et al.*, 2003; Bayoh *et al.*, 2010; Githinji *et al.*, 2020) or vector behavioural shifts (Geissbühler *et al.*, 2007; Moiroux *et al.*, 2012; Bamou *et al.*, 2018), leading to increased residual transmission (Bamou *et al.*, 2021). Since ivermectin treatment makes the vertebrate host lethal to *Anopheles* regardless of feeding location or time, ivermectin MDA would be an ideal vector control measure to directly target residual malaria transmission. As *Anopheles* vectors in South-east Asia and South America typically display outdoor-feeding and early-biting behaviours ivermectin MDA holds much promise in these regions as well.

In Latin America, ivermectin MDAs successfully eradicated 11 of 13 onchocerciasis transmission foci by escalating MDAs from one to four times per year (Cupp *et al.*, 2004; Cupp and Cupp, 2005). This indicates that ivermectin MDAs can be effectively coordinated and are a possibility for malaria control in this region. The key Amazonian vector *Anopheles darlingi* is susceptible to ivermectin at human-relevant concentrations with a 7-day-LC_{50} of 43.2 [95% CI 37.5–48.6] ng/ml (Kobylinski *et al.*, 2017a). As *An. darlingi* tends to be exophagic and exophilic (Moreno *et al.*, 2007; Girod *et al.*, 2008; Moutinho *et al.*, 2011), ivermectin MDA is an attractive vector control. Ivermectin pharmacokinetic modelling (Kobylinski *et al.*, 2017b) suggests that the 400 µg/kg dose is the ideal minimal MDA dose to target *An. darlingi*. Ivermectin at the 400 µg/kg dose is safe (Kobylinski *et al.*, 2017b) and is recommended for LF MDAs (Dembele *et al.*, 2010).

Repeated doses every 2 weeks for 12 weeks were shown to be well tolerated in a trial in Brazil (Dreyer *et al.*, 1996) and Sri Lanka (Ismail *et al.*, 1996). Another South American vector *Anopheles aquasalis* is also susceptible to ivermectin at human-relevant concentrations with a 5-day-LC_{50} of 47.0 [95% CI 44.7–49.4] ng/ml (Sampaio *et al.*, 2016). The Central American vector *Anopheles albimanus* had an excessively high ivermectin 4-day-LC_{50} of 1,468.0 [95% CI 1,153.5–1965.5] ng/ml when using the colonized *An. albimanus* STECL strain (Dreyer *et al.*, 2018). However, subsequent studies with wild-caught *An. albimanus* fed on cattle injected with ivermectin (200 µg/kg) demonstrated that *An. albimanus* are susceptible to ivermectin concentrations in treated cattle (Dreyer *et al.*, 2019), suggesting that the high LC_{50} found with colonized *An. albimanus* (STECL strain) (Dreyer *et al.*, 2018) may not be relevant to field settings (Dreyer *et al.*, 2022) and that ivermectin could be a viable control measure against this vector. Both *An. albimanus* and *An. aquasalis* feed more frequently on cattle than humans, suggesting that mass ITL may be more optimal for malaria control than human MDA. To date, no ivermectin MDA or mass ITL field trials for malaria control have been initiated in Latin America.

In South-east Asia, ivermectin use in humans remains limited with registration for strongyloidiasis treatment in Thailand and limited off-label use throughout the region. However, with the recent expansion of ivermectin, diethylcarbamazine and albendazole MDA for LF elimination efforts in the Asia and Oceania regions (e.g. India, Malaysia, Timor Leste, Papua New Guinea, American Samoa, Fiji and Tuvalu; https://www.msdresponsibility.com/access-to-health/affordability/mectizan-donation-program/), ivermectin will likely become more common in the region soon. *Anopheles* vectors in South-east Asia tend to feed outdoors more than indoors and often times earlier in the night (Trung *et al.*, 2005; Van Bortel *et al.*, 2010; Sinka *et al.*, 1984, 2011; Elyazar *et al.*, 2013), making ivermectin MDA a particularly attractive strategy to target these vectors. Within the Greater Mekong Subregion (GMS), some of the key malaria vectors have been assessed for ivermectin susceptibility, listed here by most to least susceptible: *Anopheles minimus* s.s. 7-day-LC_{50} of 16.3 [95% CI 11.6–19.4] ng/ml > *Anopheles campestris* 7-day-LC_{50} of 26.4 [95% CI 21.9–30.5] ng/ml = *Anopheles sawadwongporni* 7-day-LC_{50} of 26.9 [95% CI 24.8–28.8] ng/ml > *Anopheles dirus* 7-day-LC_{50} of 55.6 [95% CI 52.3–59.1] ng/ml. Ivermectin pharmacokinetic modelling suggests that the 400 µg/kg dose is ideal to target *An. dirus* s.s. (Kobylinski *et al.*, 2017b). Both *An. dirus* s.s. and *An. minimus* s.s. are key malaria vectors in the GMS that frequently feed on humans. However, the GMS has many other potential malaria vectors that feed more frequently on livestock (Sinka *et al.*, 2011). Thus, ivermectin MDA to humans or mass ITL could be effective vector control tools to reduce transmission in the GMS. Indeed, a small-scale mass ITL trial is under way in Vietnam, and preliminary results indicate that ivermectin-treated (200 µg/kg) cattle are lethal to colonized *An. dirus* s.s. and *Anopheles epiroticus*

(Cramer *et al.*, 2021). Outside mainland South-east Asia, ivermectin has been shown to be effective against *An. farauti* (Foley *et al.*, 2000; Kositz *et al.*, 2017; Pasay *et al.*, 2019) and *An. punctulatus* (Bockarie *et al.*, 1999), which occur in eastern Indonesia, Papua New Guinea and Oceania. It has been suggested that mass ITL in pigs could be used for control in Papua New Guinea (Foley *et al.*, 2000; Pasay *et al.*, 2019). However, the recent introduction of African Swine Fever has decimated pig populations in much of Indonesia and Papua New Guinea, making this an unlikely possibility. A pilot project on Sumba Island in Indonesia, which has potentially 10 or more *Anopheles* malaria vectors (Syafruddin *et al.*, 2020), will assess the impact of standard and long-lasting ivermectin-treated cattle and buffalo on wild *Anopheles* survival (http://www.mesamalaria.org/mesa-track/sumba-livestock-ivermectin-malaria-control-slim-project).

Impact of Sublethal Ivermectin on Vectorial Capacity

In addition to direct mortality, there are several sublethal effects of ivermectin that can impact components of vectorial capacity, such as time to re-feed, sporogony, vector density and fecundity. These sublethal ivermectin effects likely compound the impact of ivermectin MDA or mass ITL on *Plasmodium* transmission.

Ivermectin inhibits the development of the malaria parasite in the vector (i.e. sporogony) in several *Anopheles-Plasmodium* systems. The development of the malaria parasite in the vector can influence two variables of vectorial capacity, including vector competence, which is the likelihood of parasite development in the vector (b) and the extrinsic incubation period, which is duration of time that it takes for the parasite to be ingested by the mosquito, develop and be ready to transmit as sporozoites (n). It was first shown that ivermectin could inhibit vector competence (b) (i.e. sporontocidal) of *P. falciparum* in *An. gambiae*. Ivermectin compound mixed in human blood with cultured *P. falciparum* (NF54) gametocytes inhibited the proportion (i.e. prevalence) of *An. gambiae* s.s. (G3 strain) that developed oocysts or sporozoites by roughly 50% when gametocytes were co-ingested with ivermectin at the LC_{25} concentration, but no effect observed at the LC_5 concentration. Interestingly, the number of oocysts per infected mosquito (i.e. intensity) was not reduced. Additionally, the development rate of sporozoites from day 12 to day 14 was similar between control and ivermectin groups indicating that ivermectin did not delay the extrinsic incubation period (n) (Kobylinski *et al.*, 2012). Sporontocidal effect was observed when *An. dirus* and *An. minimus* co-ingested ivermectin compound and *Plasmodium vivax* collected from malaria patients with significant reductions in both oocyst prevalence and intensity at each species respective ivermectin LC_{25} and LC_5 concentrations (Kobylinski *et al.*, 2017b). There were modest but significant reductions in *An. darlingi* oocyst prevalence but not intensity when *P. vivax* from patients was co-ingested with ivermectin compound at the LC_{50} and LC_{25} concentrations (Kobylinski *et al.*, 2017a). But when plasma from ivermectin-treated volunteers was mixed with *P. vivax* infected red blood cells and fed to *An. darlingi*, there were reductions in oocyst prevalence and intensity at lower ivermectin concentrations. Indeed, in *An. aquasalis* when fed the same ivermectin-treated plasma samples, the ivermectin concentration that inhibits 50% of *P. vivax* oocyst prevalence was 5-fold lower compared to parent compound, suggesting that there are ivermectin metabolites with sporontocidal properties (Pinilla *et al.*, 2018). It should be noted that in a clinical trial where *An. gambiae* s.s. were fed blood from *Plasmodium falciparum*-infected volunteers, no sporontocidal effect was observed, however, the study was underpowered to detect this effect and half of the mosquito samples were lost before they could be analysed (Ouédraogo *et al.*, 2015). The sporotontocidal mode of action of ivermectin remains to be determined but a few possibilities have been suggested. Ivermectin delays the development and thickness of the peritrophic matrix in *Ae. aegypti* (Mahmood *et al.*, 1991) and delays blood meal digestion in *An. gambiae* (Kobylinski *et al.*, 2010) with subsequent upregulation of pertitrophic gene expression (Seaman *et al.*, 2015). As there is an intimate interaction between the ookinete and the peritrophic matrix, it is possible that alteration of the peritrophic matrix by ivermectin could impact *Plasmodium* sporogony. Ivermectin can have direct effects on *P. falciparum* blood-stage development at μM concentrations (Panchal *et al.*, 2014), so it is possible that *Plasmodium* mosquito stages are more susceptible to ivermectin. As ivermectin has been shown to affect development of some bacteria (Lim *et al.*, 2013), so it is possible that ivermectin alters the *Anopheles* midgut microbiota. Alteration of the midgut microbiota can alter the peritrophic matrix (Rodgers *et al.*, 2017) or mosquito immune response, potentially limiting *Plasmodium* development (Bai *et al.*, 2019; Sharma *et al.*, 2020).

Another critical variable in the vectorial capacity equation is the daily probability that an *Anopheles* feeds on a human (a) which is impacted by the vector feeding frequency and the human blood meal index (i.e. how likely a vector is to feed on humans). It is not known if ivermectin MDA or mass ITL could shift *Anopheles* host feeding preference. In the laboratory, it has been shown that ivermectin ingestion delays the time to re-feed on a human host. This was demonstrated in *An. gambiae* s.s. at ivermectin concentrations predicted in humans at 12 and 36 hour after ingestion of 150 μg/kg (Kobylinski *et al.*, 2010) and for *An. darlingi* at concentrations predicted in humans at 4 and 12 hour after ingestion of 200 μg/kg (Kobylinski *et al.*, 2017a). There are a few possible explanations for ivermectin induced delayed re-feeding. Sublethal ivermectin concentrations delay blood meal digestion and defecation in *An. gambiae* s.s. (Kobylinski *et al.*, 2010), which may delay time to re-feed. Sublethal ivermectin concentrations decreased *An. aquasalis* locomotor activity (Sampaio *et al.*, 2017). Adult dung beetles that ingested sublethal amounts of ivermectin in their larval dung diet had reduced chemosensory and locomotor capacity (Verdú *et al.*, 2015). Furthermore, it is likely that *Anopheles* with reduced locomotor function following sublethal ivermectin concentrations would experience higher rates of mortality in the wild due to starvation, reduced ability to avoid predation and abiotic factors.

In regard to ivermectin, vector density (m) is the least characterized variable of vectorial capacity. While reductions in vector survival following ivermectin MDA or mass ITL will occur, no field trials to date have adequately assessed vector density. Fecundity was reduced following ingestion of sublethal ivermectin compound concentrations or feeding on an ivermectin-treated host for *An. gambiae* s.s. (Fritz *et al.*, 2009; Derua *et al.*, 2015), *An. arabiensis* (Fritz *et al.*, 2009; Mekuriaw *et al.*, 2019), *An. coluzzii* (Pooda *et al.*, 2015), *An. farauti* (Pasay *et al.*, 2019) and *An. aquasalis* (Sampaio *et al.*, 2016). A reduction in vector fecundity following ivermectin MDA or mass ITL would likely contribute to reduced adult vector density. However, mathematical modelling illustrates that the impact of ivermectin on vector density is one of the least impactful points for *Plasmodium* transmission. Therefore, population age structure remains the more critical component to assess in field trials (Slater *et al.*, 2014, Slater *et al.*, 2020).

Control Efforts Using Ivermectin MDA in Humans

Ivermectin is not beneficial for the treatment of malaria as asexual blood-stage parasites were not susceptible to ivermectin at human-relevant concentrations *in vitro* (Panchal *et al.*, 2014) and ivermectin was not effective at reducing malaria burden in falciparum-infected patients co-infected with onchocerciasis (Lariviere *et al.*, 1989). However, ivermectin may have liver-stage efficacy against *Plasmodium*, raising the possibility for its use as a chemoprophylactic agent to prevent blood-stage malaria infection. In *Plasmodium berghei* infected mice it was demonstrated that ivermectin (1–10 mg/kg) orally administered 24 and 12 hour before and 12 hour after *P. berghei* sporozoite inoculation by mosquito bite in C57BL/6 mice reduced liver-stage burden at rates comparable to primaquine (10 mg/kg) (Mendes *et al.*, 2017). Intriguingly, field evidence from the RIMDAMAL trial performed in Burkina Faso demonstrated that children in the 1–5 year-old age group that weighed 15 kg or more and received ivermectin (200 µg/kg) every 3 weeks were less likely to develop clinical falciparum malaria compared to untreated counterparts, suggestive of a chemoprophylactic effect (Foy *et al.*, 2019). However, 7 days of ivermectin (0.3, 0.6 or 1.2 mg/kg) with or without chloroquine given orally to Rhesus macaques failed to prevent *Plasmodium cynomolgi* blood-stage infection or eliminate hypnozoites. Admittedly, the number of sporozoites inoculated into the macaque (1 million) was artificially high (Vanachayangkul *et al.*, 2024). A clinical trial in eight healthy adults failed to demonstrate liver-stage inhibition following a single oral dose of ivermectin (400 µg/kg) and intravenous injection of *P. falciparum* NF54 strain sporozoites (3200) (Metzger *et al.*, 2020). It may be possible that ivermectin combined with antimalarial drugs could possess liver-stage inhibition, but this remains to be explored.

The sections above discuss the potential effects of ivermectin MDA alone on *Plasmodium* transmission. However, ivermectin MDA readily lends itself to co-administration with antimalarial drugs suitable for human MDA. Ivermectin MDA could be combined with a blood-stage antimalarial drug MDA wherein the strategy would be to reduce transmission by the extant vector population while simultaneously clearing the human hosts of their *Plasmodium* parasites, so that when the next round of *Anopheles* emerges, there will be no infected humans to continue onwards transmission. In highly seasonal areas of Africa, ivermectin MDA could be combined with Seasonal Malaria Chemoprevention (SMC) with sulfadoxine-pyrimethamine plus amodiaquine (SP-AQ) in children 5 years old and younger. Ivermectin MDA could also be combined with a hypnozoiticidal drug (e.g. primaquine or tafenoquine) to clear the liver-stage reservoir of *P. vivax*. To date, human clinical trials have investigated combinations of ivermectin and the antimalarial drugs, artemether-lumefantrine (AL) (Ouédraogo *et al.*, 2015), dihydroartemisinin-piperaquine (DP) (Smit *et al.*, 2018; Kobylinski *et al.*, 2020) and primaquine (Kobylinski *et al.*, 2020).

In Burkina Faso, single or double doses of ivermectin (200 µg/kg) with AL were administered to 117 *P. falciparum*-infected patients. AL was administered six times over 3 days and ivermectin or placebo was given with the first and fifth dose of AL in three treatment groups: AL alone, AL plus ivermectin on day one and AL plus ivermectin on days 1 and 3. Blood was drawn on days 1, 3 and 7, membrane fed to *An. gambiae* s.s. and *An. funestus*, and their survival was monitored for 10 days after the blood meal. Blood collected 24 hour after an ivermectin administration was lethal to *An. gambiae* s.s. (60–70% mortality) and *An. funestus* (40–50% mortality), with the double dose regimen being slightly more impactful. For *An. gambiae* s.s., blood collected 72 and 120 hour, but not 168 hour after, an ivermectin administration caused a slight but significant reduction in mortality, and similarly for *An. funestus* with blood collected at 72 hour but was not assessed at 120 or 168 hour (Ouédraogo *et al.*, 2015). It was somewhat surprising that mosquito mortality results were not more impactful based on field MDA evidence demonstrating clear survival reductions for *An. gambiae* s.s. for 6 days post-MDA (Sylla *et al.*, 2010). AL may limit ivermectin mosquito-lethal effect by reducing parent or metabolite concentrations but could not be determined from these analyses. Interestingly, female patients had significantly higher ivermectin concentrations and mosquito-lethal effect compared to male patients. Ivermectin co-administration did not affect lumefantrine concentrations. None of the adverse events were definitively related to ivermectin (Ouédraogo *et al.*, 2015). While ivermectin and AL were safe and well tolerated, AL is not ideal for MDA, as it requires six doses over 3 days.

In Kenya, ivermectin at 0, 300 or 600 µg/kg was co-administered with DP for three consecutive days in 141 *P. falciparum*-infected patients. Venous blood was collected on days 0, 2 (4 h after last dose), 7, 14, 21 and 28 days and membrane fed to *An. gambiae* s.s. Substantial mosquito mortality results were obtained, with significant reduction in mosquito survival observed when mosquitoes were fed blood from day 21 for the ivermectin 300 µg/kg group and from day 28 for the ivermectin 600 µg/kg. For both ivermectin 300 and 600 µg/kg groups, the median mosquito survival time remained below 10 days with blood collected up to 14 days and below 14 days with blood collected from patients on days 21 and 28, meaning that *An. gambiae* would be unlikely to live long enough to complete the upper and lower bounds of the falciparum extrinsic incubation period, respectively. Again, it was observed that ivermectin-treated female patients were more lethal to mosquitoes than were male patients. There was no effect of ivermectin on piperaquine concentrations. DP is known to prolong the QT interval, which is the time period of heart muscle contraction to relaxation between beats. Overall, ivermectin co-administration did not further prolong the QT interval; however, there was one patient in the ivermectin 300 µg/kg group who had a QT interval greater than 510 ms, which is a severe adverse event. Anaphylaxis occurred in one patient in the ivermectin 600 µg/kg group. Transient minor visual disturbances were reported for the ivermectin 600 µg/kg group. Transient elevations in liver enzymes to grade three were observed in one patient in the ivermectin 300 µg/kg group and in two patients in the ivermectin 600 µg/kg group, but all three were at or above grade one at baseline. A recent study conducted in Kenya, compared the lethal effect of single dose ivermectin 400 to 300 µg/kg given on three consecutive days and concluded that single dose was non-inferior (Kamau *et al.*, 2024), suggesting that a single-dose approach may afford potential operational advantages by permitting a lower cost, higher feasibility, and likely higher adherence, and thus more effective in achieving targeted MDA coverage levels. As ivermectin accumulates in fat and dermal layers at higher concentrations than found in venous blood (Baraka *et al.*, 1996), it was postulated that direct skin feeding may be more lethal to *Anopheles* than venous blood (Kobylinski *et al.*, 2010). A small study (n=6) in Brazil with *An. aquasalis* suggested that direct skin feeding was more lethal than membrane feeding venous blood (Sampaio *et al.*, 2016). However, on day seven in the Kenya trial, there was a slight difference between ivermectin capillary and venous blood concentrations (1.32:1; Smit *et al.*, 2019a), and no difference in *An. gambiae* mortality following direct skin feeding compared to membrane feeding on venous blood collected (Smit *et al.*, 2019b).

In Thailand, a sequential drug trial was performed in 16 healthy volunteers treated with seven drug regimens: ivermectin (400 µg/kg) alone, DP (120/960 mg) alone, primaquine (30 mg) alone, DP plus primaquine, ivermectin plus primaquine, ivermectin plus DP and ivermectin plus DP plus primaquine. Venous blood was collected on hours 0 and 4 and days 1, 2, 3, 6 and 10 and membrane fed to *An. dirus* s.s. and *An. minimus* s.s. The mosquito-lethal pharmacodynamic effect anticipated from *in vitro* laboratory results and pharmacokinetic modelling of the ivermectin 400 μg/kg dose (Kobylinski *et al.*, 2017b) was a drastic underprediction from what was observed in the trial, with 50% mortality for *An. dirus* s.s. with blood from day 6 and 90% mortality for *An. minimus* s.s. with blood from day 10 post drug treatment. The *An. dirus* s.s. 7-day-LC_{50} of 2.86 [95% CI 2.67–3.05] ng/ml and *An. minimus* s.s. 7-day-LC_{50} of 0.42 [95% CI 0.39–0.46] ng/ml (Kobylinski *et al.*, 2020) were x20- and x35-fold lower compared to previous *in vitro* 7-day-LC_{50} results (Kobylinski *et al.*, 2017b).

This stark difference in ivermectin susceptibility between compound mixed in human blood and blood drawn from treated volunteers raises the distinct possibility that there are ivermectin metabolites that extend the duration of mosquito-lethal effect beyond that of the parent compound, which is the subject of ongoing research (Tipthara *et al.*, 2021; NCT03690453). Interestingly, male volunteers were more lethal to mosquitoes than female volunteers, but this could have been an artefact of the smaller sample size (n=16). Co-administration of ivermectin plus DP increased ivermectin concentrations by 33% and ivermectin plus DP plus primaquine by 54%, which compared to ivermectin alone led to an increased *An. dirus* mortality of 40% with blood from day 6 and 50–90% with blood from day 10 (Kobylinski *et al.*, 2020). This drug-drug interaction leading to increased ivermectin concentrations could be due to the fact that ivermectin (Zeng *et al.*, 1998) and piperaquine (Lee *et al.*, 2012) are primarily metabolized by CYP3A4. This ivermectin-DP interaction explains the exceptional *An. gambiae* mortality results observed in the Kenya trial (Smit *et al.*, 2018). Ivermectin plus DP increased peak piperaquine concentrations and ivermectin plus primaquine plus DP increased both piperaquine peak concentrations and total bioavailability. However, even with drug-drug interactions causing increased piperaquine concentrations, there were no QT interval prolongations of clinical consequence observed. One volunteer had transiently elevated liver enzyme levels of grades 3 and 4 concern, which returned to normal within 10 days following co-administration of ivermectin plus DP and again with ivermectin plus DP plus primaquine. Ivermectin and primaquine co-administration did not lead to any safety or tolerability issues, indicating that this combination could potentially be explored further for *P. vivax* radical cure in MDAs (Kobylinski *et al.*, 2020).

MDA with various antimalarial drugs have been trialled in numerous settings to varying degrees of success. Mathematical modelling suggests that the addition of ivermectin to DP MDAs in Africa will accelerate time to elimination and reduce the number of rounds of MDA needed for elimination (Slater *et al.*, 2014, Slater *et al.*, 2020). Currently, SMC with SP-AQ is being administered to millions of children at or under 5 years old in seasonal malaria settings of the African Sahel to prevent malaria infection. Modelling suggests that performing ivermectin MDAs in children over 5 years old at the same time as SMC in children at or under 5 years old would be a powerful tool to reduce malaria transmission and morbidity (Slater *et al.*, 2020). Ivermectin is currently contraindicated in children under 15 kg, excluding them from the benefits of ivermectin treatment and limiting MDA efficacy by reducing coverage. However, a recent systematic review suggests that ivermectin is just as safe in children under 15 kg as it is in persons at or above 15 kg (Jittamala *et al.*, 2021) and an ongoing multi-country, placebo-controlled, randomized, dose-escalation trial will assess the safety, pharmacokinetics and efficacy of ivermectin in scabies-infected children weighing less than 15 kg (NCT04332068).

Several large-scale cluster randomized trial to evaluate ivermectin MDAs are being implemented in various transmission settings, with various dosing regimens and in combination with other drugs and interventions. In Gambia, three MDAs with DP plus ivermectin (300 μg/kg) for 3 days is being implemented in 16 villages compared to control with no drugs in 16 villages in the Upper River Region along the Gambia river (MASSIV) (NCT03576313). In Guinea Bissau, three MDAs with DP plus ivermectin (300 μg/kg) for 3 days is being implemented in 10 clusters compared to control with DP alone in the Bijagos Islands (MATAMAL) (NCT04844905). In Burkina Faso, four MDAs with ivermectin (300 μg/kg) for 3 days in persons above or at 15 kg and SMC of SP-AQ in children at or under 5 years old, with controls consisting of SMC with SP-AQ alone is being implemented in the Sahel (RIMDAMAL II) (NCT03967054). In Thailand, three MDAs with single-dose ivermectin (400 μg/kg) alone in ten clusters compared to controls of no intervention in ten clusters is being implemented in rubber plantations in southern Thailand (https://mesamalaria.org/mesa-track/novel-vector-control-measure-combat-spread-artemisinin-resistance-greater-mekong). In Mozambique and Kenya, MDAs with single-dose ivermectin (400 μg/kg) will be combined with mass ITL of ivermectin (200 μg/kg) will be compared to controls with ivermectin MDA alone (BOHEMIA) (Chaccour *et al.*, 2023).

Control Efforts Using Ivermectin Treatment of Livestock

Several small-scale studies have highlighted the utility of ITL with standard ivermectin, evaluating ivermectin mosquito-lethal efficacy against colonized (Fritz *et al.*, 2009; Pooda *et al.*, 2015; Pasay *et al.*, 2019; Cramer *et al.*, 2021) and wild *Anopheles* (Naz *et al.*, 2013; Poché *et al.*, 2015; Dreyer *et al.*, 2019). What these studies illustrate is that injectable ivermectin tends to make cattle lethal to blood-feeding *Anopheles* for approximately 2–3 weeks post injection (Fritz *et al.*, 2009; Naz *et al.*, 2013; Pooda *et al.*, 2015; Dreyer *et al.*, 2019; Cramer *et al.*, 2021). Some *Anopheles* species feed more frequently on livestock than humans meaning that treating livestock with ivermectin may be a more effective way to deliver an insecticide (Killeen *et al.*, 2017). Mass ITL can be useful to target outdoor-feeding *Anopheles*, which are more difficult to manage with LLINs and IRS. A small-scale mass ITL field trial is under way in Vietnam, wherein three villages will have ivermectin (200 µg/kg) applied to cattle compared to three control villages with no treatment (Cramer *et al.*, 2021). As ivermectin is widely used in livestock for the control of gut helminths and ectoparasitic diseases, there would be numerous economic benefits for the livestock owners in terms of better animal health and increased livestock weight gains.

There are a wide range of veterinary ivermectin formulations available. However, they produce different pharmacokinetic profiles based on route of administration. Oral (Gokbulut *et al.*, 2007, 2010; Saumell *et al.*, 2017; Pasay *et al.*, 2019) and topical pour-on (Gokbulut *et al.*, 2010; Pasay *et al.*, 2019) ivermectin formulations produce far lower peak concentration, total bioavailability and shorter half-lives with diminished mosquito-lethal effects (Poché *et al.*, 2015) compared to injectable ivermectin, making currently available oral and pour-on formulations not appropriate for mass ITL for vector control. There are species differences in ivermectin pharmacokinetic profiles following injectable ivermectin with livestock ranked from lowest to highest total bioavailability as follows: pigs (Lifschitz *et al.*, 1999) < goats < sheep (González Canga *et al.*, 2009) < buffalo (Anastasio *et al.*, 2002) < horses (Marriner *et al.*, 1987) < cattle. It should be noted that horses do not tolerate injectable ivermectin well, as there appear to be issues, with the vehicle leading to necrosis and sometimes secondary Clostridial bacterial infections that can cause death (Anderson, 1984), making horses less than ideal targets for mass ITL for vector control. As cattle are typically the most abundant livestock by biomass and their ivermectin pharmacokinetic profiles are superior, they are ideal candidates for mass ITL for vector control. There are differences in ivermectin pharmacokinetics between cattle breeds (Vercruysse *et al.*, 2008; Gotardo *et al.*, 2020), and it appears that cattle of Asian origin (*Bos taurus indicus*) reach nearly double the ivermectin peak concentration and total bioavailability of European origin (*Bos taurus taurus*) (Gotardo *et al.*, 2020), which means that mass ITL may be more impactful for vector control in different parts of the world based on cattle breed.

Increasing the dose of injectable ivermectin increases peak concentration but does not provide a longer mean residence time, as ivermectin follows a first-order kinetic process. Altering the propylene glycol/glycerol ratio of the vehicle delays release of the ivermectin from the injection site extending the mean residence time and persistence of ivermectin in plasma by several weeks (Lifschitz *et al.*, 2007), thus increasing potential for mosquito-lethal efficacy. Several long-lasting injectable ivermectin formulations are now available in Latin American markets, but none have been evaluated for their efficacy for vector control. Novel slow-release ivermectin implants in cattle demonstrated substantial mosquito-lethal efficacy against *An. arabiensis* for 26 weeks (Chaccour *et al.*, 2018). Of the numerous veterinary avermectins, only eprinomectin has ivermectin comparable *Anopheles* mosquito-lethal effects (Butters *et al.*, 2012; Fritz *et al.*, 2012). Other veterinary endectocidal compounds that could be used to control *Anopheles* populations include fipronil, nitisinone, fluralaner and afoxolaner. It should be kept in mind that any endectocide used in mass could have environmental implications that must be closely monitored.

Several issues arise with the use of mass ITL for vector control. Not all vectors feed uniformly on humans or livestock, and there can be wide variations within a species even from site to site (Killeen *et al.*, 2017). This means that not all of the important vectors in the area may be affected following mass ITL. Reduction in populations of less important zoophagic malaria vectors could allow more prolific vectors to flourish due to reduced larval competition, leading to species replacement. Mass ITL would likely expose *Anopheles* larva to sublethal ivermectin concentrations, which could accelerate the development of ivermectin resistance. Mass ITL may even select for vector populations that feed more frequently on humans. In light of these issues, it seems prudent to combine mass ITL with ivermectin MDA to reduce the likelihood of these scenarios from happening while accelerating malaria elimination. Indeed, the BOHEMIA trial will look at the benefit of mass ITL in addition to ivermectin MDA (Chaccour *et al.*, 2023). Mathematical models are needed to investigate these possibilities across a range of transmission scenarios and cattle to human ratios.

Mode of Action and Potential for Ivermectin Resistance in *Anopheles*

To date, no ivermectin resistance has been identified in wild *Anopheles* populations or generated in the laboratory, but admittedly, only limited investigation has been performed. While large-scale ivermectin MDA efforts have been performed for decades for onchocerciasis and LF in Africa, the MDAs are typically administered during the dry season to clear human populations of their microfilariae before the rainy season increases their respective blackfly and mosquito vector populations, thus limiting wild *Anopheles* exposure to ivermectin. Ivermectin binds at subunit interfaces of the GluCl channel, which distorts the channel from a closed to an open state (Hibbs and Gouaux, 2011), allowing chloride to continuously flow through the channel, hyperpolarizing the cell, thus leading to flaccid paralysis in nematode (Albers-Schoenberg *et al.*, 1981; Cully *et al.*, 1994, Cully *et al.*, 1996) and arthropod (Kane *et al.*, 2000) musculature. Four GluCl isoforms were identified from *An. gambiae*, and two were successfully cloned into *Xenopus laevis* oocytes to perform two-electrode voltage clamp assays. One GluCl isoform was shown to be ivermectin sensitive while the other insensitive. Thus, overexpression of the insensitive GluCl isoform could lead to resistance in wild *Anopheles* populations (Meyers *et al.*, 2015). Cross-resistance to other insecticides could lead to ivermectin resistance in *Anopheles*, as permethrin-resistant *Ae. aegypti* strains were much less susceptible to ivermectin compared to non-permethrin-resistant strains (Deus *et al.*, 2012). However, further investigation with *Anopheles* is required. It is possible that GluCl is not the only ivermectin target in *Anopheles* as there are wide-ranging effects of ivermectin (e.g. delayed blood meal digestion and reduced fecundity) in parts of the *Anopheles* with no GluCl expression (i.e. midgut and ovaries) (Meyers *et al.*, 2015). In *An. gambiae*, it was shown that ivermectin can also activate the RDL channel. Interestingly, RNA A-to-I editing in *An. gambiae* can generate RDL isoforms that are insensitive to ivermectin (Taylor-Wells *et al.*, 2018), and it has been shown that RNA A-to-I editing does occur in the *An. gambiae* GluCl channel (Taylor-Wells *et al.*, 2015). Thus, this is an additional pathway that could lead to ivermectin resistance in wild *Anopheles* populations. As with all insecticides, metabolic resistance of *Anopheles* to ivermectin is a distinct possibility that has not been evaluated. Much still remains to be characterized regarding ivermectin interaction with the *Anopheles* GluCl channel and possible mechanisms of ivermectin resistance.

Conclusions

Ivermectin MDA and mass ITL hold promise as novel vector control tools to aid global malaria elimination efforts. The ongoing MDA trials described above will hopefully provide the necessary evidence to justify use of ivermectin MDA for vector control. Several of the investigators leading these MDA trials participated in the Ivermectin Roadmap (Billingsley *et al.*, 2020) process and will provide data to the WHO for their review and potential endorsement, which is critical for uptake of novel vector control interventions for malaria.

References

Albers-Schoenberg, G., Arison, B.H., Chabala, J.C., Douglas, A.W., Eskola, P. *et al.* (1981) Avermectins. Structure Determination. *Journal of the American Chemical Society* 103(14), 4216–4221.

Ali, A. and Nayar, J.K. (1985) Activity of an avermectin insecticide, abamectin (MK-936), against mosquitoes and chironomid midges in the laboratory. *Journal of the American Mosquito Control Association* 1(3), 384–386.

Alout, H., Krajacich, B.J., Meyers, J.I., Grubaugh, N.D., Brackney, D.E. *et al.* (2014) Evaluation of ivermectin mass drug administration for malaria transmission control across different West African environments. *Malaria Journal* 13, 417.

Anastasio, A., Esposito, M., Amorena, M., Catellani, P., Serpe, L. *et al.* (2002) Residue study of ivermectin in plasma, milk, and mozzarella cheese following subcutaneous administration to buffalo (*Bubalus bubalis*). *Journal of Agricultural and Food Chemistry* 50(18), 5241–5245. DOI: 10.1021/jf020181j.

Anderson, R. (1984) The use of ivermectin in horses: Research and clinical observations. *The Compendium on Continuing Education* 6, S516–520.

Bai, L., Wang, L., Vega-Rodríguez, J., Wang, G. and Wang, S. (2019) A gut symbiotic bacterium *Serratia marcescens* renders mosquito resistance to *Plasmodium* infection through activation of mosquito immune responses. *Frontiers in Microbiology* 10, 1580. DOI: 10.3389/fmicb.2019.01580.

Bamou, R., Mbakop, L.R., Kopya, E., Ndo, C., Awono-Ambene, P. *et al.* (2018) Changes in malaria vector bionomics and transmission patterns in the equatorial forest region of Cameroon between 2000 and 2017. *Parasites and Vectors* 11(1), 464.

Bamou, R., Rono, M., Degefa, T., Midega, J., Mbogo, C. *et al.* (2021) Entomological and anthropological factors contributing to persistent malaria transmission in Kenya, Ethiopia, and Cameroon. 12Suppl. 2, *The Journal of Infectious Diseases* 223, S155–S170.

Baraka, O.Z., Mahmoud, B.M., Marschke, C.K., Geary, T.G., Homeida, M.M. *et al.* (1996) Ivermectin distribution in the plasma and tissues of patients infected with *Onchocerca volvulus*. *European journal of clinical pharmacology* 50(5), 407–410. DOI: 10.1007/s002280050131.

Bayoh, M.N., Mathias, D.K., Odiere, M.R., Mutuku, F.M., Kamau, L. *et al.* (2010) *Anopheles gambiae*: Historical population decline associated with regional distribution of insecticide-treated bed nets in western Nyanza province, Kenya. *Malaria Journal* 9, 62. DOI: 10.1186/1475-2875-9-62.

Billingsley, P., Binka, F., Chaccour, C., Foy, B., Gold, S. *et al.* (2020) A roadmap for the development of ivermectin as a complementary malaria vector control tool. *American Journal of Tropical Medicine Hygiene* 102(2s), 3–24.

Bockarie, M.J., Hii, J.L., Alexander, N.D., Bockarie, F., Dagoro, H. *et al.* (1999) Mass treatment with Ivermectin for filariasis control in Papua New Guinea: Impact on mosquito survival. *Medical and Veterinary Entomology* 13(2), 120–123.

Burg, R.W., Miller, B.M., Baker, E.E., Birnbaum, J., Currie, S.A. *et al.* (1979) Avermectins, new family of potent anthelmintic agents: Producing organism and fermentation. *Antimicrobial Agents and Chemotherapy* 15(3), 361–367.

Butters, M.P., Kobylinski, K.C., Deus, K.M., da Silva, I.M., Gray, M. *et al.* (2012) Comparative evaluation of systemic drugs for their effects against *Anopheles gambiae*. *Acta Tropica* 121(1), 34–43. DOI: 10.1016/j.actatropica.2011.10.007.

Cartel, J.L., Sechan, Y., Spiegel, A., Nguyen, L., Barbazan, P. *et al.* (1991) Cumulative mortality rates in *Aedes polynesiensis* after feeding on Polynesian *Wuchereria bancrofti* carriers treated with single doses of Ivermectin, Diethylcarbamazine and Placebo. *Tropical Medicine and Parasitology* 42(4), 343–345.

Chaccour, C., Lines, J. and Whitty, C.J. (2010) Effect of ivermectin on *Anopheles gambiae* mosquitoes fed on humans: The potential of oral insecticides in malaria control. *The Journal of Infectious Diseases* 202(1), 113–116.

Chaccour, C., Casellas, A., Hammann, F., Ruiz-Castillo, P., Nicolas, P. *et al.* (2023) BOHEMIA: Broad one health endectocide-based malaria intervention in Africa - a phase III cluster-randomized, open-label, clinical trial to study the safety and efficacy of ivermectin mass drug administration to reduce malaria transmission in two African settings. *Trials* 24(1), 128.

Chaccour, C.J., Ngha'bi, K., Abizanda, G., Barrio, A.I., Aldaz, A. *et al.* (2018) Targeting cattle for malaria elimination: Marked reduction of Anopheles arabiensis survival for over six months using a slow-release ivermectin implant formulation. *Parasites and Vectors* 11, 287.

Chandre, F. and Hougard, J.M. (1999) Systemic action of ivermectin on *Culex quinquefasciatus* and *Simulium squamosum*. *Bulletin de La Société de Pathologie Exotique* 92(1), 71–72.

Collins, K. (2004) Profitable gifts: A history of the merck mectizan donation program and its implications for international health. *Perspectives in Biology and Medicine* 47(1), 100–109.

Cramer, E.Y., Quang, N.X., Hertz, J.C., Van Nguyen, D., Quang, H.H. *et al.* (2021) Ivermectin treatment for cattle reduced the survival of two malaria vectors, *Anopheles dirus* and *Anopheles epiroticus*, under laboratory conditions in Central Vietnam. *American Journal of Tropical Medicine and Hygiene* 104(6), 2165–2168. DOI: 4269/ajtmh.20-1239.

Crump, A. and Ōmura, S. (2011) Ivermectin, 'wonder drug' from Japan: The human use perspective. *Proceedings of the Japan Academy Series B: Physical and Biological Sciences* 87(2), 13–28.

Cully, D.F., Vassilatis, D.K., Liu, K.K., Paress, P.S., Van der Ploeg, L.H. *et al.* (1994) Cloning of an avermectin-sensitive glutamate-gated chloride channel from *Caenorhabditis elegans*. *Nature* 371(6499), 707–711. DOI: 10.1038/371707a0.

Cully, D.F., Paress, P.S., Liu, K.K., Schaeffer, J.M. and Arena, JP. (1996) Identification of a *Drosophila melanogaster* glutamate-gated chloride channel sensitive to the antiparasitic agent avermectin. *Journal of Biological Chemistry* 271(33), 20187–20191. DOI: 10.1074/jbc.271.33.20187.

Cupp, E.W. and Cupp, MS. (2005) Short report: Impact of ivermectin community-level treatments on elimination of adult *Onchocerca volvulus* when individuals receive multiple treatments per year. *American Journal of Tropical Medicine and Hygiene* 73(6), 1159–1161.

Cupp, E.W., Duke, B.O., Mackenzie, C.D., Guzmán, J.R., Vieira, J.C. *et al.* (2004) The effects of long-term community level treatment with ivermectin (mectizan) on adult *Onchocerca volvulus* in latin america. *American Journal of Tropical Medicine and Hygiene* 71(5), 602–607.

de Azambuja, P., Gomes, J.E., Lopes, F. and Garcia, E.S. (1985) Efficacy of ivermectin against the bloodsucking insect, *Rhodnius prolixus* (Hemiptera, triatominae). *Memórias Do Instituto Oswaldo Cruz* 80(4), 439–442.

Dembele, B., Coulibaly, Y.I., Dolo, H., Konate, S., Coulibaly, S.Y. *et al.* (2010) Use of high-dose, twice-yearly albendazole and ivermectin to suppress *Wuchereria bancrofti* microfilarial levels. *Clin Infect Dis* 51(11), 1229–1235. DOI: 10.1086/657063.

Derua, Y.A., Kisinza, W.N. and Simonsen, PE. (2015) Differential effect of human ivermectin treatment on blood feeding *Anopheles gambiae* and *Culex quinquefasciatus*. *Parasite and Vectors* 8, 130. DOI: 10.1186/s13071-015-0735-3.

Deus, K.M., Saavedra-Rodriguez, K., Butters, M.P., Black IV, W.C. and Foy, BD. (2012) The effect of ivermectin in seven strains of *Aedes aegypti* (Diptera: Culicidae) including a genetically diverse laboratory strain and three permethrin resistant strains. *Journal of Medical Entomology* 49(2), 356–363. DOI: 10.1603/me11164.

Distelmans, W., D'Haeseleer, F. and Mortelmans, J. (1983) Efficacy of systemic administration of ivermectin against tsetse flies. *Annales de La Société Belge de Médecine Tropicale* 63(2), 119–125.

Dreyer, G., Noroes, J., Amaral, F., Nen, A., Medeiros, Z. *et al.* (1995) Direct assessment of the adulticidal efficacy of a single dose of ivermectin in Bancroftian filariasis. *Transactions of the Royal Society of Tropical Medicine and Hygiene* 89(4), 441–443.

Dreyer, G., Addiss, D., Noroes, J., Amaral, F., Rocha, A. *et al.* (1996) Ultrasonographic assessment of the adulticidal efficacy of repeat high-dose ivermectin in Bancroftian filariasis. *Tropical Medicine and International Health* 1(4), 427–432.

Dreyer, S.M., Morin, K.J. and Vaughan, J.A. (2018) Differential susceptibilities of *Anopheles albimanus* and *Anopheles stephensi* mosquitoes to ivermectin. *Malaria Journal* 17(1), 148.

Dreyer, S.M., Leiva, D., Magaña, M., Pott, M., Kay, J. *et al.* (2019) Fipronil and ivermectin treatment of cattle reduced the survival and ovarian development of field-collected *Anopheles albimanus* in a pilot trial conducted in northern Belize. *Malaria Journal* 18(1), 296.

Dreyer, S.M., Morin, K.J., Magaña, M., Pott, M., Leiva, D. *et al.* (2022) Oral susceptibility to ivermectin is over fifty times greater in a wild population of *Anopheles albimanus* mosquitoes from Belize than the STECLA laboratory reference strain of this mosquito. *Malaria Journal* 21(1), 72.

Drummond, R.O. (1984) Control of larvae of the common cattle grub (Diptera: Oestridae) with animal systemic insecticides. *Journal of Economic Entomology* 77(2), 402–406.

Drummond, R.O., Whetstone, T.M. and Miller, J.A. (1981) Control of ticks systemically with Merck MK-933, an avermectin. *Journal of Economic Entomology* 74(4), 432–436.

Duke, B.O., Zea-Flores, G., Castro, J., Cupp, E.W. and Munoz, B. (1992) Effects of three-month doses of ivermectin on adult *Onchocerca volvulus*. *American Journal of Tropical Medicine and Hygiene* 46(2), 189–194.

Egerton, J.R., Ostlind, D.A., Blair, L.S., Eary, C.H., Suhayda, D. *et al.* (1979) Avermectins, new family of potent anthelmintic agents: Efficacy of the b1a component. *Antimicrobial Agents and Chemotherapy* 15(3), 372–378.

Elyazar, I.R., Sinka, M.E., Gething, P.W., Tarmidzi, S.N., Surya, A. *et al.* (2013) The distribution and bionomics of *Anopheles* malaria vector mosquitoes in Indonesia. *Advances in Parasitology* 83, 173–266.

Focks, D.A., McLaughlin, R.E. and Linda, S.B. (1991) Effects of ivermectin (MK-933) on the reproductive rate of *Aedes aegypti* (Diptera: Culicidae). *Journal of Medical Entomology* 28(4), 501–505.

Foley, D.H., Bryan, J.H. and Lawrence, G.W. (2000) The potential of ivermectin to control the malaria vector *Anopheles farauti*. *Transactions of the Royal Society of Tropical Medicine and Hygiene* 94(6), 625–628.

Foy, B.D., Alout, H., Seaman, J.A., Rao, S., Magalhaes, T. *et al.* (2019) Efficacy and risk of harms of repeat ivermectin mass drug administrations for control of malaria (RIMDAMAL): A cluster-randomised trial. *The Lancet* 393(10180), 1517–1526.

Fritz, M.L., Walker, E.D. and Miller, JR. (2012) Lethal and sublethal effects of avermectin/milbemycin parasiticides on the African malaria vector, *Anopheles arabiensis*. *Journal of Medical Entomology* 49(2), 326–331. DOI: 10.1603/me11098.

Fritz, M.L., Siegert, P.Y., Walker, E.D., Bayoh, M.N., Vulule, J.R. *et al.* (2009) Toxicity of bloodmeals from ivermectin-treated cattle to *Anopheles gambiae* s.l. *Annals of Tropical Medicine and Parasitology* 103(6), 539–547.

Gardner, K., Meisch, M.V., Meek, C.L. and Biven, W.S. (1993) Effects of ivermectin in canine blood on *Anopheles quadrimaculatus, Aedes albopictus* and *Culex salinarius*. *Journal of the American Mosquito Control Association* 9(4), 400–402.

Geissbühler, Y., Chaki, P., Emidi, B., Govella, N.J., Shirima, R. *et al.* (2007) Interdependence of domestic malaria prevention measures and mosquito-human interactions in urban Dar es Salaam, Tanzania. *Malaria Journal* 6, 126.

Girod, R., Gaborit, P., Carinci, R., Issaly, J. and Fouque, F. (2008) *Anopheles darlingi* bionomics and transmission of *Plasmodium falciparum, Plasmodium vivax and Pasmodium malariae* in Amerindian villages of the Upper-Maroni Amazonian forest, French Guiana. *Memórias Do Instituto Oswaldo Cruz* 103(7), 702–710.

Githinji, E.K., Irungu, L.W., Ndegwa, P.N., Machani, M.G., Amito, R.O. *et al.* (2020) Impact of insecticide resistance on *P. falciparum* vectors' biting, feeding, and resting behaviour in selected clusters in Teso north and south subcounties in Busia county, Western Kenya. *Journal of Parasitology Research* 9423682.

Glaziou, P., Nyguyen, L.N., Moulia-Pelat, J.P., Cartel, J.L. and Martin, P.M. (1994) Efficacy of ivermectin for the treatment of head lice (*Pediculosis capitis*). *Tropical Medicine and Parasitology* 45(3), 253–254.

Gokbulut, C., Karademir, U. and Boyacioglu, M. (2007) Comparison of plasma pharmacokinetic profile of ivermectin following administration of subcutaneous injection (Baymec) and oral tablet (Efektin) in goats. *Journal of Veterinary Pharmacology and Therapeutics* 30(5), 489–491. DOI: 10.1111/j.1365-2885.2007.00888.x.

Gokbulut, C., Cirak, V.Y., Senlik, B., Aksit, D., Durmaz, M. *et al.* (2010) Comparative plasma disposition, bioavailability and efficacy of ivermectin following oral and pour-on administrations in horses. *Vet Parasitol* 170(1–2), 120–126. DOI: 10.1016/j.vetpar.2010.01.041.

González Canga, A., Sahagún Prieto, A.M., José Diez Liébana, M., Martínez, N.F., Vega, M.S. *et al.* (2009) The pharmacokinetics and metabolism of ivermectin in domestic animal species. *Journal of Veterinary Medicine* 179(1), 25–37. DOI: 10.1016/j.tvjl.2007.07.011.

Gotardo, A.T., Fonseca, R.L., Barreto, F., Tomaszewski, C.A., Barnet, L.S. *et al.* (2020) The effect of breed, sex, and drug concentration on the pharmacokinetic profile of ivermectin in cattle. *Journal of Veterinary Pharmacology and Therapeutics* 43(5), 477–484. DOI: 10.1111/jvp.12862.

Gustavsen, K.M., Colatrella, B.D. and McCoy, T. (2018) For as long as necessary: Examining 30 years of MSD's focus on achieving elimination of *Onchocerciasis* and *Lymphatic filariasis*. *International Health* 10, i3–i6.

Hibbs, R.E. and Gouaux, E. (2011) Principles of activation and permeation in an anion-selective Cys-loop receptor. *Nature* 474(7349), 54–60. DOI: 10.1038/nature10139.

Iakubovich, V., Zakharova, N.F., Alekseev, A.N. and Alekseev, E.A. (1989) Evaluation of the action of ivermectin on blood-sucking mosquitoes. *Meditsinskaia Parazitologiia i Parazitarnye Bolezni* 3, 60–64.

Ismail, M.M., Weil, G.J., Jayasinghe, K.S., Premaratne, U.N., Abeyewickreme, W. *et al.* (1996) Prolonged clearance of microfilaraemia in patients with Bancroftian filariasis after multiple high doses of ivermectin or diethylcarbamazine. *Transactions of the Royal Society of Tropical Medicine and Hygiene* 90(6), 684–688.

Jittamala, P., Monteiro, W., Smit, M.R., Pedrique, B., Specht, S. *et al.* (2021) A systematic review and an individual patient data meta-analysis of ivermectin use in children weighing less than fifteen kilograms: Is it time to reconsider the current contraindication? *PLOS Neglected Tropical Diseases* 15(3), e0009144. DOI: 10.1371/journal.pntd.0009144. Erratum in: *PLOS Neglected Tropical Diseases* doi: 10.1371/journal.pntd.0011053

Jones, J.W., Meisch, M.V., Meek, C.L. and Bivin, W.S. (1992) Lethal effects of ivermectin on *Anopheles quadrimaculatus*. *Journal of the American Mosquito Control Association* 8(3), 278–280.

Kamau, Y., Tuwei, M., Wanjiku, C., Ominde, K., Ngamaet, M. *et al.* (2024) Mosquitocidal efficacy and pharmacokinetics of single-dose ivermectin versus three-day dose regimen for malaria vector control in comparison with albendazole and no treatment: An open-label randomized controlled trial. *International Journal of Infectious Diseases*. Available at: https://www.sciencedirect.com/science/article/pii/S1201971224003072

Kanamitie, J.N., Ahorlu, C.S., Otchere, J., Aboagye-Antwi, F., Kwansa-Bentum, B. *et al.* (2017) Twelve-month longitudinal parasitological assessment of lymphatic filariasis-positive individuals: Impact of a biannual treatment with Ivermectin and Albendazole. *Tropical Medicine and International Health* 22(11), 1451–1456.

Kane, N.S., Hirschberg, B., Qian, S., Hunt, D., Thomas, B. *et al.* (2000) Drug-resistant Drosophila indicate glutamate-gated chloride channels are targets for the Antiparasitics Nodulisporic acid and ivermectin. *Proceedings of the National Academy of Sciences USA* 97(25), 13949–13954. DOI: 10.1073/pnas.240464697.

Killeen, G.F., Tatarsky, A., Diabate, A., Chaccour, C.J., Marshall, J.M. *et al.* (2017) Developing an expanded vector control toolbox for malaria elimination. *BMJ Global Health* 2(2), e000211. DOI: 10.1136/bmjgh-2016-000211.

Kobylinski, K.C., Foy, B.D. and Richardson, J.H. (2012) Ivermectin inhibits the sporogony of *Plasmodium falciparum* in *Anopheles gambiae*. *Malaria Journal* 11, 381. DOI: 10.1186/1475-2875-11-381.

Kobylinski, K.C., Jittamala, P., Hanboonkunupakarn, B., Pukrittayakamee, S., Pantuwatana, K. *et al.* (2020) Pharmacokinetics, and mosquito-lethal effects of Ivermectin in combination with dihydroartemisinin-piperaquine and primaquine in healthy adult Thai subjects. *Clinical pharmacology and therapeutics* 107(5), 1221–1230. DOI: 10.1002/cpt.1716.

Kobylinski, K.C., Deus, K.M., Butters, M.P., Hongyu, T., Gray, M. *et al.* (2010) The effect of oral anthelmintics on the survivorship and re-feeding frequency of Anthropophilic mosquito disease vectors. *Acta Tropica* 116(2), 119–126.

Kobylinski, K.C., Sylla, M., Chapman, P.L., Sarr, M.D. and Foy, B.D. (2011) Ivermectin mass drug administration to humans disrupts malaria parasite transmission in Senegalese villages. *American Journal of Tropical Medicine and Hygiene* 85(1), 3–5.

Kobylinski, K.C., Escobedo-Vargas, K.S., López-Sifuentes, V.M., Durand, S., Smith, E.S. *et al.* (2017a) Ivermectin susceptibility, sporontocidal effect, and inhibition of time to re-feed in the Amazonian malaria vector. *Malaria Journal* 16(1), 474.

Kobylinski, K.C., Ubalee, R., Ponlawat, A., Nitatsukprasert, C., Phasomkulsolsil, S. *et al.* (2017b) Ivermectin susceptibility and sporontocidal effect in Greater Mekong Subregion *Anopheles*. *Malaria Journal* 16(1), 280.

Kositz, C., Talina, J., Diau, J., Asugeni, R., Whitehorn, C. *et al.* (2017) Incidental mosquitocidal effect of an ivermectin mass drug administration on Anopheles farauti conducted for scabies control in the Solomon Islands. *Royal Society of Tropical Medicine and Hygiene* 111(3), 97–101. DOI: 10.1093/trstmh/trx025.

Lariviere, M., Beauvais, B., Aziz, M., Garin, Y.J.F., Abeloos, J. *et al.* (1989) Study in Ivory Coast (1985–1987) of the efficacy and tolerance of Ivermectin (Mectizan) in human onchocerciasis. 1. A double-blind comparative study of 200 patients treated with a single oral dose of 100 mcg/kg, 150 mcg/kg or 200 mcg/kg. *Bulletin De La Societe De Pathologie Exotique* 82, 35–47.

Lee, T.M., Huang, L., Johnson, M.K., Lizak, P., Kroetz, D. *et al.* (2012) *In vitro* metabolism of piperaquine is primarily mediated by CYP3A4. *Xenobiotica* 42(11), 1088–1095. DOI: 10.3109/00498254.2012.693972.

Lifschitz, A., Pis, A., Alvarez, L., Virkel, G., Sanchez, S. *et al.* (1999) Bioequivalence of ivermectin formulations in pigs and cattle. *Journal of Veterinary Pharmacology and Therapeutics* 22(1), 27–34. DOI: 10.1046/j.1365-2885.1999.00172.x.

Lifschitz, A., Virkel, G., Ballent, M., Sallovitz, J., Imperiale, F. *et al.* (2007) Ivermectin (3.15%) long-acting formulations in cattle: Absorption pattern and pharmacokinetic considerations. *Veterinary parasitology* 147(3–4), 303–310. DOI: 10.1016/j.vetpar.2007.04.009.

Lim, L.E., Vilchèze, C., Ng, C., Jacobs, W.R., Jr, Ramón-García, S. *et al.* (2013) Anthelmintic avermectins kill *Mycobacterium tuberculosis*, including multidrug-resistant clinical strains. *Antimicrobial Agents and Chemotherapy* 57(2), 1040–1046.

Mahmood, F., Walters, L.L., Guzman, H. and Tesh, R.B. (1991) Effect of ivermectin on the ovarian development of *Aedes aegypti* (Diptera: Culicidae). *Journal of Medical Entomology* 28(5), 701–707.

Marriner, S.E., McKinnon, I. and Bogan, JA. (1987) The pharmacokinetics of ivermectin after oral and subcutaneous administration to sheep and horses. *Journal of Veterinary Pharmacology and Therapeutics* 10(2), 175–179. DOI: 10.1111/j.1365-2885.1987.tb00097.x.

Mascari, T.M. and Foil, L.D. (2010) Oral treatment of rodents with ivermectin for the control of *Phlebotomus papatasi* (Diptera: Psychodidae) under laboratory conditions. *Veterinary Parasitology* 171(1–2), 130–135.

Mascari, T.M., Mitchell, M.A., Rowton, E.D. and Foil, L.D. (2008) Ivermectin as a rodent feed-through insecticide for control of immature sand flies (Diptera: Psychodidae). *Journal of the American Mosquito Control Association* 24(2), 323–326. DOI: 10.2987/5678.1.

MDP (2020) *Investing for Impact: 2020 Annual Highlights*. Mectizan Donation Program.

MDP (2017) 35 Years: The Mectizan® Donation Program. Available at: https://www.msd.com/stories/mectizan/#:~:text=In%20November%202017%2C%20in%20support,the%20MDP%20speak%20for%20themselves (accessed 4 December 2022).

Mekuriaw, W., Balkew, M., Messenger, L.A., Yewhalaw, D., Woyessa, A. *et al.* (2019) The effect of ivermectin® on fertility, fecundity and mortality of *Anopheles arabiensis* fed on treated men in Ethiopia. *Malaria Journal* 18(1), 357. DOI: 10.1186/s12936-019-2988-3.

Mendes, A.M., Albuquerque, I.S., Machado, M., Pissarra, J., Meireles, P. *et al.* (2017) Inhibition of Plasmodium Liver Infection by Ivermectin. *Antimicrob Agents Chemother* 61(2), e02005-16. DOI: 10.1128/AAC.02005-16.

Metzger, W.G., Theurer, A., Pfleiderer, A., Molnar, Z., Maihöfer-Braatting, D. *et al.* (2020) Ivermectin for causal malaria prophylaxis: A randomised controlled human infection trial. *Tropical medicine & international health* 25(3), 380–386. DOI: 10.1111/tmi.13357.

Meyers, J.I., Gray, M., Kuklinski, W., Johnson, L.B., Snow, C.D. *et al.* (2015) Characterization of the target of ivermectin, the glutamate-gated chloride channel, from *Anopheles gambiae*. *Journal of Experimental Biology* 218(Pt 10), 1478–1486. DOI: 10.1242/jeb.118570.

Miller, J.A., Oehler, D.D., Siebenaler, A.J. and Kunz, S.E. (1986) Effect of ivermectin on survival and fecundity of horn flies and stable flies (Diptera: Muscidae). *Journal of Economic Entomology* 79(6), 1564–1569.

Miller, T.W., Chaiet, L., Cole, D.J., Cole, L.J., Flor, J.E. *et al.* (1979) Avermectins, new family of potent anthelmintic agents: Isolation and chromatographic properties. *Antimicrobial Agents and Chemotherapy* 15(3), 368–371.

Moiroux, N., Gomez, M.B., Pennetier, C., Elanga, E., Djènontin, A. *et al.* (2012) Changes in *Anopheles funestus* biting behavior following universal coverage of long-lasting insecticidal nets in Benin. *The Journal of Infectious Diseases* 206, 1622–1629.

Moreno, J.E., Rubio-Palis, Y., Páez, E., Pérez, E. and Sánchez, V. (2007) Abundance, biting behaviour and parous rate of Anopheline mosquito species in relation to malaria incidence in gold-mining areas of southern Venezuela. *Medical and Veterinary Entomology* 21(4), 339–349.

Mounsey, K.E., Holt, D.C., McCarthy, J., Currie, B.J. and Walton, S.F. (2008) Scabies: Molecular perspectives and therapeutic implications in the face of emerging drug resistance. *Future Microbiology* 3(1), 57–66.

Moutinho, P.R., Gil, L.H., Cruz, R.B. and Ribolla, P.E. (2011) Population dynamics, structure and behavior of *Anopheles darlingi* in a rural settlement in the Amazon rainforest of Acre, Brazil. *Malaria Journal* 10, 174.

Mwangangi, J.M., Mbogo, C.M., Nzovu, J.G., Githure, J.I., Yan, G. *et al.* (2003) Blood-meal analysis for anopheline mosquitoes sampled along the Kenyan coast. *Journal of the American Mosquito Control Association* 19(4), 371–375.

Naz, S., Maqbool, A., Ahmad, M. and Zaman, S. (2013) Efficacy of ivermectin for control of zoophilic malaria vectors in Pakistan. *Pakistan Journal of Zoology* 45(6), 1585–1591. DOI: 0030-9923/2013/0006-1585.

Omitola, O.O., Umunnakwe, C.U., Bayegun, A.A., Anifowose, S.A., Mogaji, H.O. *et al.* (2021) Impacts of ivermectin mass drug administration for onchocerciasis on mosquito populations of Ogun state, Nigeria. *Parasite & Vectors* 14(1), 212. DOI: 10.1186/s13071-021-04716-3.

Omura, S. and Crump, A. (2004) The life and times of ivermectin – a success story. *Nature Reviews Microbiology* 2(12), 984–989.

Ouédraogo, A.L., Bastiaens, G.J., Tiono, A.B., Guelbéogo, W.M., Kobylinski, K.C. *et al.* (2015) Efficacy and safety of the mosquitocidal drug ivermectin to prevent malaria transmission after treatment: A double-blind, randomized, clinical trial. *Clinical Infectious Diseases* 60(3), 357–365. DOI: 10.1093/cid/ciu797.

Pampiglione, S., Majori, G., Petrangeli, G. and Romi, R. (1985) Avermectins, MK-933 and MK-936, for mosquito control. *Transactions of the Royal Society of Tropical Medicine and Hygiene* 79(6), 797–799.

Panchal, M., Rawat, K., Kumar, G., Kibria, K.M., Singh, S. *et al.* (2014) *Plasmodium falciparum* signal recognition particle components and anti-parasitic effect of ivermectin in blocking nucleo-cytoplasmic shuttling of SRP. *Cell Death & Disease* 5(1), e994. DOI: 10.1038/cddis.2013.521.

Pasay, C.J., Yakob, L., Meredith, H.R., Stewart, R., Mills, P.C. *et al.* (2019) Treatment of pigs with endectocides as a complementary tool for combating malaria transmission by *Anopheles farauti* (s.s.) in Papua New Guinea. *Parasite and Vectors* 12(1), 124. DOI: 10.1186/s13071-019-3392-0.

Pinilla, Y.T., C P Lopes, S., S Sampaio, V., Andrade, F.S., Melo, G.C. *et al.* (2018) Promising approach to reducing Malaria transmission by ivermectin: Sporontocidal effect against *Plasmodium vivax* in the South American vectors *Anopheles aquasalis* and *Anopheles darlingi*. *PLOS Neglected Tropical Diseases* 12(2), e0006221. DOI: 10.1371/journal.pntd.0006221.

Pitterna, T., Cassayre, J., Hüter, O.F., Jung, P.M., Maienfisch, P. *et al.* (2009) New ventures in the chemistry of avermectins. *Bioorganic and Medicinal Chemistry* 17(12), 4085–4095.

Poché, R.M., Burruss, D., Polyakova, L., Poché, D.M. and Garlapati, RB. (2015) Treatment of livestock with systemic insecticides for control of *Anopheles arabiensis* in western Kenya. *Malaria Journal* 14, 351. DOI: 10.1186/s12936-015-0883-0.

Pooda, H.S., Rayaisse, J.B., Hien, D.F., Lefèvre, T., Yerbanga, S.R. *et al.* (2015) Administration of ivermectin to peridomestic cattle: A promising approach to target the residual transmission of human malaria. *Malaria Journal* 13(Suppl 1), 496. DOI: 10.1186/s12936-015-1001-z.

Rodgers, F.H., Gendrin, M., Wyer, C.A.S. and Christophides, GK. (2017) Microbiota-induced peritrophic matrix regulates midgut homeostasis and prevents systemic infection of malaria vector mosquitoes. *PLOS Pathogens* 13(5), e1006391. DOI: 10.1371/journal.ppat.1006391.

Sampaio, V.S., Beltrán, T.P., Kobylinski, K.C., Melo, G.C., Lima, J.B. *et al.* (2016) Filling gaps on ivermectin knowledge: Effects on the survival and reproduction of *Anopheles aquasalis*, a Latin American malaria vector. *Malaria Journal* 15(1), 491. DOI: 10.1186/s12936-016-1540-y.

Sampaio, V.S., Rivas, G., Kobylinski, K., Pinilla, Y.T., Pimenta, P.F.P. *et al.* (2017) What does not kill it makes it weaker: Effects of sub-lethal concentrations of ivermectin on the locomotor activity of *Anopheles aquasalis*. *Parasites and Vectors* 10(1), 623.

Saumell, C., Lifschitz, A., Baroni, R., Fusé, L., Bistoletti, M. *et al.* (2017) The route of administration drastically affects ivermectin activity against small strongyles in horses. *Vet Parasitol* 236, 62–67. DOI: 10.1016/j.vetpar.2017.01.025.

Seaman, J.A., Alout, H., Meyers, J.I., Stenglein, M.D., Dabiré, R.K. *et al.* (2015) Age and prior blood feeding of *Anopheles gambiae* influences their susceptibility and gene expression patterns to ivermectin-containing blood meals. *BMC Genomics* 16, 797. DOI: 10.1186/s12864-015-2029-8.

Sharma, P., Rani, J., Chauhan, C., Kumari, S., Tevatiya, S. *et al.* (2020) Altered gut microbiota and immunity defines *Plasmodium vivax* survival in *Anopheles stephensi*. *Frontiers in Immunology* 11, 609. DOI: 10.3389/fimmu.2020.00609.

Sinka, M.E., Bangs, M.J., Manguin, S., Chareonviriyaphap, T., Patil, A.P. *et al.* (1984) The dominant *Anopheles* vectors of human malaria in the Asia-pacific region: Ocurrence data, distribution maps and bionomic précis. *Parasites and Vectors* 4, 89.

Sinka, M.E., Bangs, M.J., Manguin, S., Chareonviriyaphap, T., Patil, A.P. *et al.* (2011) The dominant Anopheles vectors of human malaria in the Asia-Pacific region: Occurrence data, distribution maps and bionomic précis. *Parasite and Vectors* 4, 89. DOI: 10.1186/1756-3305-4-89.

Slater, H.C., Walker, P.G., Bousema, T., Okell, L.C. and Ghani, AC. (2014) The potential impact of adding ivermectin to a mass treatment intervention to reduce malaria transmission: A modelling study. *The Journal of infectious Diseases* 210(12), 1972–1980. DOI: 10.1093/infdis/jiu351.

Slater, H.C., Foy, B.D., Kobylinski, K., Chaccour, C., Watson, O.J. *et al.* (2020) Ivermectin as a novel complementary malaria control tool to reduce incidence and prevalence: A modelling study. *The Lancet. Infectious Diseases* 20(4), 498–508. DOI: 10.1016/S1473-3099(19)30633-4.

Smit, M.R., Ochomo, E.O., Aljayyoussi, G., Kwambai, T.K., Abong'o, B.O. *et al.* (2018) Safety and mosquitocidal efficacy of high-dose ivermectin when co-administered with dihydroartemisinin-piperaquine in Kenyan adults with uncomplicated malaria (IVERMAL): A randomised, double-blind, placebo-controlled trial. *Lancet Infect Dis* 18(6), 615–626.

Smit, M.R., Ochomo, E.O., Waterhouse, D., Kwambai, T.K., Abong'o, B.O. *et al.* (2019a) Pharmacokinetics-pharmacodynamics of high-dose ivermectin with dihydroartemisinin-piperaquine on mosquitocidal activity and QT-prolongation (IVERMAL). *Clinical Pharmacology and Therapeutics* 105(2), 388–401. DOI: 10.1002/cpt.1219.

Smit, M.R., Ochomo, E.O., Aljayyoussi, G., Kwambai, T.K., Abong'o, B.O. *et al.* (2019b) Human direct skin feeding versus membrane feeding to assess the mosquitocidal efficacy of high-dose ivermectin (IVERMAL trial). *Clinical Infectious Diseases* 69(7), 1112–1119. DOI: 10.1093/cid/ciy1063.

Standfast, H.A., Muller, M.J. and Wilson, D.D. (1984) Mortality of *Culicoides brevitarsis* (Diptera: Ceratopogonidae) fed on cattle treated with ivermectin. *Journal of Economic Entomology* 77(2), 419–421.

Stapley, E.O. and Woodruff, H.B. (1982) Avermectins, antiparasitic lactones produced by *Streptomyces avermitilis* isolated from a soil in Japan. In: *Proceedings of an International Conference on Trends in Antibiotic Research: Genetics, Biosyntheses, Actions and New Substances*, Japan Antibiotic Research Association, pp. 154–170.

Syafruddin, D., Asih, P.B.S., Rozi, I.E., Permana, D.H., Nur Hidayati, A.P. *et al.* (2020) Efficacy of a spatial repellent for control of malaria in Indonesia: A cluster-randomized controlled trial. *American Journal of Tropical Medicine and Hygiene* 103(1), 344–358. DOI: 10.4269/ajtmh.19-0554.

Sylla, M., Kobylinski, K.C., Gray, M., Chapman, P.L., Sarr, M.D. *et al.* (2010) Mass drug administration of ivermectin in South-eastern Senegal reduces the survivorship of wild-caught, blood fed malaria vectors. *Malaria Journal* 9, 365.

Taylor-Wells, J., Brooke, B.D., Bermudez, I. and Jones, A.K. (2015) The neonicotinoid imidacloprid, and the pyrethroid deltamethrin, are antagonists of the insect Rdl GABA receptor. *Journal of Neurochemistry* 135(4), 705–713. DOI: 10.1111/jnc.13290.

Taylor-Wells, J., Senan, A., Bermudez, I. and Jones, A.K. (2018) Species specific RNA A-to-I editing of mosquito RDL modulates GABA potency and influences agonistic, potentiating and antagonistic actions of ivermectin. *Insect Biochemistry and Molecular Biology* 93, 1–11. DOI: 10.1016/j.ibmb.2017.12.001.

Tesh, R.B. and Guzman, H. (1990) Mortality and infertility in adult mosquitoes after the ingestion of blood containing ivermectin. *American Journal of Tropical Medicine and Hygiene* 43(3), 229–233.

Tipthara, P., Kobylinski, K.C., Godejohann, M., Hanboonkunupakarn, B., Roth, A. *et al.* (2021) Identification of the metabolites of ivermectin in humans. *Pharmacology Research & Perspectives* 9(1), e00712. DOI: 10.1002/prp2.712.

Trung, H.D., Bortel, W.V., Sochantha, T., Keokenchanh, K., Briët, O.J. *et al.* (2005) Behavioural heterogeneity of *Anopheles* species in ecologically different localities in southeast Asia: A challenge for vector control. *Tropical Medicine and International Health* 10(3), 251–262.

Vanachayangkul, P., Kodchakorn, C., Ta-Aksorn, W., Im-Erbsin, R., Tungtaeng, A. *et al.* (2024) Safety, pharmacokinetics, and potential neurological interactions of ivermectin, tafenoquine, and chloroquine in *Rhesus macaques*. *Antimicrob Agents Chemother* e0018124. DOI: 10.1128/aac.00181-24.

Van Bortel, W., Trung, H.D., Hoi Le, X., Van Ham, N., Van Chut, N. *et al.* (2010) Malaria transmission and vector behaviour in a forested malaria focus in central Vietnam and the implications for vector control. *Malaria Journal* 9, 373.

Vercruysse, J., Deprez, P., Everaert, D., Bassissi, F. and Alvinerie, M. (2008) Breed differences in the pharmacokinetics of ivermectin administered subcutaneously to Holstein and Belgian Blue calves. *Veterinary Parasitology* 152(1–2), 136–140. DOI: 10.1016/j.vetpar.2007.11.021.

Verdú, J.R., Cortez, V., Ortiz, A.J., González-Rodríguez, E., Martinez-Pinna, J. *et al.* (2015) Low doses of ivermectin cause sensory and locomotor disorders in dung beetles. *Scientific Reports* 5, 13912. DOI: 10.1038/srep13912.

Wei, L., Wei, G., Zhang, H., Wang, P.G. and Du, Y. (2005) Synthesis of new, potent avermectin-like insecticidal agents. *Carbohydrate Research* 340(9), 1583–1590.

Weil, G.J., Jacobson, J.A. and King, J.D. (2020) A triple-drug treatment regimen to accelerate elimination of lymphatic filariasis: From conception to delivery. *International Health* 13(Suppl 1), S60–S64. DOI: 10.1093/inthealth/ihaa046.

WHO (2018) World Malaria Report 2018. Geneva, Switzerland: World Health Organization.

Yoon, Y.J., Kim, E.S., Hwang, Y.S. and Choi, C.Y. (2004) Avermectin: Biochemical and molecular basis of its biosynthesis and regulation. *Applied Microbiology and Biotechnology* 63(6), 626–634.

Zeng, Z., Andrew, N.W., Arison, B.H., Luffer-Atlas, D. and Wang, R.W. (1998) Identification of cytochrome P4503A4 as the major enzyme responsible for the metabolism of ivermectin by human liver microsomes. *Xenobiotica* 28(3), 313–321. DOI: 10.1080/004982598239597.

8

House Modification for Malaria Control

J. D. Charlwood*

Global Health and Tropical Medicine, Instituto de Hygiene e Medicina Tropical, Lisbon, Portugal

Abstract

If there is one intervention that leads to improvements in people's health in sub-Saharan Africa, it is likely to be changes to their houses. These include replacing the traditional roof to one of corrugated iron and changing walls to be cement block or brick. These changes are being undertaken without any professional involvement, even though climate-friendly alternatives might be designed, and even though tin roofs may turn houses into 'ovens'. The role of the professional should be to provide advice on the most biodiversity-friendly 'do-it-yourself' improvements and their cost. In this chapter, simple additions to these modifications (such as adding netting screens to windows) are described. It is argued that using pressed, rather than burnt, bricks would reduce deforestation.

> '*Good quality housing is part of the basic health infrastructure required for development to take place.*'
>
> – The malERA consultative group on vector control (2011)

> '*Food and shelter are both essential for human life but food is anything from a bowl of rice a day to some exquisite mouthful for a moment's pleasure. Junk food is somewhere in between, but so too is just the right amount of nutrition our bodies needs. It's the same with shelter. 'We have bread buildings that fill, cake buildings that thrill, and junk buildings that make us want more. All misfits' wants is a nutritious architecture that does the shelter thing well, makes us feel good because it is good for us, doesn't cost the earth, or cost us the earth.*'
>
> – Misfit Architects Manifesto

*Corresponding author: jdcharlwood@gmail.com

DOI: 10.1079/9781789248845.0008

'The improvement of homes by the application of mosquito-proofing, . . . is probably the most practical, economical, and effective malarial control measure available'

– Kiker (1941)

Introduction

The early malariologist Lewis Wendell Hackett famously observed, 'Malaria, like chess, is a game of a small number of pieces but with an infinite set of possibilities' (Smith and Schapira, 2012). These possibilities include transmission by mosquitoes biting outside in the early evening. It is a truism, however, to say that most malaria transmission occurs inside houses and most of it inside houses in Africa. Indeed, even mosquitoes that may bite outside during one or the other gonotrophic cycle are still likely to enter houses during other cycles (Charlwood *et al.*, 1986; Killeen *et al.*, 2016). Making people's houses more difficult to access for disease vectors, particularly malaria vectors, while keeping the internal environment within the 'comfort zone', would help reduce transmission drastically.

The Comfort Zone

The physical comfort zone is that combination of temperature and humidity in which a person feels 'comfortable'. This generally falls within the blue-shaded portion of the figure.

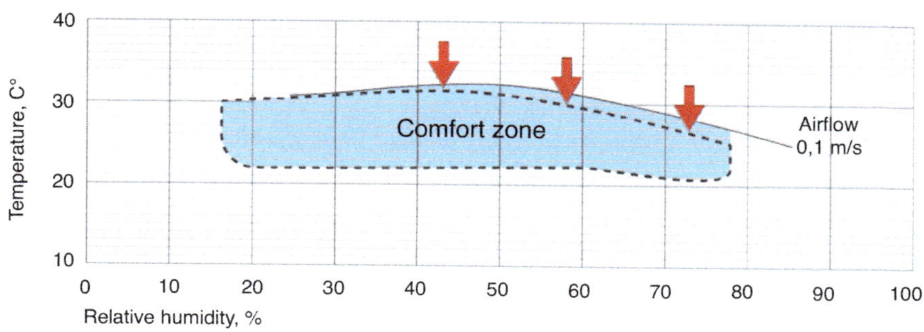

Airflow extends the comfort zone. The red arrows indicate the reduction in the comfort zone if airflow is decreased. At the upper edge of the comfort zone, even a subtle reduction in airflow can result in a climate that is intolerable for most people (from von Seidlein *et al.*, 2012).

Metabolic rate, clothing, air temperature, air speed and relative humidity as well as psychological factors all have an influence on how comfortable we feel. There may be considerable differences between different individuals. People are highly sensitive to even small differences in environmental temperature. At 24°C, a difference of 0.38°C can be detected between the temperature of two rooms. A perceived ability to control one's thermal environment tends to widen one's range of tolerable temperatures and natural ventilation enhances acceptance of a wider range of temperatures. Females tend to be more sensitive to temperatures, but males tend to be more sensitive to the other major contributor to comfort levels – relative humidity. Enhanced humidity means that heating is less effective, leading to a perception of discomfort.

Poor housing is an independent risk factor for malaria separate from house location, and it remains to be seen whether malaria transmission will remain a major problem once transmission inside houses is completely controlled (Gillies, 1988).

People have developed a number of solutions based on the way they built their houses that suited how they wanted to live, using whatever materials were available and all things considered did the shelter thing rather well. Although shelter may be the predominant factor in designing and building houses, people since ancient times, have adopted techniques to avoid the 'slings and arrows' of mosquito bites. The Greek historian Herodotus in 2500 BP described how people in Egypt 'sleep at night on raised structures, which is of great benefit to them because the gnats are prevented by the wind from flying high' (Herodotus, 1966). This was probably a species of *Mansonia*, since they are a noted painful biter. In São Tomé, people built their houses on stilts to both avoid floods and mosquitoes (in this case, *Anopheles coluzzii*). They were right to do so, since studies in the 1990s indicated a significant reduction in mosquito numbers in such houses compared to those at ground level (Charlwood *et al.*, 2003). In The Gambia entry of *An. gambiae* into houses declined with increasing height, with a hut at 3m reducing entry by 84% when compared with huts on the ground (Carrasco-Tenezaca *et al.*, 2021).

This effect may not occur in areas where *Anopheles dirus* is the vector. In Cambodia, this mosquito entered houses built more than 2m off the ground (Charlwood *et al.*, 2017). This is perhaps because, historically, it bit monkeys that lived in trees and so developed the ability to fly upwards, even in the absence of a wall to guide it. Whether it uses the post supporting the house (that acts like a tree trunk) as a reference or whether it flies upwards by directly following the host-odour stream is not known. House construction affected malaria transmission in Vietnam in the mid-1970s. The hill people in the north of the country built their houses on stilts to avoid *Anopheles minimus*, which seldom fly more than 2–3m above the ground. The anophelines fed on the animals kept under the houses and avoided the people. When people from the lowlands moved into the hills, they brought their culture with them. Their houses were built on the ground, and animals were stabled alongside. As a result, *An. minimus*, which will feed on cattle but prefers humans, became an active vector of malaria (Laderman, 1975).

How vectors enter houses is contentious. In what was a short, but classic, paper on house entry behaviour by African anophelines, Snow (1987) showed that if openings were at ground level, a variety of species entered an experimental hut, but as the height of the opening increased, the number of species entering the hut was reduced, until it was only malaria vectors that entered when the opening was close to the roof. This either means that the insects were able to fly up to the opening by following the odour plume, or they used the wall as a guide. In Papua New Guinea, *Anopheles farauti* appeared unable to follow an odour plume that was merely 50 cm off the ground (Charlwood *et al.*, 1984). The diminution in numbers of mosquitoes entering houses built off the ground in São Tomé also indicates that the vectors, unlike the anthropophilic descendant of an ornithophilic mosquito *Culex quinquefasciatus*, were unable to fly up an odour plume (Charlwood *et al.*, 2003). More recently, however, in a semifield situation, using video to track the flight paths of mosquitoes around the eaves of an experimental hut, Spitzen *et al.* (2016) found no clear indications that mosquitoes arrived at low altitude and moved upwards along the wall towards the eave. Rather, they described *Anopheles gambiae* flying directly upwards following the odour plume. Semifield setups may reduce wind speeds, which may affect flight height in mosquitoes.

Given that, however they get there, most mosquito entry is via the gables or eaves, the simplest modification might be to close the eave, or gable, gap. This, however, reduces airflow and makes the house hotter and more uncomfortable. If the eaves are closed, smaller alternative openings also become the source of all odours leaving the house, making them more attractive to mosquitoes. A better, and even simpler, solution is to close the gap with old mosquito netting. This allows for air and light to get in while allowing odour out so that mosquitoes continue to be attracted to this opening rather than to others. In Mozambique, the use of netting over the gaps between roof and walls has been shown to reduce the numbers of malaria vectors entering houses (Kampango *et al.*, 2013) and does no harm. Once in place, the householder does not need to do anything. It can be installed on a DIY basis or could be done for a fee by local installers (Fig. 8.1).

It also is a good use of old nets. Over 2.3 billion LLINs have been produced and distributed globally at the rate of 300–400 million per year (Lorenz *et al.*, 2020). They are supposed to have a useful life of 3 years, although studies from a number of countries on net attrition rates indicate that they may be much less than that. The disposal of old nets, equivalent to 216,000 metric tonnes of homogeneous polymer, is now recognized as a problem (Ramanantsoa *et al.*, 2017). Obviously, the long-term solution is to either produce more durable nets or to produce nets that are biodegradable, such as those that might be made from cellulose (Brake *et al.*, 2022). But this does not address the problem of how best to dispose of the millions of nets that are already in use and needing replacement. A more recent development of this technique used hessian infused with transfluthrin, a spatial repellent (Mmbando *et al.*, 2018; Kaindoa *et al.*, 2021). These also reduced entry rates, as much probably due to the effect of the repellent as the physical barrier provided by the hessian. The effect of the transfluthrin lasted for 6 months, which, although considerable, is a lot shorter than the life of the house or the material. The use of old nets, however,

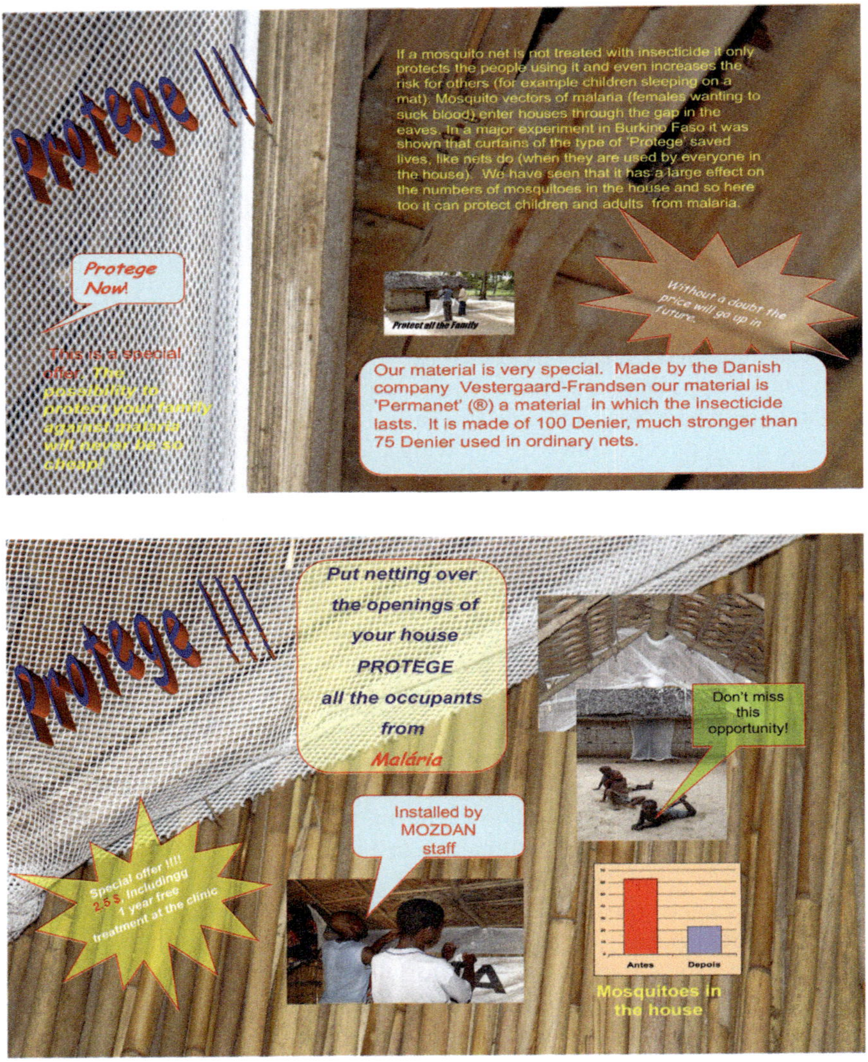

Fig. 8.1. The brochure (translated here into English) used to advertise the use of netting over the eave gap in houses from Furvela. (Photo: J.D. Charlwood.).

has a number of advantages over hessian. It does not reduce airflow to the same extent and allows light into the room, and it helps recycle old nets. By not discouraging mosquitoes from attempting to enter, they may be diverted less often to other non-treated houses.

Even with the best protection around the roof, badly fitting doors remain a problem. As long ago as 1901, Celli showed that screened porches (Fig. 8.2) constructed around the front doors and the chimneys closed with

Fig. 8.2. A veranda attached to a house in the Italian marshes circa 1900 from Celli (1901).

mesh reduced malaria among rail workers' living in an endemic area in Italy (Lindsay *et al.*, 2002). In the 2 year study, there were 96% fewer cases of malaria in the protected group than in the control group. Only 10 out of the 207 railway workers contracted malaria in the protected group, compared with nearly everyone in unscreened control houses, where mosquitoes were found to enter continuously, and 89% (204/229) of residents were afflicted with malaria (Celli, 1901).

A Celli-like solution of a netting veranda in front of the door might be one solution applicable in Africa. Village houses are built on an individual basis, not in estates by building companies. There are a number of options that might be available (screen doors, for example), according to people's budgets. The use of pressed clay bricks, rather than burnt ones, would have an enormous impact on the environment. This would be the most environmentally friendly technique for building improved houses. It would be nice to think that organizations like the Gates Foundation would subsidize their distribution across Africa. A 'Ferrari' house, built to the best specifications available, might serve as both an office for these products and a kind of 'lived-in' display house. Pressed bricks would be a cheap, basic (possibly subsidized) option and the first option to be recommended to potential house builders.

A 'one-size-fits-all' veranda could be purchased in a hardware shop and could be used on different houses. Again, it could be installed by local entrepreneurs for a modest fee or could be installed on a DIY basis. In areas where the house construction is suitable, the same applies to the installation of eave tubes.

Eave Tubes

Eave tubes are a recently developed system in which the eaves of local houses are closed, and a few cylindrical holes are made in the walls, into which plastic tubes with insecticide-laden electrostatic netting are inserted. Where houses already have blocked eaves, these cylindrical holes are drilled, and the tubes inserted. Despite the promising results obtained using this technique (Sternberg *et al.*, 2016), there remain a number of caveats surrounding its wide-scale implementation, namely: (i) there is a need for extensive house modifications before eave tubes are inserted; (ii) the tubes are only suitable for well-built houses, which means the poorest and highest-risk households living in housing structures not amenable to eave tubes cannot be assisted; and (iii) there are poor synergies when eave tubes are combined with LLINs or IRS in the same households. Given that placing old netting over the eave gap in even the poorest house is a relatively simple procedure (Kampango *et al.*, 2013) that mimics the effects of eave tubes (at much lower cost and without the added inconvenience of reduced ventilation due to blocking the eave gap), it is difficult to warrant such a system compared to the simpler alternative. Nevertheless, eave tubes above windows (which are closed with mosquito-proof shutters at night) might be useful for maintaining airflow, especially if the tubes were located at opposite ends of the house.

Some changes to houses, which are likely to affect mosquito populations, will occur without outside 'professional' intervention. For example, people tend to install a tin (corrugated iron) roof for their houses, if they can afford one. These houses are hotter, and mosquitoes are more likely to leave them in an attempt to find a more suitable resting site. Houses with traditional thatch roofs are cooler in the warm season and warmer in the cooler months than outdoors. The habit of resting inside houses, before the advent of IRS, resulted in a higher survival rate in mosquitoes. In Africa, this habit must have developed early in their association with humans and is perhaps one reason why African mosquitoes are such good vectors. It remains to be seen if the trend for such roofs reverses if houses become unbearably hot or if, as is likely, the installation of windows, which help with ventilation, becomes an increasing trend.

If the roof is made from metal sheeting in order to reduce the likelihood of the roof being carried off by gusts of wind that lift the overhanging section, many householders close the gap between the eaves and the roof, limiting mosquito entry. Brick houses tend to have windows (while smaller houses made with traditional material are often windowless). In areas where dengue is a problem, windows should be screened in order to prevent mosquito entry. This applies particularly to preventing access to *Aedes aegypti*. In many places, windows are often closed at night with burglar-proof, and hopefully mosquito-proof, shutters. Such windows may not need screens, since screens will, to some extent, reduce airflow. Indeed, an area that remains not well explored is the effect of screens of different mesh sizes on airflow.

A further improvement to houses that will reduce exposure to disease vectors is the installation of a ceiling. Ceilings are typically made of wood, cardboard, palm leaves, cloth or sacking installed below the eave space. Ceilings prevent mosquitoes that have entered through the eaves from accessing hosts in the room. In The Gambia, simple corrugated cardboard sheets (of the sort that are used to transport domestic electrical items such as fridges, which are considered 'rubbish' by shop owners) tacked on to a wooden frame offered protection against *An. gambiae* s.l. Twice as many rooms without ceilings contained malaria vectors compared to rooms with ceilings (Lindsay *et al.*, 1990). Mesh ceilings were also more effective than plywood ceilings at reducing the number of mosquitoes inside rooms (70–80% reduction in mosquito numbers vs 52–59% reduction) (Kirby, 2013). It is also worth pointing out that mesh treatments did not raise the indoor temperature compared to the control.

In Dar es Salaam, a reinterpretation of the data obtained in the larviciding project described by Maheu-Giroux and Castro (2013) showed that most of the empirically fitted downward time trend in *Plasmodium falciparum* malaria prevalence over the course of the study could be plausibly explained by association with an upward trend in city-wide window screening coverage (Killeen *et al.*, 2019).

Compared to unscreened houses, houses with either treated or untreated screens in Ifakara, Tanzania, almost entirely excluded *Anopheles arabiensis* (Msoffe *et al.*, 2022). On Bioko Island, the prevalence of malaria infection in 2- to 14-year-old children living in houses with open eaves was 23.0%, compared to 18.8% for those living in houses with closed eaves (OR = 0.81, 95% CI 0.67–0.98), while the prevalence of infection for children in screened houses was 9.1% compared to 20.1% for those living in unscreened houses (OR = 0.44, 95% CI 0.27–0.71) (Bradley *et al.*, 2013). Thus, screening of windows makes a difference. Perhaps most critical to the acceptability of house screening are the related issues of durability, affordability, ventilation and indoor microclimate.

Sometimes, a method that seems counter-intuitive may have beneficial results, as is the case when traffic lights and road signs, and even roads, are removed from congested towns, resulting in increased flow and fewer accidents (Baker, 2009). Thus, one might think that a house with a greater number of windows would attract more mosquitoes, but in The Gambia, installing two or more screened windows in houses, each 0.5 m^2 in area, significantly reduced the number of mosquitoes entering them than poorly ventilated houses (Jatta *et al.*, 2021; Mmbando *et al.*, 2022; Carrasco-Tenezaca *et al.*, 2023). Well-ventilated houses are cooler at night and more comfortable to live in. Increasing the number of windows lowered the indoor temperature at night by 0.2–0.6°C, depending on outdoor wind speed (Jatta *et al.*, 2021). It is important, however, that there is the possibility of cross-ventilation between windows. In this case, windows are positioned on opposite walls without any internal obstruction. How easy this might be in practice may be a problem.

New Constructions – Pressed Bricks Versus Fired Bricks

Modifications are one thing; starting from the ground up is another. It is the author's belief that most Africans presently living in villages may well prefer a brick house to a traditional house, even though it might be more uncomfortable. Traditional bricks, however, require a considerable amount of fuel to be manufactured (Fig. 8.3).

In most places, this means wood (although in rice-growing areas, such as Ifakara in Tanzania, rice husks may be used) (Fig. 8.4).

Traditional brick making is responsible for a large part of the deforestation that is continuing in Africa.

An alternative is already available. Houses should be made from pressed, rather than fired, bricks. The technique has been developed and the equipment is already available. These bricks (aka compressed earth blocks) are as durable as fired bricks, or concrete, and have a minimal impact on the environment compared to conventional techniques (Jayasinghe, 2007). Pressed bricks are more regular in shape than conventional bricks, so they fit together better, meaning that houses can be built with less mortar (Fig. 8.5). An economic system that encouraged the development of pressed bricks should be pursued, whether in the form of subsidy or loan, competition or even alternative currency (Lietaer *et al.*, 2012).

Cooler alternatives to bricks might require a considerable degree of persuasion on the part of malariologists for them to be appreciated by locals. One

Fig. 8.3. Traditional brick making in Kyamurwa, Muleba, in. the Kagera region of Tanzania. (Photo: J.D. Charlwood.).

Fig. 8.4. Brick making close to the Kilobero River, Ifakara, Tanzania. Clay comes from the river bed. Rice husks are used to fire these bricks. (Photo: J.D. Charlwood.).

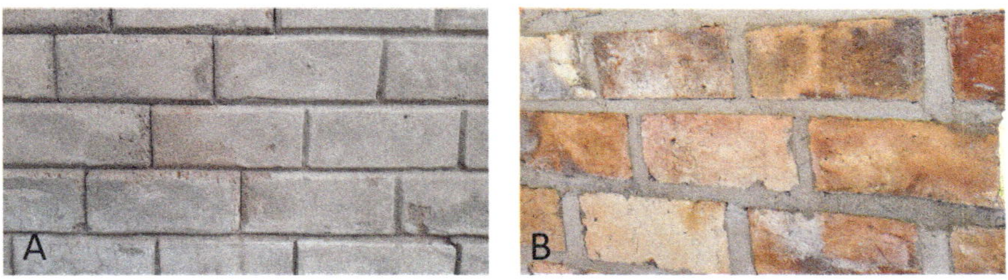

Fig. 8.5. (A) Pressed bricks and (B) fired bricks used in the construction of houses in Mozambique. A lacquer has been applied to the pressed bricks. Note the even nature of the bricks compared to traditional bricks, which means less mortar is needed to join them together. (Photo: J.D. Charlwood.).

way people might be persuaded to adopt mosquito-proofing techniques is for them to see what they look like in practice and to know how much they might cost. In many housing developments in Europe, when a company builds a large housing estate and attempts to persuade potential owners to buy them, they build a 'show house' that potential purchasers can visit. A 'show house' (that may also act as an office for potential contractors) built in an easily accessible location (perhaps next to a builder's merchants where products can be purchased) would enable people to see for themselves what might be done. Each component of the house (such as the cost of a screened window or of a veranda) can be priced so that a mixture of components might eventually be purchased according to available funds. If the effect that individual components might have on both vector numbers inside the house and the quality of the environment inside it (how well it adheres to the 'comfort zone') were available, this would help potential purchasers decide on what they might incorporate into their own houses. Some components (such as a veranda) might need a subsidy before they are widely adopted. Once techniques become more widely adopted, then, as shown by the cost of renewable energy in Europe and America, the price will almost certainly drop to a point where they become more sustainable. These improvements become permanent interventions rather than the temporary ones (like the distribution of LLINs or IRS) currently adopted. The WHO, however, is shackled from endorsing such techniques, because they have yet to show significant results from randomised control trials (Tusting *et al.*, 2016). The same attitude was initially adopted towards the endorsement of bed nets treated with insecticide, despite the fact that, even if information on the impact on disease transmission was minimal, such nets gave users a 'good night's sleep'. Thus, although the evidence for house improvement having an impact on disease transmission in the 'pure sense' is limited, it is our opinion, like that of early malariologists, that it will be the most important factor in reducing vector-borne disease transmission – and it will largely be undertaken without professional help.

In 2013, Matt Kirby described an 'ideal' house (Fig. 8.6).

Expanded Polystyrene Beads

Although it is an efficient vector of bird malaria, *Cx. quinquefasciatus* is an inefficient vector of human filariasis. It is extremely synanthropic (closely associated with humans), and both males and females spend much of their lives in and around houses. It is probably the mosquito that people in the tropics would most like to see disappear. It is a 'sit-and-wait' predator and is the mosquito that buzzes in one's ears at night before attempting to feed. A single mosquito is enough to disturb one's sleep. Males, too, are annoying, because they will swarm inside houses and can use people's heads as markers. The mosquitoes are well known for breeding in foul water, septic tanks, wet pit latrines or ditches of polluted water. A large proportion of people, both rural and urban, in Africa use pit latrines (Fig. 8.7).

A good proportion of these may contain water in which *Cx. quinquefasciatus* thrive. The control of *Cx. quinquefasciatus* should be largely through control of the larvae rather than the adults. This is because adult mortality, even in the absence of any intervention, is low. Typically, 80% of a collection of adults have newly emerged with only 20% being parous (compared to the opposite situation for Anopheline vectors of malaria) (Charlwood, 1979). Thus, killing the adults will have little perceived effect on the numbers biting. Larval densities can be very high. For example, 5 dips with a 9-cm-diameter cup in one

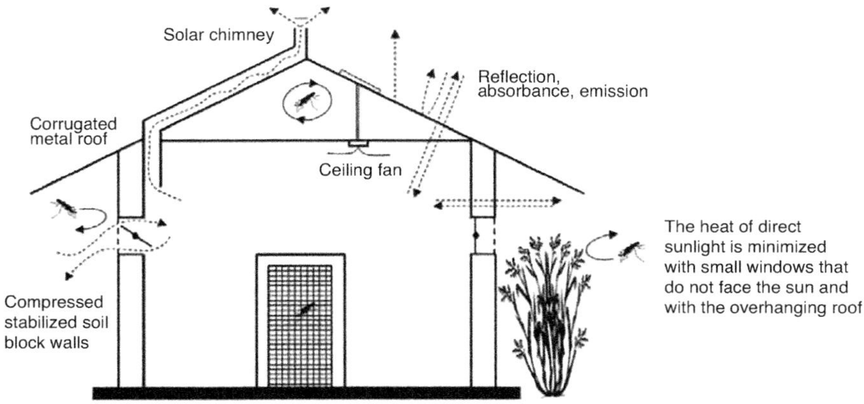

Fig. 8.6. The ideal house? Mosquitoes approaching doors and windows are prevented from entering by screening material over the openings and further discouraged by repellent plants. Solar-powered fans and a solar chimney improve air movement indoors. The ceiling prevents any mosquitoes in the roof space from reaching the room and should absorb some of the heat from the metal roof. Carefully chosen construction materials improve the durability of the house. A solar-powered Suna trap placed outside close to the walls might also help as might the addition of eave tubes over the windows. (Figure from Kirby, 2013.)

3-m-diameter septic tank in Ifakara yielded 2213, 2405, 3002, 1982 and 2137 larvae (Charlwood, 1994).

Expanded polystyrene beads, which float and form an impenetrable barrier to larval, pupal and ovipositing female mosquitoes, but not to faeces, are one tool that can be used to control *Cx. quinquefasciatus* breeding in wet pit latrines (Reiter, 1978; Curtis and Minjas, 1985). When manufactured, the beads are similar in size to grains of sand with diameter of 0.6–2.5 mm. They contain a gas which, when heated in steam or boiling water, expands. Thus, the beads increase in size to about 35–40 times of their original volume and decrease in density so that they float. When compressed together, the beads are commonly used as packing material (material which often ends up polluting the sea). Use of such beads was effective in controlling filariasis in coastal villages on the island of Zanzibar (Maxwell *et al.*, 1990). They were later adopted as a control technique in Dar es Salaam.

> Open breeding sites such as areas of flooded land and blocked drains were treated with pyriproxyfen (an insect growth regulator) at a concentration of 0.1 ppm. Emergence of *Cx. quinquefasciatus* adults from these sites was inhibited for 4 weeks during the rainy season and for up to 11 weeks during the dry season. The problem of mosquito breeding sites caused by bathroom sullage water was addressed through a combination of health education and indirect pressure from the Urban Malaria Control Project (UMCP) via local community leaders.

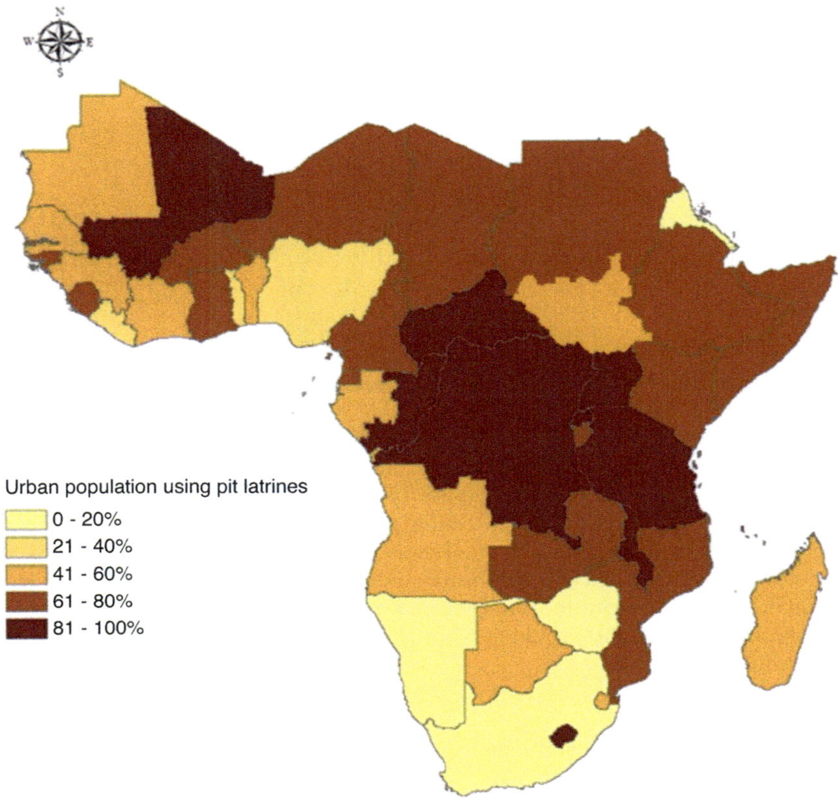

Fig. 8.7. Map of urban users of pit latrines in Africa. (From Nakagiri *et al.*, 2016, used under license CC BY 4.0.).

> Households responsible for these sites were encouraged to eliminate them by diverting the water into an enclosed drainage structure, usually a pit latrine. After two weekly visits, 64.7% of households had complied and 93.4% had complied after five visits. Densities of *Cx. quinquefasciatus* adults dropped by 76.7% in one site and by 46.2% in another following intervention but increased by 84.9 and 25.6% in two untreated comparison areas.
>
> (Chavasse *et al.*, 1995)

The technique remains underused. One reason may be that although the technique can theoretically be applied on an individual basis (or by local entrepreneurs; Charlwood, 2008), it is not possible to obtain small amounts of the beads sufficient to treat a single latrine. The gas in the unexpanded beads will eventually diffuse away, and so the beads need to be used a relatively short time after manufacture. This also impedes their use on a small scale. Vacuum-packed quantities sufficient to treat a standard latrine might be produced, which might make them more easily available and enable longer storage.

VIP Latrines

One technique that can, however, be used to control mosquitoes and flies that are attracted to pit latrines is by building ventilated improved pit latrines (VIP latrines) (WHO and UNICEF, 2014). They work on the principle that hot air rises (Fig. 8.8).

The pipe on the outside of the latrine heats up during the day, and because hot air rises, air is drawn into the latrine rather than out. This reduces any odour that might help gravid females to find it. Mosquitoes and flies are also less likely to enter the latrine in the first place compared to a standard latrine (because it is dark, and because the odour that might attract them leaves the latrine via the ventilation pipe), and if they do, then they will not easily leave (because they will attempt to do so via the light coming from the ventilation pipe that is covered with a mesh that prevents them leaving). The best size of mesh is 1.5 mm × 1.2 mm. Larger holes allow flies to get through, while smaller holes restrict the airflow.

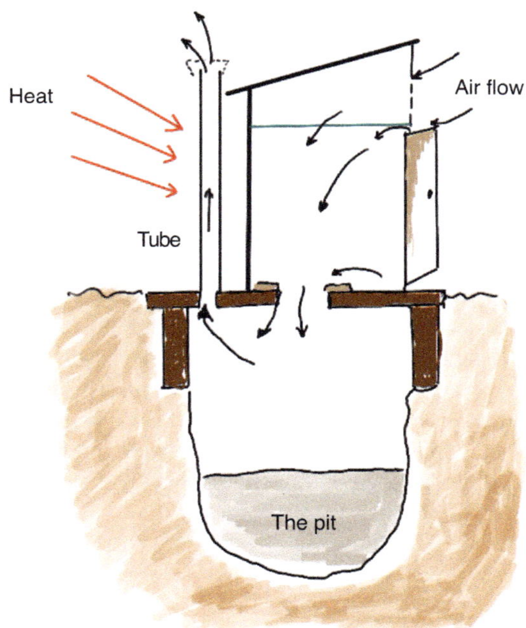

Fig. 8.8. Diagram of a VIP latrine. The air in the external pipe is heated by the sun which then generates the airflow that means that the odour that would otherwise attract flies is dispersed away from the source. The seat is optional. (Figure redrawn from Nwaneri *et al.*, 2008).

Ventilation of a VIP latrine is most effective when the entry is facing the direction of the prevailing wind and the ventilation pipe exposed to the sun (so the air in the pipe heats up and rises). The idea is a simple and good one, but such VIP latrines are not commonly built, perhaps because of a lack of understanding of the principle behind their design.

Another even simpler technique that can be applied by individual householders with septic tanks is to make sure that there are no openings available for mosquitoes to either enter or leave. Very often, septic tanks develop small openings as they settle after construction. Even very small ones can be discovered by *Cx. quinquefasciatus* (the odour is, after all, rather potent) who will enter to oviposit. As Leonard Cohen sings, 'There is a crack, a crack in everything. That's how the light gets in' (and the smell gets out). Mosquitoes use the light to enable them to exit such damaged tanks.

Finding holes in tanks and filling them and covering any vents with mesh to prevent mosquito passage is the simplest of control techniques. Once again, it is probably through ignorance that the status quo is maintained. Simple advertising/information campaigns might make a difference, but since the mosquito rarely transmits any disease, no funding agency is likely to support such an initiative.

The control of larvae by an individual householder may have a limited impact on the number of adults biting that same householder because of immigrant mosquitoes from nearby untreated sites. Nevertheless, there is evidence that because of the limited flight range of the mosquito, it might have an effect. For example, in Recife, Brazil, Medeiros *et al.* (1992) found that a house with an adjacent latrine treated with expanded polystyrene beads (to control the larvae) had a very low adult population, despite there being many other untreated pits in the neighbourhood. Similarly, in Ifakara, Tanzania, despite there being a marked effect in the hospital wards (with numbers declining to 10% of the original control collections), a house 200m from a treated septic tank in the hospital grounds showed no evidence of an effect (Charlwood, 1994). More information on the effect of the impact of individual actions on community benefits would be useful. Once these actions (be it house improvement for malaria control or interventions against *Culex* larvae) reach a threshold, then a 'mass effect' will ensue such that even people not undertaking such measures will be partially protected, and the protection for those that do will be improved.

References

Baker, L. (2009) Removing roads and traffic lights speeds urban travel. *Scientific American Magazine* 300, 2.

Bradley, J., Rehman, A.M., Schwabe, C., Vargas, D., Monti, F. *et al.* (2013) Reduced prevalence of malaria infection in children living in houses with window screening or closed eaves on Bioko island, Equatorial Guinea. *PLoS ONE* 8, e80626. DOI: 10.1371/journal.pone.0080626.

Brake, S., Gomez-Maldonado, D., Hummel, M., Zohdy, S. and Peresin, M.S. (2022) Understanding the current state-of-the-art of long-lasting insecticide nets and potential for sustainable alternatives. *Current Research in Parasitology and Vector Borne Disease* 2, 100101.

Carrasco-Tenezaca, M., Abdi, M.Y., Bradley, J., Brittain, O.S., Ceesay, S. *et al.* (2021) The relationship between house height and mosquito entry: An experimental study in rural Gambia. *Journal of the Royal Society Interface* 18, 178.

Carrasco-Tenezaca, M., Jawara, M., Lee, D.S., Holmes, M.S., Ceesay, S. *et al.* (2023) Effect of passive and active ventilation on malaria mosquito house entry and human comfort: An experimental study in rural Gambia. *Journal of the Royal Society Interface* 20, 201. DOI: 10.1098/rsif.2022.0794.

Celli, A. (1901) The new preventative treatment of malaria in Latium. In: *Collected Papers on Malaria. Angelo Celli.* London School of Hygiene and Tropical Medicine, London, pp. 1–12.

Charlwood, J.D., Pinto, J., Ferrara, P., Sousa, C.A., Ferreira, C. *et al.* (2003) Raised houses reduce mosquito bites. *Malaria Journal* 2, 45.

Charlwood, J.D. (1979) Estudos sobre a biologia e habitos de *Culex quinquefasciatus* say em Manaus, Amazonas, Brasil. *Acta Amazonica* 9, 271–278.

Charlwood, J.D. (1994) The control of *Culex quinquefasciatus* breeding in septic tanks using expanded polystyrene beads in southern Tanzania. *Transactions of the Royal Society of Tropical Medicine and Hygiene* 88, 380.

Charlwood, J.D. (2008) Women's groups and the marketing of health interventions a Tanzanian experience. *PNG Medical Journal* 51, 102–104.

Charlwood, J.D., Paru, R. and Dagaro, H. (1984) Raised platforms reduce mosquito bites. *Transactions of the Royal Society of Tropical Medicine & Hygiene* 78, 141–142.

Charlwood, J.D., Birley, M.H. and Graves, P.M. (1986) Capture-recapture studies of females of the *Anopheles punctulatus* group of mosquitoes (Diptera, culicidae) from Papua New Guinea. *Bulletin of Entomological Research* 76, 211–227.

Charlwood, J.D., Hall, T., Nenhep, S., Rippon, E., Branca-Lopes, A. *et al.* (2017) Spatial repellents and malaria transmission in an endemic area of Cambodia with high mosquito net usage. *Malaria World Journal* 8, 11.

Chavasse, D.C., Lines, J.D., Ichimori, K., Majala, A.R., Minjas, J.N. *et al.* (1995) Mosquito control in Dar es Salaam. II. impact of expanded polystyrene beads and pyriproxyfen treatment of breeding sites on *Culex quinquefasciatus* densities. *Medical and Veterinary Entomology* 9, 147–154. DOI: 10.1111/j.1365-2915.1995.tb00171.x.

Curtis, C.F. and Minjas, J. (1985) Expanded polystyrene for mosquito control. *Parasitology Today* 1(1), 36. DOI: 10.1016/0169-4758(85)90106-1.

Gillies, M.T. (1988) Anopheline mosquitos: Vector behaviour and bionomics. In: Wernsdorfer, W.H. and McGregor, S.I. (eds) *Malaria – Principles and Practice of Malariology*, Vol. 1. Churchill Livingstone, Edinburgh, UK, pp. 453–485.

Herodotus (1966) Translated by A. de Sélincourt. *The Histories*. Penguin Books, London.

Jatta, E., Carrasco-Tenezaca, M., Jawara, M., Bradley, J., Ceesay, S.D. *et al.* (2021) Impact of increased ventilation on indoor temperature and malaria mosquito density: An experimental study in the Gambia. *Journal of the Royal Society Interface* 18(178), 20201030. DOI: 10.1098/rsif.2020.1030.

Jayasinghe, C. (2007) Comparative performance of burnt clay bricks and compressed stabilized earth bricks and block. *The Engineer* 3, 33–40.

Kaindoa, E.W., Mmbando, A.S., Shirima, R., Hape, E.E. and Okumu, F.O. (2021) Insecticide-treated eave ribbons for malaria vector control in low-income communities. *Malaria Journal* 20, 415. DOI: 10.1186/s12936-021-03945-2.

Kampango, A., Bragança, M., de Sousa, B. and Charlwood, J.D. (2013) Netting barriers to prevent mosquito entry into houses in Southern mozambique: A pilot study. *Malaria Journal* 12, 99.

Kiker, C.C. (1941) Housing with special reference to mosquito-proofing for malaria control. *Symposium on Human Malaria, American Association for the Advancement of Science* 308–314.

Killeen, G.F., Govella, N.J., Lwetoijera, D.W. and Okumu, F.O. (2016) Most outdoor malaria transmission by behaviourally-resistant *Anopheles arabiensis* is mediated by mosquitoes that have previously been inside houses. *Malaria Journal* 15, 225.

Killeen, G.F., Govella, N.J., Mlacha, Y.P. and Chaki, P.P. (2019) Suppression of malaria vector densities and human infection prevalence associated with scale-up of mosquito-proofed housing in Dar es Salaam, Tanzania: Re-analysis of an observational series of parasitological and entomological surveys. *Lancet Planet Health* 3(3), e132–e143. DOI: 10.1016/S2542-5196(19)30035-X.

Kirby, M.J. (2013) House screening. In: Cameron, M.M. and Lorenz, L.M. (eds) *Biological and Environmental Control of Disease Vectors*. CAB International, Wallingford, UK, pp. 117–143.

Laderman, C. (1975) Malaria and progress: Some historical and ecological considerations. *Social Science and Medicine* 9, 587–594.

Lietaer, B., Arnsperger, C., Goerner, S. and Brunnhuber, S. (2012) *Money and Sustainability: The Missing Link*. Triarchy Press, Charmouth, UK.

Lindsay, S.W., Campbell, H., Adiamah, J.H., Greenwood, A.M., Bangali, J.E. *et al.* (1990) Malaria in a peri-urban area of the Gambia. *Annals of Tropical Medicine & Parasitology* 84(6), 553–562. DOI: 10.1080/00034983.1990.11812510.

Lindsay, S.W., Emerson, M.P. and Charlwood, J.D. (2002) Professor Angelo Celli's forgotten experiments: Reducing malaria by mosquito-proofing homes. *Trends in Parasitology* 18, 510–514.

Lorenz, L.M., Bradley, J., Yukich, J., Massue, D.J., Mageni Mboma, Z.P. *et al.* (2020) Comparative functional survival and equivalent annual cost of 3 long-lasting insecticidal net (LLIN) products in Tanzania: A randomised trial with 3-year follow up. *PLoS Medicine* 10, 1371.

Maheu-Giroux, M. and Castro, M.C. (2013) Impact of community-based larviciding on the prevalence of malaria infection in Dar es Salaam, Tanzania. *PLoS One* 8(8), e71638. DOI: 10.1371/journal.pone.0071638.

malERA consultative group on vector control (2011) A research agenda for malaria eradication: Vector control. *PLoS Medicine* 8(1), e1000401. DOI: 10.1371/journal.pmed.1000401.

Maxwell, C.A., Curtis, C.F., Haji, H., Kisumuku, S., Thalib, A.I. *et al.* (1990) Control of Bancroftian filariasis by integrating therapy with vector control using polystyrene beads in wet pit latrines. *Transactions of the Royal Society of Tropical Medicine and Hygiene* 84, 709–714.

Medeiros, Z., Dreyer, G., Andrade, L. and Pires, M.L. (1992) A preliminary study of vector control with polystyrene beads in Recife, Brazil. *Memorias Do Instituto Oswaldo Cruz* 87, 165–166.

Mmbando, A.S., Ngowo, H., Limwagu, A., Kilalangongono, M., Kifungo, K. *et al.* (2018) Eave ribbons treated with the spatial repellent, transfluthrin, can effectively protect against indoor-biting and outdoor-biting malaria mosquitoes. *Malaria Journal* 17, 368. DOI: 10.1186/s12936-018-2520-1.

Mmbando, A.S., Bradley, J., Kazimbaya, D., Kasubiri, R., Knudsen, J. *et al.* (2022) The effect of light and ventilation on house entry by *Anopheles arabiensis* sampled using light traps in Tanzania: An experimental hut study. *Malaria Journal* 21, 36. DOI: 10.1186/s12936-022-04063-3.

Msoffe, R., Hewitt, M., Masalu, J.P., Finda, M., Kavishe, D.R. *et al.* (2022) Participatory development of practical, affordable, insecticide-treated mosquito proofing for a range of housing designs in rural Southern Tanzania. *Malaria Journal* 21, 318. DOI: 10.1186/s12936-022-04333-0.

Nakagiri, A., Niwagaba, C.B., Nyenje, P.M., Kulabako, R.N., Tumuhairwe, J.B. *et al.* (2016) Are pit latrines in urban areas of sub-Saharan Africa performing? A review of usage, filling, insects and odour nuisances. *BMC Public Health* 16, 120.

Nwaneri, C., Foxon, K., Bakare, B. and Buckley, C. (2008) Biological degradation processes within a pit latrine. In: *WISA 2008 Conference (Sun City)*. Available at: https://www.researchgate.net/publication/267715964_Biological_degradation_processes_within_a_pit_latrine (accessed 23 September 2023).

Ramanantsoa, A., Wilson-Barthes, M., Rahenintsoa, R., Hoibak, S., Ranaivoharimina, H. *et al.* (2017) Can the collection of expired long-lasting insecticidal nets reduce their coverage and use? Sociocultural aspects related to LLIN life cycle management and use in four districts in Madagascar. *Malaria Journal* 16, 404.

Reiter, P. (1978) Expanded polystyrene beads: An idea for mosquito control. *Annals of Tropical Medicine and Parasitology* 72, 595–596.

Smith, T.A. and Schapira, A. (2012) Reproduction numbers in malaria and their implications. *Trends in Parasitology* 28, 3–8.

Snow, W.F. (1987) Studies of house-entering habits of mosquitoes in the Gambia, West Africa: Experiments with prefabricated huts with various wall apertures. *Medical and Veterinary Entomology* 1, 9–21.

Spitzen, J., Koelewijn, T., Mukabana, W.R. and Takken, W. (2016) Visualization of house-entry behaviour of malaria mosquitoes. *Malaria Journal* 15, 233. DOI: 10.1186/s12936-016-1293-7.

Sternberg, E.D., Ng'habi, K.R., Lyimo, I.N., Kessy, S.T., Farenhorst, M. *et al.* (2016) Eave tubes for malaria control in Africa: Initial development and semi-field evaluations in Tanzania. *Malaria Journal* 15, 447.

Tusting, L.S., Willey, B. and Lines, J. (2016) Building malaria out: Improving health in the home. *Malaria Journal* 15, 320. DOI: 10.1186/s12936-016-1349-8.

von Seidlein, L., Ikonomidis, K., Bruun, R., Jawara, M., Pinder, M. *et al.* (2012) Airflow attenuation and bed net utilization: Observations from Africa and Asia. *Malaria Journal* 11, 200. DOI: 10.1186/1475-2875-11-200.

WHO and UNICEF (2014) Country files – WHO / UNICEF joint monitoring programme for water supply and sanitation; estimates on the use of water sources and sanitation facilities. WHO, UNICEF, Geneva, Switzerland.

9

Imagining the Future of Bed Nets for Malaria Control in Africa

FREDROS OKUMU[1,2*] AND EMMANUEL KAINDOA[1]

[1]*Environmental Health & Ecological Sciences, Ifakara Health Institute, Ifakara, Tanzania*
[2]*School of Biodiversity, One Health and Veterinary Medicine, University of Glasgow, Glasgow, Scotland*

Abstract

Insecticide-treated nets (ITNs) are today the most dominant tool for malaria prevention. By 2021, more than 2.2 billion units had been delivered since the start of the Roll Back Malaria Partnership in 1998. Most ITNs are still treated using pyrethroids, but the threat of insecticide resistance in malaria vectors is driving the development of new ITN brands incorporating enzyme synergists, such as piperonyl butoxide (PBO), and non-pyrethroid active ingredients. While the functionality of an ITN constitutes both a physical barrier and chemical bio-efficacy, prevailing conversations overwhelmingly imply only the latter. The resultant drive for new insecticides has, to a large extent, overshadowed other important attributes of bed nets, such as their physical integrity, design and aesthetics, durability and continuous availability. Beyond these aspects, manufacturing of ITNs in Africa remains marginal, and even countries with the highest malaria burden still rely primarily on imports. Moreover, the private sector market for mosquito control products appears to be substantial but remains largely ignored, ceding ground to a public sector market powered largely by international financing and pooled procurement. Despite African manufacturers regularly producing high-quality apparel for global export markets, their involvement in bed net production remains limited, in part because many of them lack the technology for incorporating insecticides to nets. As a result, they are often deprioritized in favour of international manufacturers. This paper examines the plausible futures of bed nets for malaria control in Africa and highlights opportunities for reinvention beyond just the insecticidal properties and pooled international procurement. The central argument

*Corresponding author: fredros@ihi.or.tz

 Practical Control of Mosquito Disease Vectors (ed J.D. Charlwood)
DOI: 10.1079/9781789248845.0009

is that bed net industry should focus on enhancing durability, accessibility, aesthetic appeal, biodegradability and local production while ensuring competitive quality and pricing, without placing excessive emphasis on insecticidal properties.

Keywords: Malaria, ITNs, Ifakara Health Institute, Insecticide-Treated Nets

Background

The years since 2000 have heralded extraordinary initiatives to combat malaria, an infectious disease that still affects ~250 million and kills over 600,000 people annually (World Health Organization, 2023). Perhaps the most important was the Abuja meeting of the African heads of state and government, which first proposed wide-scale distribution of malaria interventions to at-risk households (World Health Organization, 2000). The Roll Back Malaria Partnership (RBM), which was created in 1998, took the lead role in the campaign to scale up ITN distribution, starting initially with conventionally treated nets which required retreatment every ~6 months (RBM, 2002). The Global Fund for HIV, Tuberculosis and Malaria, which was created in 2002 and faced a series of challenges in the beginning, is now one of the leading mechanisms and the top financier of public health initiatives in the world (Kapp, 2002, 2003; Ramsay, 2002). In 2005, the US President's Malaria Initiative was formed with the mission to reduce deaths from malaria by 50% by scaling up WHO-approved interventions, notably ITNs (USAID/PMI, 2023). The invention of long-lasting insecticide-treated net technology (LLIN), which was first approved by WHO in 2000, enabled extraordinary scalability and inspired opportunities to achieve universal coverage as envisioned by the WHO Global Malaria Action Plan (GMAP) of 2008 (World Health Organization, 2008). By 2020, more than two billion insecticide-treated nets (TTNs) had been delivered to households in malaria-endemic countries, under the auspices of the RBM and the partners (Roll Back Malaria, 2020). In their 2023 report, PMI alone reported they had delivered ~37 million ITNs, protecting nearly 74 million people (USAID/PMI, 2023).

A Cochrane review completed in 2004 showed that ITNs reduce malaria cases by 39–62% and overall child mortality by 14–29% (Lengeler, 2004). A 2018 update taking into consideration the rise of insecticide resistance concluded that ITNs reduced child mortality by approximately 17%, uncomplicated malaria by approximately 17% and severe malaria incidence by about 44% relative to no nets (Pryce *et al.*, 2018). Compared to untreated nets, ITNs reduced uncomplicated malaria by a tenth (Pryce *et al.*, 2018). Costing just US$1.5–2.5 per unit every 2–4 years, ITNs are arguably also among the most cost-effective public health interventions after childhood immunization. One comprehensive analysis of use and impact in Africa indicated that between 2000 and 2015, ITNs alone contributed nearly 68% of all gains made against malaria, making them the top buy for control programmes (Bhatt *et al.*, 2015).

The impact of ITNs is derived from both the physical barrier effects, providing users direct personal protection, and the killing effects on mosquitoes of the insecticide, an attribute which guarantees additional benefits not just to the users but also to neighbouring non-users (Fig. 9.1). In some villages in East Africa, ITNs were so effective that populations of *Anopheles gambiae* s.s., which feed and rest indoors, were either locally eliminated or reduced to insignificance (Bayoh *et al.*, 2010; Russell *et al.*, 2010; Lwetoijera *et al.*, 2014; Msugupakulya *et al.*, 2023). In rural south-eastern Tanzania, the roll-out of ITNs resulted in a more than tenfold reduction in malaria transmission intensities, subsequently resulting in a significant reduction in malaria incidence and prevalence rates (Russell *et al.*, 2010).

The ITN-dominated strategies, though extensively impactful, have in recent years started to reach saturation (World Health Organization, 2023). While progress against deaths has continued (exceeding a 60% decline relative to 2000), malaria incidence and case counts have plateaued or are increasing in some countries (World Health Organization, 2023) for multiple reasons beyond the functionality of ITNs themselves. The most commonly cited challenges include the rise of insecticide resistance (Ranson and Lissenden, 2016) and inadequate financing for regular redistribution of ITNs. Other factors

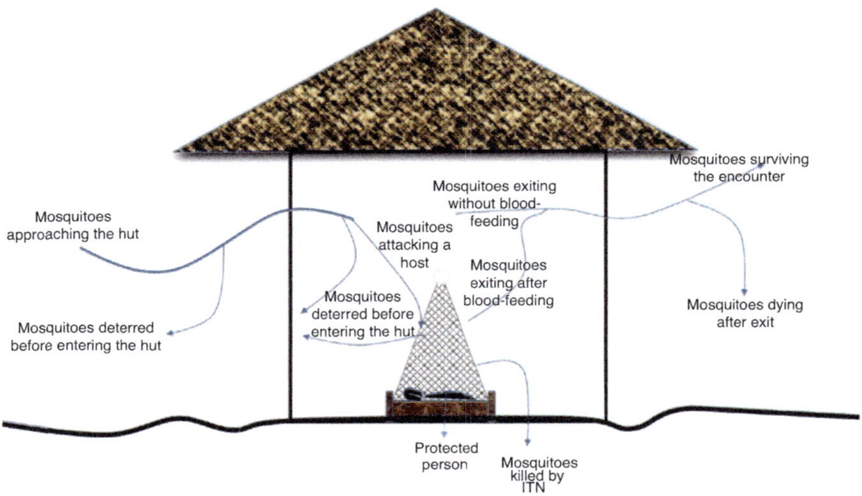

Fig. 9.1. Diagrammatic representation of various effects of mosquito nets (ITNs) on mosquitoes that enter or attempt to enter houses. Mosquitoes can be deterred and diverted before they enter houses, killed by the insecticides used on ITNs inside houses or irritated so that they exit houses earlier than normal. This exit may occur before or after the mosquitoes have blood-fed, but both fed and unfed mosquitoes may subsequently die after exiting. The nets may also reduce the mosquitoes' ability to successfully bite and transmit disease. (Adapted from Okumu and Moore, 2011, USED under license CC BY 2.0.)

include shifts in mosquito behaviours or replacement of populations by those that readily bite outdoors and those that bite non-human vertebrates and thus are not directly impacted by ITNs (Durnez and Coosemans, 2013; Sherrard-Smith *et al.*, 2019). Another emerging concern is the limited durability of the commercial ITNs, which have been shown to last shorter durations than the standard claim of 3 years (Lorenz *et al.*, 2020; Bertozzi-Villa *et al.*, 2021).

Today, multiple efforts are under way to address specific challenges associated with ITNs, and to restore the expected epidemiological impact. Most of these constitute testing and incorporation of new chemical compounds for the ITNs to remain substantially mosquitocidal, despite pyrethroid resistance, as well as the incorporation of piperonyl butoxide (PBO) synergists alongside pyrethroids. There are also nets that incorporate insect-growth regulators, namely pyriproxyfen, to reduce reproductive capacities of the adult *Anopheles*. WHO has now approved all these ITN types, and the consolidated guidelines suggest that some of these can be deployed specifically as a response to resistance (www.who.int/pq-vector-control/pre-qualified-lists/en). The Innovative Vector Control Consortium (IVCC), created in 2005 (Hemingway *et al.*, 2006), has rejuvenated the search for new actives, building multiple partnerships with industries in an attempt to beat the resistance treadmill (www.ivcc.com). *It remains to be seen whether this strategy is actually feasible in the long term.*

An important question therefore is whether the dominant pathway for improving ITNs – the search for new actives and/or synergists – is the most appropriate to guarantee the efficacy of an ITN programme. Related questions include whether there can be room and incentives for alternative pathways, whether these bed nets have to contain insecticides and what the long-term future of bed nets as a malaria control tool could be. This paper, therefore, examines possible future scenarios of bed nets for malaria control in Africa and highlights opportunities for reinvention beyond just the insecticidal content, pooled international procurement and opportunities for local manufacturing in endemic countries where the nets are most required. It imagines a target product profile for bed nets that will be in use in Africa in the years ahead.

Impact of Pyrethroid Resistance on Nets is Hard to Measure Directly

While it has been conclusively demonstrated that insecticide resistance reduces the effectiveness of IRS (Mabaso *et al.*, 2004; Wondji *et al.*, 2012; Coetzee *et al.*, 2013), the evidence for effects on ITNs remains implied rather than direct. The updated Cochrane review on the epidemiological impact of ITNs explicitly explored the possible effects of insecticide resistance on transmission. The conclusion was that 'although there is some evidence that insecticide resistance frequency has some effects on mosquito mortality, it is unclear how quantitatively important resistance is' (Pryce *et al.*, 2018).

Laboratory studies have clearly demonstrated that bio-efficacy of ITNs wanes in the face of resistance, but multiple field studies examining the impact on disease have generally failed to verify these failures (Wondji *et al.*, 2012; Strode *et al.*, 2014; Lindblade *et al.*, 2015; Kleinschmidt *et al.*, 2018). Multiple reasons have been cited for this situation, one of them being that ITNs may still have sublethal effects, especially on older mosquitoes (Viana *et al.*, 2016), which are the most important in malaria epidemiology (Garrett-Jones and Shidrawi, 1969). The analysis by Viana *et al.* (2016), which fed laboratory data into Bayesian models, predicted that ITNs would still substantially prevent malaria transmission even under pyrethroid resistance. However, follow-up studies in Burkina Faso demonstrated that pyrethroid resistance was so strong that such expected sublethal benefits were essentially nil (Noble *et al.*, 2019).

Another possible reason is that the ITNs provide the barrier protection, and at high coverage, this protection accounts for a significant proportion of the communal benefits previously afforded by the insecticidal effects in the days when the mosquito populations were susceptible (Hawley *et al.*, 2003). Therefore, where most people use pyrethroid-treated or untreated nets consistently, the combined personal protection may begin to approximate the communal benefits, regardless of resistance status of the prevailing *Anopheles* populations. The Cochrane review on ITNs determined that compared with untreated nets, ITNs probably reduce malaria prevalence by just one-tenth (Lengeler, 2004). Later, a large-scale WHO-backed trial designed to investigate this question in five countries (Kleinschmidt *et al.*, 2015) had concluded that the ITNs performed equally despite resistance in the vectors but that people without nets were less protected than those with nets (Kleinschmidt *et al.*, 2018).

One study in north-eastern Tanzania, which is often quoted as having yielded conclusive evidence, was a four-group cluster randomized trial designed to evaluate performance of PBO Nets compared with standard LLINs, PBO Nets plus IRS and standard LLINs plus IRS (Protopopoff *et al.*, 2018). The investigators observed that adding IRS improved protection relative to LLINs alone and that the PBO Nets performed better than standard LLINs. The authors argued that the greater performance of PBO Nets compared to standard nets together with the absence of effect when standard LLINs were initially distributed in the area suggested indirectly that insecticide resistance was compromising the effectiveness of standard pyrethroid nets in the area (Protopopoff *et al.*, 2018). However, this indirect evidence does not rule out that the additional benefits were due to increased killing of the resistant mosquitoes rather than reversal of the resistance. Moreover there was no direct comparison of the nets between sites with and without resistance. More recently, however, additional studies with PBO and dual active nets, containing a pyrethroid and a pyrrole, have shown far greater benefits of these net types in terms of reduced malaria incidence compared to pyrethroid-only nets in an area where insecticide resistance is widespread (Mosha *et al.*, 2022). This suggests that this extra layer of insecticidal activity can provide greater protection, effectively countering any resistance, but it remains to be determined whether this can be a long-term strategy in itself, how long it will take before the mosquitoes become resistant again and whether the insecticidal efficacy can last the full lifetime of the ITNs.

A comprehensive analysis of this subject was provided by Strode *et al.*, who concluded that evidence was limited for impact of resistance on ITNs (2014). In line with the hypothesis put forth a decade earlier by Ranson *et al.* (2010), a recent analysis attributed the difficulties in quantifying impacts of resistance to multiple factors, notably: (i) the multiplicity of resistance mechanisms; (ii) difficulty in finding a resistance-free control area; (iii) variations in intensities of resistance and (iv) the reluctance to roll-out untreated nets in field trials as a comparator (Okumu, 2020). It is clear that the mosquitocidal impacts that previously yielded communal benefits of ITNs (Hawley *et al.*, 2003; Abdulla *et al.*, 2005) are lost when ITNs no longer kill large numbers of mosquitoes. However, the evidence for resistance impacting overall epidemiological impact of nets remains inconclusive.

Notwithstanding the inconclusive data, efforts to address the resistance challenge are primarily based on the assumption that insecticide resistance has negative epidemiological impact on ITNs and that ignoring it will exacerbate the situation. As a consequence, the responses are mostly focused on insecticidal content of the ITNs, even though nets have multiple modes of actions. It can be argued that while resistance has depleted the bio-efficacy of nets, it has not impacted the physical barrier effect, which if expanded, could address the current gaps in both personal and communal protections.

Too Much Focus on Insecticides Impedes Other Opportunities for ITN Development

To accelerate innovation around bed net technology, it is important that the processes are transparent and comprehensive to cover as many opportunities as possible. Similarly, while pyrethroid-treated nets are still desirable in some settings, it is possible that adding insecticides to nets is not an absolute necessity in all settings. For this reason, developments around the bed net technology should also examine other characteristics of nets, such as durability, coverage, access and even acceptability.

As illustrated in Fig. 9.1, mosquito nets work by preventing mosquito bites or by killing the mosquitoes. The former is mostly dependent on the physical durability and integrity of the net, how well users can maintain them intact and how consistently or properly the nets are used. The killing function, on the other hand, is dependent on the chemical content of the nets and the degree to which the local mosquito populations are susceptible to whichever insecticides are used on them. To investigate the relevance of insecticidal content, a paper in 2020 examined different attributes of untreated or pyrethroid-treated nets alongside findings of multiple experimental hut and epidemiological studies (Okumu, 2020). The main findings were that the widespread resistance in Afro-tropical *Anopheles* mosquitoes has greatly weakened the historical mass-killing effects that ITNs used to have and which had contributed the extensive communal benefits to users and non-users (Hawley *et al.*, 2003). However, because the pyrethroid-ITNs still elicited substantial epidemiological impact in multiple settings with pyrethroid resistance, it was likely that physical integrity, consistent use and population-level coverage were becoming the key variables, perhaps more important than the insecticidal content.

To achieve effectiveness, the ITNs must be scaled to large populations and used properly and consistently. Decision makers must therefore ensure barrier-free access to the target users. Similarly, the users must consistently and appropriately use the nets, and the nets must be sufficiently functional for the entire duration of use. Identifying a new insecticide to treat the nets may be one solution to the plateauing impact of malaria control, but there is a chance that society may achieve equal or more control by improving on the other attributes. This includes acknowledging and evaluating the potential and cost-effectiveness of durable non-insecticidal nets across settings. It has been argued that longer-lasting ITNs are more cost-effective and that both WHO and National Malaria Control Programs should institute programmes to monitor the physical quality of the ITNs post marketing (Lorenz *et al.*, 2020; Bertozzi-Villa *et al.*, 2021). For African countries and companies without the technology to incorporate insecticides on to nets, it may be reasonable to consider the option of durable nets and ensure that these nets are widely acceptable and widely available.

If the WHO or individual country's jurisdictions could prequalify some bed nets based solely on personal protection and bite prevention, it could stimulate the industry and promote the development of durable, potentially biodegradable nets that can be locally manufactured. Currently, the excessive and biased focus on insecticides may be limiting innovation and the sustainable development of new, competitively priced bed net types and designs.

Given the demand-supply dynamics and the international pooled-procurement models, ITN distribution systems today take minimal consideration of people's choices or preferences. Instead, consignments of nets are procured and distributed, usually on a top-down approach at scale. These mass distribution approaches, together with other net distribution methods (e.g. through antenatal clinics (Kramer *et al.*, 2017) or through schools (Yukich *et al.*, 2020)), have demonstrated extraordinary coverage and reduced equity gaps across communities (Noor *et al.*, 2007). However, they lack sustainability at community level. The restrictions are partly due to the need to have pesticides on nets. By extension, it is also conceivable that a vibrant ITN market could produce more fashionable nets with greater aesthetic appeal than the homogenous ones currently circulating and make these more accessible, thus creating more demand and increasing acceptability (Fig. 9.2).

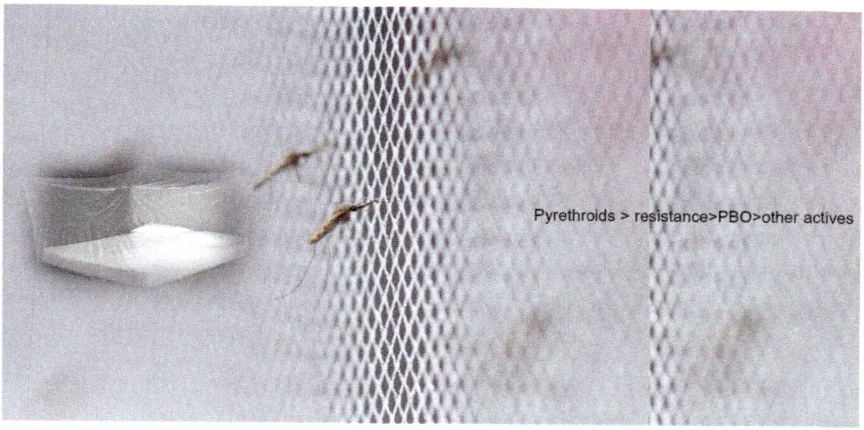

Fig. 9.2. The race to beat resistance has yielded multiple new bed net types incorporating both pyrethroids and synergists like PBO or pyrethroids and other non-pyrethroid actives. An important question is whether this approach is sustainable and whether there are other attributes of the net which may yield equally significant gains at scale.

The Private Sector Market for ITNs in Endemic Countries is Substantial but Largely Ignored

The mass distribution channels have been a great success but are inadequate to sustaining access and coverage in communities, especially in between the national campaigns. Fortunately, there is a substantial, albeit largely ignored, private sector market in many endemic countries that could be relied upon to support ITN distributions. This consists of many outlets selling different mosquito control products such as topical repellents, mosquito coils and bed nets, as well as aerosol sprays such as those containing non-residual pyrethroids. A recent audit in western Kenya showed that nearly three-quarters of such commercial outlets regularly stocked mosquito coils, and about 40% stocked topical repellents, while 14% stocked ITNs (Oria *et al.*, 2021). Another interesting observation from the Kenyan study was that the retailers made profits and were, at the time of the audit, planning to continue stocking the mosquito control products (Oria *et al.*, 2021). The retail system has already been widely relied upon to support delivery of essential medicines for malaria. In Tanzania, for example, Alba *et al.* (2010) demonstrated improvements in access to artemisinin combination treatments following the introduction of accredited drug dispensing outlets.

The private sector for mosquito control products could be specifically harnessed to serve the middle class, thus freeing resources for low-income households. With that approach, nets may carry certain specific features desirable and affordable to the middle class. One discrete choice experiment in Ghana has recently highlighted opportunities where the private sector outlets could help improve access to ITNs for the middle class (Alfonso *et al.*, 2020). Here, researchers offered household representatives different paired combinations of LLINs with different features and prices and asked them to select which one of each pair they preferred and whether they would purchase it if they had the cash. They also offered the participants a certain amount of cash, which the participants could either keep or use to buy the selected ITNs immediately. Under these conditions, the researchers determined that household probability of purchasing an LLIN with special features was over 40%. Even more interestingly, the average willingness to pay for an ITN with special features was US$18.5 (in this case, the selected features included zipper entry, queen size, conical or rectangular shapes) (Alfonso *et al.*, 2020). Studies such as this highlight another opportunity to stratify resource allocation by socio-economic status rather than geography. For example, in the Ghana study, the implication was that letting the middle class purchase their own ITNs would relieve the public sector of nearly 40% of the budget, while generating 8.1 million USD revenue for the distributors, retailers and importers (Alfonso *et al.*, 2020).

The WHO has today pre-qualified and listed over 20 ITN brands (www.who.int/pq-vector-control/pre-qualified-lists/en; World Health Organization, 2024), of which only three have any manufacturing within Africa. The rest are all manufactured outside endemic countries, which then have to import. As previously argued, any efforts to encourage local manufacturing must be in such a way that does not lower the quality of the ITNs. We have previously noted that the countries with most malaria in the world, including Nigeria, which has a quarter of the burden, still rely primarily on importations of ITNs (Okumu, 2020). Tanzania has been an exception, where the Sumitomo Ltd has provided a royalty-free tech transfer to A to Z Textile in Tanzania to manufacture ITNs incorporating permethrin (Masum *et al.*, 2010). The company already has capacity to produce in excess of 30 million ITNs in response to international tenders. Other countries pursuing this approach include Rwanda, but the majority of countries still do not have local manufacturing. This is another opportunity for private sector involvement in malaria-endemic countries.

Several reasons have been given for the lack of local manufacturing, the most common being the difficulty in ensuring good quality and competitive pricing relative to ITNs manufactured elsewhere, such as in China, India or Vietnam. A related challenge is that companies in endemic countries are inadequately equipped to compete for the international pooled-procurement programmes that currently permeate the ITN initiatives. Another reason is that most African manufacturers, despite some of them being major producers of apparel for export market, do not have the technology to incorporate the active insecticide into the ITN fabric, and as a result, they may be unable to manufacture insecticidal nets even when they can manufacture untreated nets. There have been attempts for cut-and-sew operations in Africa, where pretreated fabric is imported and the rest of the manufacturing is done locally, but these too have been marginal (www.tananetting.com/blogs/empowering-people-nigeria-dawaplus-made-africa-project; Tanna Netting, 2017). Other countries such as Rwanda have opted to invite companies with products already pre-qualified by WHO so that these pre-qualified ITNs are produced locally. Tanzania was able to start large-scale operations only after the unique royalty-free technology transfer, which enabled the Arusha-based A to Z Textile to begin producing Olyset Nets in partnership with Sumitomo chemicals (Masum *et al.*, 2010). Given these circumstances, we have previously argued that removing the absolute requirement for the insecticide component, as a universal requirement in ITNs, may allow more local manufacturers to get involved with the ITN manufacturing.

A related challenge here is the lack of competitiveness at local level and, as a consequence, the minimal choice options that users have. Today, ITNs being distributed meet the basic minimum standards, partly because of the competitive pricing that manufacturers are required to meet when there are only so few buyers. More than 30 African countries procure ITNs every year for the public sector. However, their procurement choices are tightly controlled to meet the budgetary restrictions by their domestic coffers and the major external financiers, such as the Global Fund, USAID-PMI and UNICEF. As a result, manufacturers cannot introduce any design features or enhancements, such as making ITNs more aesthetically appealing, of varied sizes (including larger nets and baby-sized nets) and with easier opening and closing (e.g. using zippers) rather than the standard-sized and homogenously coloured ITNs currently being distributed. Any such additional design features would increase manufacturing costs, yet they would also increase the likelihood that people purchase more nets or use nets more. There needs to be a balance between meeting competitive design features, while remaining competitive on pricing. A vibrant private sector would allow this and allow user households to get the nets they need and to use those nets.

Insecticide-Treated Nets May Become Supplementary Rather Than Core Interventions

The WHO Global Technical Strategy for Malaria Elimination 2016–2030 (GTS) included three pillars, the first of which was the scale up of core interventions, including ITNs and IRS, alongside effective case management (World Health Organization, 2015). Alongside this guidance, malaria-endemic countries in Africa now include ITNs and/or IRS as the core vector control methods, with everything else being considered supplementary.

The main reason that ITNs are considered core is that they are the most scalable and have been the most cost-effective. They are also applicable across all socio-economic groups, from the poorest households to the wealthiest. ITNs alone are argued to have contributed nearly 70% of all the gains made against malaria (Bhatt *et al.*, 2015). However, as with the other commodities, ITNs must be replaced every few years, usually 3–4 years. Without constant replenishment, the gains achieved by ITNs may not be sustainable. Though this commoditization challenge is now prevalent in malaria control, it is recognized that these approaches will be inadequate for malaria control. Yet, in the absence of a better technological commodity, ITNs are currently considered by most as the lead intervention.

Before DDT was synthesized, malaria control mostly relied on environmental management and housing improvement, complemented with mass drug administration using drugs such as quinine (Wilson *et al.*, 2020). However, highly efficacious tools such as insecticides, and more recently ITNs and medicines, resulted in a decline of innovations for malaria control, and with it a decline of the pursuit of other interventions perceived to be more expensive. In particular, environmental management and house improvement have been widely neglected, despite the early evidence of effectiveness.

Insecticides, medicines and nets can make significant progress, reducing malaria transmission significantly, but real progress requires much more holistic approaches. To achieve sustainable control, a transitional approach is therefore necessary for malaria prevention, involving judicious use of current tools like ITNs while gradually building long-term resilience against mosquitoes and transmission (Okumu and Finda, 2021). This approach could also allow countries to carefully transition from insecticide-based to non-insecticidal approaches without losing the gains. Even with the new forms of ITNs, the challenge of regular replenishments remains unaddressed.

A more sustainable approach, therefore, is to establish programmes that build long-term resilience to sustain the accrued gains and prevent transmission rebounds. Sustaining malaria control will be possible only once vector control approaches are permanent. These require multi-sectorial initiatives such as environmental sanitation to reduce vectors as well as improved housing to prevent mosquito bites indoors. Once integrated into the wider public health context, this approach could contain ongoing malaria transmission, reduce over-reliance on insecticide-based tools and minimize transmission rebounds even in poor communities (Okumu and Finda, 2021). It is therefore conceivable that once most households have sufficient housing improvements, and if these potentially transformative innovations are successful, interventions such as ITNs and IRS will be supplementary rather than being core interventions for malaria control.

Another reason that ITNs may eventually become secondary is if some of the potentially transformational technologies such as gene drive-modified mosquitoes become available (Nolan, 2021). These genetically modified mosquitoes are designed to reduce malaria infections, either by significantly suppressing malaria vector populations or modifying them to be refractory to infections. These tools, though currently being discussed as complementary, may eventually be so impactful that endemic countries begin to consider them the primary interventions.

Financing of ITNs Should Transition From External to Domestic

The estimated annual budget for malaria control was US$5.6 billion in 2019, of which just $3.1 billion was actually invested in vector control (World Health Organization, 2020). The WHO World Malaria Report of 2020 suggested that starting 2000, a total of US$39 billion had been spent but also that only one-third of this had been contributed by endemic countries themselves, the remaining US$26 billion being from external donors (World Health Organization, 2020). A significant proportion of the available malaria funding is used for procurement of commodities, including ITNs. Indeed, it can be argued that most publicly available ITNs distributed in Africa today are externally financed.

The current financing model may be appropriate now, but long-term sustainability of malaria control requires that endemic countries build up sustainable domestic financing, either through the public sector or some combination of public and private sector financing. Since population sizes are increasing and the net types are changing to more expensive ones, it is inevitable that procurement costs will rise as well (Cohen *et al.*, 2022). Therefore, if the financing basket does not expand, as has been the case since 2015, the increased costs and greater demands could result in much lower quantities of ITNs available. While innovative financing is beyond the scope of this paper, individual countries may explore different opportunities. As discussed above, the rejuvenation of the private sector may offer one opportunity for some households, especially the middle-to-high income, or other forms of local financing to take up the costs of ITNs eventually. There may be other innovative financing systems that countries could select.

Environmental Concerns Will Necessitate a Change of Net Designs

In 2014, WHO provided five recommendations on proper management of old ITNs (World Health Organization, 2014). The recommendations were as follows: (i) Old ITNs should continue to be used even if they have holes until they are replaced with new ones; (ii) the users should be advised not to dispose the old ITNs in water bodies; (iii) national malaria programmes should only collect the old ITNs for disposal; (iv) the collected ITNs and their packages should be disposed of in high-temperature incineration and should not be burned in open air; and (v) malaria control programmes should work with national environment authorities to ensure the recommendations are taken into consideration by formulating local guidance and regulations (World Health Organization, 2014).

It is unlikely that many malaria-endemic countries, with their understaffed NMCPs, will have the full capacity to follow these guidelines. Yet, data on net use suggest very high levels of attrition already. In Tanzania, for example, the average ITN lasts 2.5 years, compared to the 3 years specified by the manufacturers (Lorenz *et al.*, 2020), and these durability estimates are far lower in several other countries (Bertozzi-Villa *et al.*, 2021).

Without formal waste management systems, the billions of nets already out there, most of them polyethylene or polyester, are destined to remain unaccounted for. Multiple alternative uses of retired ITNs (Fig. 9.3), such as crop protection, child games, house screening, fencing and others, have been reported in several parts of Africa, many of which may lead to environmental harm (Santos *et al.*, 2020). This widespread alternative of ITNs puts in focus the challenges of environmental protection in the context of ongoing malaria elimination efforts. For instance, insecticides like permethrin embedded in the ITNs may be toxic to aquatic life. Thus, any forms of repurposing should consider protecting such life as stipulated in the SDGs 14 and 15. While these issues are not currently prioritized, the future ITN discussions must integrate environmental and public health concerns.

While new tools and approaches are being developed to overcome insecticide resistance, new strategies for environmental protection as the result of vector control are equally important and should also be developed and evaluated. With the current technological advancements in Africa, development of environmentally friendly nets should be possible. One of the oldest industries in Africa is fabric making, relying on organic fabric like cotton. The potential of local industries to produce organic bed nets needs to be explored and encouraged in order to improve access and ownership of bed nets and to reduce environmental harm caused by current nets (Okumu, 2020). Further considerations should ensure the integrity and other desirable features of environmentally friendly bed nets.

Most ITNs today are either polyethylene or polyester and are often distributed in plastic packaging either singly or in bales. The use of these materials

Fig. 9.3. Examples of old ITNs used for crop protection in Kilombero, Tanzania.

for manufacturing was meant to enhance the longevity and durability of ITNs. However, such physical strength can also be achieved by other means such as improved knitting patterns. In recent years, several countries have banned single-use plastics, including polyethylene bags regardless of their thickness, but ITNs appear to be exempted. Thus, development of environmentally friendly bags for ITNs should be explored to adhere to the current and future legislation specified by endemic countries.

Many governments have already adopted approaches to reduce the harmful effects of plastic bags, including bans. However, limited measures exist for ITN bag disposal in Africa. Health authorities warn that reused ITN packaging can cause pesticide poisoning, soil and groundwater pollution and release of dangerous toxins from open burning. It can be expected that these concerns are expected to grow, necessitating new production and packaging methods for bed nets that protect environmental, human and animal health.

Conclusions

This paper has examined plausible futures of bed nets for malaria control in Africa. It observes that ITNs have multiple modes of action, but their innovations are focused almost entirely on bio-efficacy; that the ITNs are mostly imported in countries that require them most and that the local private sector is minimal, and the nets are made of non-biodegradable material. It argues that the ITN technology requires significant rethink to allow reinvention beyond just the insecticide actives and that a rejuvenated private sector market and possibly high-quality local manufacturing are necessary to guarantee regular supplies at competitive pricing to households in malaria-endemic countries. The vector control industry should therefore focus on enhancing durability, increasing accessibility, promoting aesthetic appeal and ensuring environmental sustainability through the use of biodegradable materials on bed nets. Local manufacturing should also be actively supported to produce high-quality nets at competitive prices. Most importantly, reducing the current overemphasis on insecticidal properties will pave the way for innovation in other aspects, improving both the effectiveness and attractiveness of bed nets overall.

References

Abdulla, S., Gemperli, A., Mukasa, O., Armstrong Schellenberg, J.R., Lengeler, C. *et al.* (2005) Spatial effects of the social marketing of insecticide-treated nets on malaria morbidity. *Tropical Medicine and International Health* 10, 11–18.

Alba, S., Hetzel, M.W., Goodman, C., Dillip, A., Liana, J. *et al.* (2010) Improvements in access to malaria treatment in Tanzania after switch to artemisinin combination therapy and the introduction of accredited drug dispensing outlets - a provider perspective. *Malaria Journal* 9, 1–15.

Alfonso, Y.N., Lynch, M., Mensah, E., Piccinini, D. and Bishai, D. (2020) Willingness-to-pay for long-lasting insecticide-treated bed nets: A discrete choice experiment with real payment in Ghana. *Malaria Journal* 19, 1–13.

Bayoh, M.N., Mathias, D.K., Odiere, M.R., Mutuku, F.M., Kamau, L. *et al.* (2010) *Anopheles gambiae*: Historical population decline associated with regional distribution of insecticide-treated bed nets in Western Nyanza Province, Kenya. *Malaria Journal* 9, 62.

Bertozzi-Villa, A., Bever, C.A., Koenker, H., Weiss, D.J., Vargas-Ruiz, C. *et al.* (2021) Maps and metrics of insecticide-treated net access, use, and nets-per-capita in Africa from 2000–2020. *Nature Communications* 12, 1–12.

Bhatt, S., Weiss, D., Cameron, E., Bisanzio, D., Mappin, B. *et al.* (2015) The effect of malaria control on *Plasmodium falciparum* in Africa between 2000 and 2015. *Nature* 526, 207–211.

Coetzee, M., Kruger, P., Hunt, R., Durrheim, D., Urbach, J. *et al.* (2013) Malaria in South Africa: 110 years of learning to control the disease. *SAMJ: South African Medical Journal* 103, 770–778.

Cohen, J.M., Okumu, F. and Moonen, B. (2022) The fight against malaria: Diminishing gains and growing challenges. *Science Translational Medicine* 14, eabn3256.

Durnez, L. and Coosemans, M. (2013) Residual transmission of malaria: An old issue for new approaches. In: Manguin, S. (ed.) *Anopheles Mosquitoes – New Insights into Malaria Vectors*. InTech, pp. 671–704. DOI: 10.5772/3392.

Garrett-Jones, C. and Shidrawi, G.R. (1969) Malaria vectorial capacity of a population of *Anopheles gambiae*. *Bulletin of the World Health Organization* 40, 531–545.

Hawley, W.A., Phillips-Howard, P.A., Ter Kuile, F.O., Terlouw, D.J., Vulule, J.M. *et al.* (2003) Community-wide effects of permethrin-treated bed nets on child mortality and malaria morbidity in Western Kenya. *American Journal of Tropical Medicine and Hygiene* 68, 121–127.

Hemingway, J., Beaty, B.J., Rowland, M., Scott, T.W. and Sharp, B.L. (2006) The innovative vector control consortium: Improved control of mosquito-borne diseases. *Trends in Parasitology* 22, 308–312.

Kapp, C. (2002) Global fund on AIDS, tuberculosis, and malaria holds first board meeting. *The Lancet* 359, 414.

Kapp, C. (2003) Struggling global fund approves grants for HIV, TB, and malaria. But board meeting is marred by warnings about shortfalls and accusations of political manoeuvering. *The Lancet* 362, 1381.

Kleinschmidt, I., Mnzava, A.P., Kafy, H.T., Mbogo, C., Bashir, A.I. *et al.* (2015) Design of a study to determine the impact of insecticide resistance on malaria vector control: A multi-country investigation. *Malaria Journal* 14, 282.

Kleinschmidt, I., Bradley, J., Knox, T.B., Mnzava, A.P., Kafy, H.T. *et al.* (2018) Implications of insecticide resistance for malaria vector control with long-lasting insecticidal nets: A WHO-coordinated, prospective, international, observational cohort study. *The Lancet Infectious Diseases* 18, 640–649.

Kramer, K., Mandike, R., Nathan, R., Mohamed, A., Lynch, M. *et al.* (2017) Effectiveness and equity of the Tanzania national voucher scheme for mosquito nets over 10 years of implementation. *Malaria Journal* 16, 255.

Lengeler, C. (2004) Insecticide-treated bed nets and curtains for preventing malaria. *Cochrane Database of Systematic Reviews* 2004(2), CD000363. DOI: 10.1002/14651858.CD000363.pub2.

Lindblade, K.A., Mwandama, D., Mzilahowa, T., Steinhardt, L., Gimnig, J. *et al.* (2015) A cohort study of the effectiveness of insecticide-treated bed nets to prevent malaria in an area of moderate pyrethroid resistance, Malawi. *Malaria Journal* 14, 31.

Lorenz, L.M., Bradley, J., Yukich, J., Massue, D.J., Mageni Mboma, Z. *et al.* (2020) Comparative functional survival and equivalent annual cost of 3 long-lasting insecticidal net (LLIN) products in Tanzania: A randomised trial with 3-year follow up. *PLoS Medicine* 17, e1003248.

Lwetoijera, D.W., Harris, C., Kiware, S.S., Dongus, S., Devine, G.J. *et al.* (2014) Increasing role of *Anopheles funestus* and *Anopheles arabiensis* in malaria transmission in the Kilombero valley, Tanzania. *Malaria Journal* 13, 331.

Mabaso, M.L., Sharp, B. and Lengeler, C. (2004) Historical review of malarial control in Southern African with emphasis on the use of indoor residual house-spraying. *Tropical Medicine and International Health* 9, 846–856.

Masum, H., Shah, R., Schroeder, K., Daar, A.S. and Singer, P.A. (2010) Africa's largest long-lasting insecticide-treated net producer: Lessons from A to Z textiles. *BMC International Health and Human Rights* 10, S6.

Mosha, J.F., Kulkarni, M.A., Lukole, E., Matowo, N.S., Pitt, C. *et al.* (2022) Effectiveness and cost-effectiveness against malaria of three types of dual-active-ingredient long-lasting insecticidal nets (LLINs) compared with pyrethroid-only LLINs in Tanzania: A four-arm, cluster-randomised trial. *The Lancet* 399, 1227–1241.

Msugupakulya, B.J., Urio, N.H., Jumanne, M., Ngowo, H.S., Selvaraj, P. *et al.* (2023) Changes in contributions of different *Anopheles* vector species to malaria transmission in East and Southern Africa from 2000 to 2022. *Parasites and Vectors* 16, 408.

Noble, C., Min, J., Olejarz, J., Buchthal, J., Chavez, A. *et al.* (2019) Daisy-chain gene drives for the alteration of local populations. *Proceedings of the National Academy of Sciences* 116, 8275–8282.

Nolan, T. (2021) Control of malaria-transmitting mosquitoes using gene drives. *Philosophical Transactions of the Royal Society B* 376, 20190803.

Noor, A.M., Amin, A.A., Akhwale, W.S. and Snow, R.W. (2007) Increasing coverage and decreasing inequity in insecticide-treated bed net use among rural Kenyan children. *PLoS Medicine* 4, e255.

Okumu, F. (2020) The fabric of life: What if mosquito nets were durable and widely available but insecticide-free?. *Malaria Journal* 19.

Okumu, F.O. and Moore, S.J. (2011) Combining indoor residual spraying and insecticide-treated nets for malaria control in Africa: A review of possible outcomes and an outline of suggestions for the future. *Malaria Journal* 10, 208.

Okumu, F.O. and Finda, M.F. (2021) Creating long-term resilience against malaria vectors while addressing the immediate need to suppress pathogen transmission. In: Koenraadt, C.J.M., Spitzen, J. and Takken, W. (eds) *Innovative Strategies for Vector Control*, Vol. 6. Wageningen Academic Publishers, Wageningen, Netherlands, pp. 33–57. DOI: 10.3920/978-90-8686-895-7_3.

Oria, P.A., Moshi, V., Odero, J.I., Ekodir, S., Monroe, A. *et al.* (2021) A retail audit of mosquito control products in Busia County, Western Kenya. *Malaria Journal* 20, 1–12.

Protopopoff, N., Mosha, J.F., Lukole, E., Charlwood, J.D., Wright, A. *et al.* (2018) Effectiveness of a long-lasting piperonyl butoxide-treated insecticidal net and indoor residual spray interventions, separately and together, against malaria transmitted by pyrethroid-resistant mosquitoes: A cluster, randomised controlled, two-by-two factorial design trial. *The Lancet* 391, 1577–1588.

Pryce, J., Richardson, M. and Lengeler, C. (2018) Insecticide-treated nets for preventing malaria. *Cochrane Database Systematic* 11(11), CD000363. DOI: 10.1002/14651858.CD000363.pub3.

Ramsay, S. (2002) Global fund makes historic first round of payments. *The Lancet* 359, 1581–1582.

Ranson, H. and Lissenden, N. (2016) Insecticide resistance in African *Anopheles* mosquitoes: A worsening situation that needs urgent action to maintain malaria control. *Trends in Parasitology* 32, 187–196.

Ranson, H., N'Guessan, R., Lines, J., Moiroux, N., Nkuni, Z. *et al.* (2010) Pyrethroid resistance in African *Anopheline* mosquitoes: What are the implications for malaria control? *Trends in Parasitology* 27(2), 91–98. DOI: 10.1016/j.pt.2010.08.004.

RBM (2002) *Scaling Up Insecticide Treated Netting Programmes in Africa: A Strategic Framework for Coordinated National Action.* World Health Organization, Geneva, Switzerland.

Roll Back Malaria (2020) 2 billion mosquito nets delivered worldwide since 2004 [Online]. RBM Partnership. Available at: https://endmalaria.org/news/2-billion-mosquito-nets-delivered-worldwide-2004 (accessed 20 February 2020).

Russell, T.L., Lwetoijera, D.W., Maliti, D., Chipwaza, B., Kihonda, J. *et al.* (2010) Impact of promoting longer-lasting insecticide treatment of bed nets upon malaria transmission in a rural Tanzanian setting with pre-existing high coverage of untreated nets. *Malaria Journal* 9, 187.

Santos, E.M., Coalson, J.E., Munga, S., Agawo, M., Jacobs, E.T. *et al.* (2020) 'After those nets are torn, most people use them for other purposes': An examination of alternative bed net use in western Kenya. *Malaria Journal* 19, 1–11.

Sherrard-Smith, E., Skarp, J.E., Beale, A.D., Fornadel, C., Norris, L.C. *et al.* (2019) Mosquito feeding behavior and how it influences residual malaria transmission across Africa. *Proceedings of the National Academy of Sciences* 116, 15086–15095.

Strode, C., Donegan, S., Garner, P., Enayati, A.A. and Hemingway, J. (2014) The impact of pyrethroid resistance on the efficacy of insecticide-treated bed nets against African *Anopheline* mosquitoes: Systematic review and meta-analysis. *PLoS Medicine* 11, e1001619.

Tanna Netting (2017) Empowering the people of Nigeria through DawaPlus Made in Africa program [Online]. Available at: https://www.tananetting.com/blogs/empowering-people-nigeria-dawaplus-made-africa-project/ (accessed 19 January 2019).

USAID/PMI (2023) 18th Annual Report to Congress USAID/PMI.

Viana, M., Hughes, A., Matthiopoulos, J., Ranson, H. and Ferguson, H.M. (2016) Delayed mortality effects cut the malaria transmission potential of insecticide-resistant mosquitoes. *Proceedings of the National Academy of Sciences* 113, 8975–8980.

Wilson, A.L., Courtenay, O., Kelly-Hope, L.A., Scott, T.W., Takken, W. *et al.* (2020) The importance of vector control for the control and elimination of vector-borne diseases. *PLoS Neglected Tropical Diseases* 14, e0007831.

Wondji, C.S., Coleman, M., Kleinschmidt, I., Mzilahowa, T., Irving, H. *et al.* (2012) Impact of pyrethroid resistance on operational malaria control in Malawi. *Proceedings of the National Academy of Sciences* 109, 19063–19070.

World Health Organization (2000) The Abuja declaration and the plan of action. An extract from the African Summit on Roll Back Malaria, Abuja, WHO/CDS/RBM/2000.

World Health Organization (2008) *Global Malaria Action Plan.* World Health Organization, Roll Back Malaria Partnership, Geneva, Switzerland.

World Health Organization (2014) *Recommendations on the Sound Management of Old Long-Lasting Insecticidal Nets.* WHO, Geneva, Switzerland.

World Health Organization (2015) *Global Technical Strategy for Malaria 2016–2030.* WHO, Geneva, Switzerland.

World Health Organization (2020) *World Malaria Report 2020.* WHO, Geneva, Switzerland.

World Health Organization (2023) *Malaria Report 2023.* WHO, Geneva, Switzerland.

World Health Organization (2024) *Prequalification Vector Control: Prequalified Lists of Vector Control Products* [Online]. WHO, Geneva, Switzerland. Available at: https://extranet.who.int/pqweb/vector-control-products/prequalified-product-list (accessed 1 April 2024).

Yukich, J., Stuck, L., Scates, S., Wisniewski, J., Chacky, F. *et al.* (2020) Sustaining LLIN coverage with continuous distribution: The school net programme in Tanzania. *Malaria Journal* 19, 1–12.

Attractive Toxic Sugar Bait (ATSB®)

10

Amir Galili* and Meir Morag*

Westham Co., Hod Hasharon, Israel

Background

The challenges facing mosquito control are enormous. Recent studies have shown alarming spikes in worldwide mosquito populations. The number of outbreaks of mosquito-borne illnesses documented by researchers in recent years has ballooned, including Yellow Fever, Dengue, Chikungunya and Zika (Kraemer *et al.*, 2019). More specifically, it was estimated that by 2050, two key disease-spreading mosquitoes – *Aedes aegypti* and *Aedes albopictus* – will significantly expand their range, posing a threat to 49% of the global human population. In the US alone, mosquito populations have increased as much as tenfold, and mosquito populations have become two- to fourfold larger over the last five decades (Rochlin *et al.*, 2016). Similar trends have been identified in Europe and Asia (Caminade *et al.*, 2019; Ryan *et al.*, 2019). The assumed drivers behind these increases are primarily global warming, the decay of DDT and accelerated urbanization.

Mosquito control is used in battling both nuisance mosquitoes and vector mosquitoes. The two different categories share similar preventive methods such as removal of mosquito habitats, structure barriers, pesticides and non-pesticidal approaches. Non-pesticidal management includes biological and genetic methods, natural mosquito predators and entomopathogenic fungi. All these are more environmentally friendly but have so far shown only modest effects vis-à-vis dramatically decimating mosquito populations and are thus considerably limited (Chaudhry *et al.*, 2019).

*Corresponding authors: Amir@westhamco.com and Meir.morag@westhamco.com

 Practical Control of Mosquito Disease Vectors (ed J.D. Charlwood)

DOI: 10.1079/9781789248845.0010

Chemical insecticides, which have proven efficacy in mosquito control, are therefore key factors in various indoor and outdoor control methods. Nevertheless, they are known for having substantial environmental and health drawbacks. These substances were found to contaminate soil and water, as well as reduce biodiversity. Their poisoning effects on human health include mutagenicity and carcinogenicity, reproductive damage and behavioural disorders. A critical drawback is that mosquitoes have developed resistance to commonly used chemical insecticides. The mechanisms through which these pests have developed resistance are diverse and include mutation-induced changes in target sites, enhanced metabolic detoxification (which also causes cross-resistance relative to different chemical insecticides with similar chemical structures), the development of a thick cuticle layer that reduces the penetration of insecticide into the body and gene overexpression. These resistance mechanisms are developing as a result of insecticides in both indoor and outdoor use (Balabanidou *et al.*, 2016; Bamou *et al.*, 2018).

Most protective strategies for vector control target host-seeking and blood-feeding mosquitoes. They rely on potential human hosts for attracting the vectors to the vicinity of the insecticide exclusively used inside residences. For the past 20 years, long-lasting insecticide nets (LLINs) and Indoor Residual Spray (IRS), singly or in combination, have been the dominant indoor control methods for fighting mosquito-borne diseases such as malaria. However, by definition these methods are only applicable indoors.

In addition, although widely accepted, challenges involved in using LLINs lead to their restricted use. Only about one-third of households own LLIN as protection against mosquito bite (Gobena *et al.*, 2012). Non-persistent LLIN usage was associated with unintended uses, misperceptions about malaria, insufficient education and inadequate follow-up regarding LLINs utilization (Oliveira Sousa *et al.*, 2019).

Similarly, the successful use of IRS is hindered by the substantial investments of time, personnel, insecticides and the incorporation of quality control systems which are needed if IRS are to be efficient and effective (Knapp *et al.*, 2015). These limitations threaten the long-term sustainability of IRS programmes and necessitate increased focus on updated and integrated mosquito control tools and techniques that leverage advances in innovation, targeting and training.

The Need

In view of the rapid global increase in mosquito populations and the limitations of current methods, the challenges facing both nuisance and vector control are immense. Even modest improvements in current tools and techniques could help limit the evident expansion of both nuisance and arboviral diseases and prevent eroding the recent gains in reducing the global mosquito-related burden (Knapp *et al.*, 2015). While mosquito control remains central to eradicating mosquito-borne diseases, existing drawbacks have highlighted the necessity to find additional complementary methods to be utilized in integrative vector management (IVM) for mosquito population control (Revay *et al.*, 2015; Benelli and Beier, 2017).

There is undoubtedly a need to think beyond the norms, employ new paradigms and develop complementary tools that can narrow the existing disparities (Alonso *et al.*, 2011). In addition to control efficacy, various factors such as demography, ecology, education and suitability of treatment, as well as many other factors have to be taken into consideration when weighing new methods and tools. None the less, finding a balance between cost and effectiveness in resource-poor settings is certainly a decisive issue (Tenywa *et al.*, 2017).

Introducing Attractive Toxic Sugar Bait (ATSB®)

Female mosquitoes feed on sugar derived mainly from plants in order to obtain energy for survival, flight and fecundity (Yuval, 1992). They quest for sugar before and after obtaining a blood meal so as to maintain their fitness throughout their life. In order to meet their energetic demands, both male and female mosquitoes ingest sugar and are very selective in choosing their source (Schlein and Müller, 2006; Schlein and Müller, 2008). Such sources can include fruit, floral nectar or honeydew. In searching for an alternative control strategy, several studies have explored the efficacy of a control method that is based on this specific mosquito feeding behaviour as opposed to the search for blood meals (Barredo and DeGennaro, 2020).

The newly introduced method is novel to mosquito control, being based on the concept of toxic bait. The bait is a mix of sugar and attractant (either natural or a mix of sugar with attractive odorant) and a gut toxin. Mosquitoes are attracted to the bait mixture, residual sugar elicits a feeding response, and the active ingredient is delivered by oral ingestion. This bypasses resistance mechanisms that involve cuticle or wax layer modifications.

ATSB® spray

Testing the validity of the attractive toxic sugar baits (ATSB®) concept, initial experiments were conducted where a mixture of sugar and oral insecticide dissolved in water was sprayed on plants known to be highly attractive to local mosquito populations. A study comparing the effects of ATSB® presented in different ways to isolated populations of *Phlebotomus papatasi* Scopoli found that spraying ATSB® reduced the population to about 5% of the control area population. Barrier ATSB® coated fences had a similar effect, decreasing the population to about 12% of the concurrent catch in the control site (Müller and Schlein, 2011). In another study, at a small, isolated oasis, a spray of sugar and food-dye solution on the few flowering *Acacia raddiana* trees saw dye labelling in 80–90% of *Anopheles sergentii*. At a similar oasis, this spray with the addition of oral insecticide virtually eliminated the local mosquitoes (Schlein and Müller, 2006).

A study was conducted that sought to validate the impact of availability of sugar sources on regulating the dynamics of mosquito populations and their vector potential. It was found that the estimated vectorial capacity, a standard measure of malaria transmission potential, was over 250 times higher in the sugar-rich oasis than in a site with poor sugar availability. These findings further support the potency of the sugar-based approach to controlling vector mosquitoes (Gu *et al.*, 2011). The impressive effect of ATSB® spray in decimating local mosquito populations was further demonstrated in a study using guava and honey melons as Attractive Sugar Bait (ASB) for *Anopheles gambiae* and ATSB® that additionally contained boric acid as an oral insecticide (Müller *et al.*, 2010).

Although experiments with ATSB® spray showed significant controlling effects, three challenges remain to be considered in order to use it extensively in outdoor conditions. The first is that vegetation sprayed with ATSB® may be accessed by non-target insects, such as bees and other pollinators. The second is that ATSB® spray may not be durable enough in harsh environmental conditions, as it can be washed out by rain or dried out in hot climates. The third challenge is to avoid contamination of soil and water by the active (toxic) ingredients even though only minuscule amounts are used. These challenges can be mitigated by (i) spraying the bait on non-flowering vegetation; (ii) spraying the vegetation again after every rain; and (iii) using natural active ingredients such as eugenol or garlic oil. A blanket spray of ATSB® can be used in regions suffering from a massive mosquito burden. However, another form of ATSB® can be highly effective in residential areas.

ATSB® station

A novel implementation of the ATSB® concept can take the form of suspended removable bait stations. These devices should protect the bait from harsh climate conditions while avoiding environmental contamination. The immediate benefits of this implementation are versatility of use (indoor/outdoor), together with the potential impact for both nuisance and vector control and for addressable global territories.

Early studies with ATSB® stations emulating a very basic prototype comprised cotton wool pads soaked with bait and placed on plinths showed significant reduction in mosquito populations (Stewart *et al.*, 2013). A study using ATSB® indoor bait stations conducted in five villages in Mali produced significant reductions of *An. gambiae* at 90% in female and 93% in male populations. A 3.8-fold decrease in the proportion of females that had undergone four or more gonotrophic cycles was recorded at the experimental village, compared to a 1.2-fold increase at the control village (Qualls *et al.*, 2015b).

The effect of ATSB stations using sucrose, nectarine juice, slow-release substances, preservatives, red food-dye marker and the oral insecticide Spinosad on *An. sergentii* and *Aedes caspius* populations was also studied at a small oasis in Israel's southern desert. Feeding on non-insecticidal baits was monitored as a control in a similar neighbouring oasis. The insecticide caused a drastic decrease in the number of mosquitoes. Compared to the control site, the *An. sergentii* population was reduced to less than a tenth and that of *Ae. caspius* declined to a third (Müller *et al.*, 2008).

Bait core

BLOOD BAIT CORE. Leveraging on the ATSB principle, a possible solution for indoor control of blood-seeking pyrethroid-resistant female mosquitoes would utilize a similar bait station containing modified bait core. Blood feeding by females is a requirement for anautogenous mosquito reproduction, as blood is an exclusive source of proteins and amino acids. Bait for blood-seeking mosquitoes would utilize kairomone attractants to supplement host CO_2 cues, along with feeding cues (phagostimulants) required to induce engorging responses in the target vectors. The modified station would be installed on bed nets, exploiting CO_2 generated by the potential blood host to attract mosquitoes to the vicinity of the station, while the added attractants would lure the mosquitoes at short distances. The edible bait can incorporate a variety of gut toxins, which are proposed to be much more efficient against resistant mosquitoes, due to both alternative mode of action, and the (oral) exposure route that bypasses some resistance mechanisms. Although this type of bait station has shown promising results in preliminary trials, further work is required to validate its efficacy against resistant mosquitoes under various conditions and for prolonged periods of use.

Active ingredients

Most of the trials testing ATSB® efficacy as a vector control strategy employed one of three compounds (Spinosad, boric acid, Dinotefuran) as oral insecticides (Müller *et al.*, 2008; Khallaayoune *et al.*, 2013; Naranjo *et al.*, 2013). In some other cases, synthetic substances such as chlorfenapyr and tolfenpyrad have also been used (Stewart *et al.*, 2013). Ecological and environmental concerns have prompted studies that tested alternative natural substances, such as eugenol and garlic oil, which were shown in several studies to possess effective insecticidal properties (Qualls *et al.*, 2014; Karunaratne *et al.*, 2018).

Resistance to some insecticides can develop very quickly, more so if used repeatedly. An effective resistance combat strategy is to use substances in rotation, or in combination with toxins from a different chemical group. Having a wide range of oral toxins allows for reducing selection pressure for specific toxins. The ATSB®'s active ingredients are used in minuscule amounts and are therefore readily available for such resistance-free strategy.

Requirements for a Novel Bait Station

Based on numerous indoor and outdoor ATSB® studies across different countries and targeting different mosquito types, the accumulated data and experience allow for designing a bait station that will present worldwide mosquito control effectiveness. The following are the basic requirements enabling widespread bait station usage:

Efficacy A solution that targets decimation of mosquito populations should demonstrate a significant eradication impact whether used alone or together with other methods.

Resistance free Given the major problem of biological resistance in contemporary pesticide approaches, the solution should provide an effective way to overcome this challenge.

Affordability Low cost of production is a mandatory prerequisite for any widely used solution, especially for vector control. This implicates both the materials used and a scalable production design.

Durability Outdoor solutions in particular, must withstand harsh environmental conditions (temperature fluctuations and extremes, UV radiation, rain, wind, etc.) for prolonged periods of time.

Environmentally friendly The two most important ecological aspects that the solution should meet are (i) active ingredients must be non-hazardous to the environment and human health; and (ii) toxic materials, if used, must be inaccessible to non-targets such as bees.

Simple to distribute and easy to use Making the solution available to millions of users, the solution must be easy to distribute and simple to operate.

Following an extensive research process that employed wide-scale experimentation and numerous laboratory and field tests, a novel bait station has been developed to meet the above often conflicting requirements.

Description of a Novel Bait Station

The bait station has a 2D sheet-like structure, is the size of an A4 page and contains several layers.

Back layer. The back layer is a polymer-based surface, thermoformed with an array of 16 (8 × 2) cells. Each cell's dimension is 85 × 23 mm. The cells constitute the bait containers.

Bait. The bait is a syrup-like solution made of date, a natural sugar-rich and odorant liquid, mixed with an active ingredient such as boric acid, Spinosad or Dinotefuran.

Membrane. A thin polymer-based sheet covers the bait-containing cells. The membrane is semipermeable, allowing bate volatiles (attractants) to evaporate and permit mosquitoes to puncture it with their proboscis and access the bait.

Fig. 10.1. Front of a bait station comprising 8 × 2 cells covered by a perforated black membrane.

Following experiments, it was found that mosquito feeding behaviour can be greatly enhanced if the membrane is perforated with miniature holes, enhancing access to the bait. However, perforation may result in both inward penetration of moisture such as rain and other precipitations (in outdoor use), and leakage of the liquid bait through the perforations to the outside of the membrane.

Bait station improvements. In order to maintain high feeding behaviour using a perforated membrane, but without compromising on water penetration and bait leakage, two measures were taken:

Supportive layer. A pierceable support layer, made of a non-woven fabric or similar material, to prevent both inward water penetration and outward bait leakage via the perforations. A hydrophobic material can be added to further enhance the support layer's functionality.

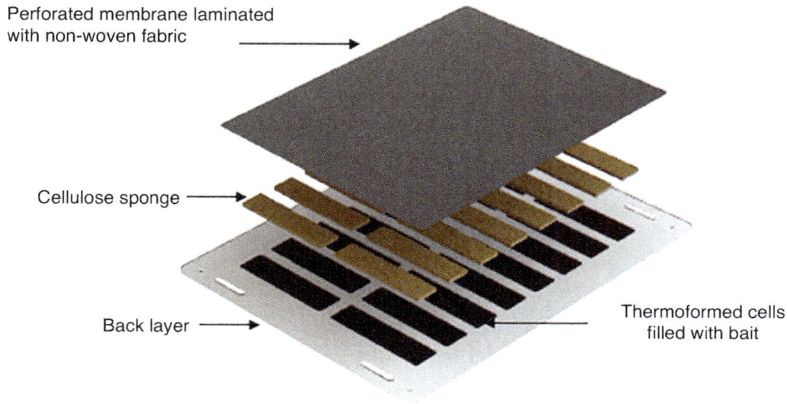

Fig. 10.2. 3D view of a disassembled bait station. The back layer is a plastic sheet that shapes and stabilizes the station and is formed with a matrix of cells that contain the bait. Each cell is fitted with a cellulose insert that absorbs the liquid bait and restrains its fluidity. A perforated membrane laminated with a non-woven fabric seals the bait cells, allows attractant volatiles to diffuse and provides mosquitoes with easy access to the bait.

Absorbent material An absorbent material which provides a reservoir for bait station liquids that simultaneously restrains bait fluidity.

Field Trials in Africa

The first ATSB® field study with the novel bait station was conducted by a team of researchers in Mali, Africa (Traore *et al.*, 2020). The field trial aimed to evaluate the efficacy of attractive toxic sugar baits (ATSB®) in Mali, where sustained malaria transmission occurs despite the use of LLINs. ATSB® stations were deployed in 7 of 14 similar villages participating in the study and where LLINs were already in widespread use. The combined use of ATSB® and LLINs was tested to see if it would substantially reduce parasite transmission by *An. gambiae* beyond the use of LLINs alone.

Testing the malaria vector feeding rate on natural sugar vs ASB stations was carried out in a 2-day field experiment in several villages. The number of mosquitoes feeding on natural sugar was counted vs mosquitoes feeding on bait stations containing Attractive Sugar Bait (without toxin: ASB) but with food dye. The experiment was conducted monthly for nine consecutive months in seven random villages. In the following year, in seven treatment villages, two ATSB® stations containing the insecticide Dinotefuran were placed on the outer walls of each building in the selected villages. Vector population density was evaluated monthly by CDC-UV light traps, malaise traps, Pyrethrum Spray

(PSCs) and Human Landing Catches (HLCs). Female samples of the catch were tested for age by examination of the ovarioles in dissected ovaries and identification of *Plasmodium falciparum* sporozoite infection by ELISA. Entomological Inoculation Rates (EIR) were calculated, and reductions between treated and untreated villages were determined.

In the 2 day experiment with ASB conducted each month, there was a lower number of male and female mosquitoes feeding on the natural sugar sources than on the ASB. ATSB® deployment reduced CDC-UV trap female catches in September, when catches were highest, by 57.4% compared to catches in control sites. Similarly, malaise trap catches showed a 44.3% reduction of females in August and PSC catches of females were reduced by 48.7% in September. Reduction of females in HLCs were lower by 19.8% indoors and 26.3% outdoors in September. The high reduction seen in the rainy season was similar for males. Reduction in population density for both males and females was over 70% during the dry season. Reductions of females with ≥ 3 gonotrophic cycles were recorded every month, amounting to 97.1% in October and 100% in December. Reductions in monthly EIRs ranged from 77.76 to 100% indoors and 84.95 to 100% outdoors. The number of sporozoite-infected females from traps was reduced by 97.83% in treated villages compared to controls.

This large field trial in Mali shows that ATSB® is competitive with local sugar sources and its use has significant impacts on population density, age group proportions, incidence of malaria infection in mosquitoes and EIR. It

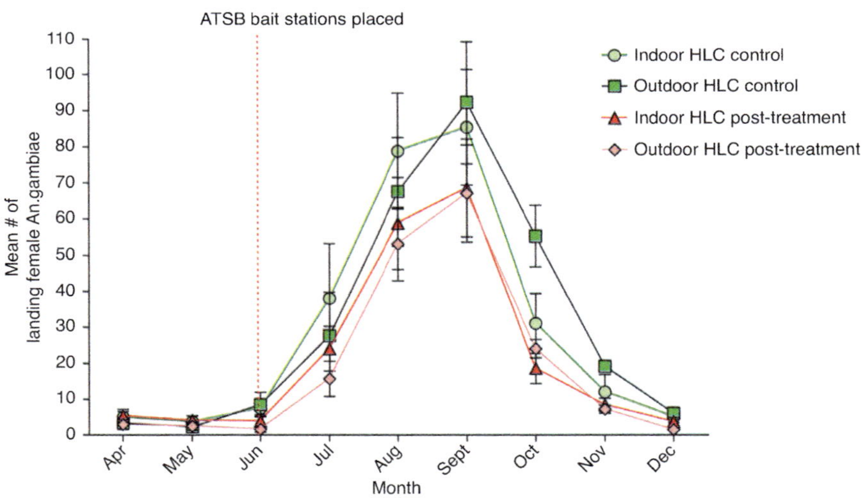

Fig. 10.3. Landing/biting by *An. gambiae* females in HLC observations after placement of ATSB® stations. Indoor and outdoor data collected from ATSB® treated villages and from control villages were grouped and the mean number of landing/biting per volunteer per night was calculated for each month.

also demonstrates that the use of ATSB® stations could virtually obstruct transmission in the dry season and could be an effective new class of tool to combat malaria.

Forward-Looking Vision

The current ATSB® station is now being tested in entomological, and later on epidemiological, trials in three countries in Africa. Preliminary encouraging results have prompted preparations for scaling up production of the ATSB® stations to supply a demand of close to one million stations for the 2-year trials. In view of their current performance and the likelihood of wide-scale acceptance of ATSB® as a complementary effective tool for mosquito control, consideration can already be given to some future directions for ATSB®.

Sophisticated attractants

Since attractants are critical for mosquito feeding behaviour, it is important to gain a better understanding of their olfactory mechanism. A recent study has demonstrated the selective response of different species – three *Aedes* and one *Anopheles* – to different plant-based odorants (Nyasembe *et al.*, 2018). Assessing the exact attractant effects on different insect species may therefore greatly optimize and improve the overall controlling effect of ATSB®. In addition, the potency of natural attractants currently used in ATSB® tends to decline over time. Biochemical research may lead to identifying certain

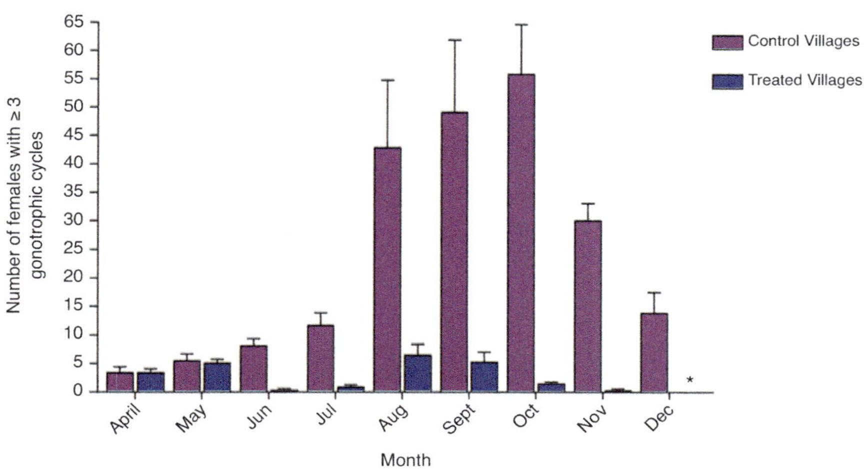

Fig. 10.4. Mean number per village and month (±SE) of female *An. gambiae* collected from all trapping methods and found infected with sporozoites in control villages compared to treated villages.

synthetic compounds that can serve as more powerful attractants assisting to enhance and/or preserve the fragrance over a longer period of time.

Novel active ingredients

The introduction of gut toxins as insecticides opens a wide range of opportunities for using compounds that can help combat biological resistance while presenting less of a hazard to the environment and human health. One option introduces compounds with novel modes of action. Rather than developing and synthesizing completely novel compounds, a more economic and promising route is to screen existing pesticides, predominantly those developed for use in agriculture, and repurpose the applicable ones to mosquito control. This method is supported by a recent study. Using a laboratory bioassay testing cascade against adult female *An. gambiae* mosquitoes, 11 compounds of interest were identified as vector control agents from a long list of representative candidates (Lees *et al.*, 2019). Further research in this direction is of paramount importance. In addition to synthetic pesticides, natural active ingredients such as Spinosad, eugenol and garlic oil have been successfully used in ATSB®. However, their lethal effect was lower than synthetic toxics such as Dinotefuran. The ecological and health implications for using natural ingredients are obvious, especially for ATSB® spray. Careful systematic research is needed to identify additional potent natural ingredients.

Blood bait core

Leveraging the ATSB® principle, a powerful solution with a focus on indoor control can be established by replacing the sugar bait core with a blood-feeding stimulant (BFS) for inducing biting mosquitoes to feed. This tool exploits the blood-feeding behaviour of female mosquitoes. Blood-feeding by females is a requirement for anautogenous mosquito reproduction, as blood is a good source of proteins and amino acids. A BFS in a liquid form can easily replace the sugar-based core as effective bait using the station's novel structure. This station variant seems particularly relevant for indoor use and can be applied as a standalone unit or as a patch attached to the LLINs. Although all benefits of a bait station apply to this novel implementation, further research is needed to evaluate its efficacy.

Beyond mosquitoes

The very same ATSB® mechanism that works for mosquitoes can be applied to other blood-seeking and sugar-feeding vector insects such as the sand fly and black fly genera. These insects are the primary vectors for serious diseases:

leishmaniasis and pappataci fever by sand fly, and onchocerciasis and mansonellosis by black fly. Previous studies have already indicated that ATSB® is effective in decimating sand fly populations (Saghafipour *et al.*, 2011; Qualls *et al.*, 2015a).

Acknowledgements

ATSB® method as well as the ATSB® station, developed by Westham Co., based on successful research by Prof Schlein and Dr Müller with support from Bill and Melinda Gates Foundation.

References

Alonso, P.L., Brown, G., Arevalo-Herrera, M., Binka, F., Chitnis, C. *et al.* (2011) A research agenda for malaria eradication – vector control. *PLOS Medicine* 8(1):e1000406. DOI: 10.1371/journal.pmed.1000406

Balabanidou, V., Kampouraki, A., MacLean, M., Blomquist, G.J., Tittiger, C. *et al.* (2016) Cytochrome P450 associated with insecticide resistance catalyzes cuticular hydrocarbon production in *Anopheles gambiae*. *Proceedings of the National Academy of Sciences of the United States of America*. 113(33), 9268–9273. https://doi.org/10.1073/pnas.1608295113

Bamou, R., Mbakop, L.R., Kopya, E., Ndo, C., Awono-Ambene, P. *et al.* (2018) Changes in malaria vector bionomics and transmission patterns in the equatorial forest region of Cameroon between 2000 and 2017. *Parasites and Vectors* 11, 464.

Barredo, E. and DeGennaro, M. (2020) Not just from blood: Mosquito nutrient acquisition from nectar sources. *Trends in Parasitology* 36(5), 473–484. https://doi.org/10.1016/j.pt.2020.02.003

Benelli, G. and Beier, J.C. (2017) Current vector control challenges in the fight against malaria. *Acta Tropica* 174, 91–96.

Caminade, C., McIntyre, M. and Jones, A.E. (2019) Impact of recent and future climate change on vector-borne diseases. *New York Academy of Sciences* 1436, 157–173.

Chaudhry, A., Jabeen, R., Sarfraz, B. and Mazhar, S. (2019) Mosquito control methods and their limitations. *Pure and Applied Biology* 8(4), 2389–2398.

Gobena, T., Berhane, Y. and Worku, A. (2012) Low long-lasting insecticide nets (LLINs) use among household members for protection against mosquito bite in Kersa, Eastern Ethiopia. *Public Health* 12, 914.

Gu, W., Muller, G., Schlein, Y., Novak, R.J. and Beier, J.C. (2011) Natural plant sugar sources of *Anopheles* mosquitoes strongly impact malaria transmission potential. *PLOS One* 6, 1, E15996. https://doi.org/10.1371/journal.pone.0015996

Karunaratne, P., Silva, W.A.P.P., Weeraratne, T.C. and Surendran, S.N. (2018) Insecticide resistance in mosquitoes: Development, mechanisms and monitoring. *S.H.P. Ceylon Journal of Science* 47(4), 299–309.

Khallaayoune, K., Qualls, W.A., Revay, E.E., Allan, S.A., Arheart, K.L. *et al.* (2013) Attractive toxic sugar baits: control of mosquitoes with the risk active ingredient dinotefuran and potential impacts on nontarget organisms in Morocco. *Environmental Entomology* 42(5), 1040–1045.

Knapp, J., Macdonald, M., Malone, D., Hamon, N. and Richardson, J.H. (2015) Disruptive technology for vector control – the innovative vector control consortium and the US military join forces to explore transformative insecticide application technology for mosquito control programmes. *Malaria Journal* 14, 371.

Kraemer, M.U.G., Reiner, R.C., Jr, Brady, O.J., Messina, J.P., Gilbert, M. *et al.* (2019) Past and future spread of the arbovirus vectors *Aedes aegypti* and *Aedes albopictus*. *Nature Microbiology* 854(4), 854–863.

Lees, R., Praulins, G., Davies, R., Brown, F., Parsons, G. *et al.* (2019) A testing cascade to identify repurposed insecticides for next-generation vector control tools – screening a panel of chemistries with novel modes of action against a malaria vector. *Gates Open Research* 2019(3), 1464.

Müller, G.C. and Schlein, Y. (2011) Different methods of using attractive sugar baits (ATSB®) for the control of *Phlebotomus papatasi*. *Journal of Vector Ecology* 36(1), S64–S70.

Müller, G.C., Kravchenko, V.D. and Schlein, Y. (2008) Decline of *Anopheles sergentii* and *Aedes caspius* populations following presentation of attractive toxic (spinosad) sugar bait stations in an oasis. *Journal of the American Mosquito Control Association* 24(1), 147–149.

Müller, G.C., Beier, J.C., Traore, S.F., Toure, M.B., Traore, M.M. *et al.* (2010) Successful field trial of attractive toxic sugar bait (ATSB®) plant-spraying methods against malaria vectors in the *Anopheles gambiae* complex in Mali, West Africa. *Malaria Journal* 9, 210.

Naranjo, D.P., Qualls, W.A., Müller, G.C., Samson, D.M., Roque, D. *et al.* (2013) Evaluation of boric acid sugar baits against *Aedes albopictus* (Diptera: Culicidae) in tropical environments. *Parasitology Research* 112, 1583–1587.

Nyasembe, V.O., Tchouassi, D.P., Pirk, C.W.W., Sole, C.L. and Torto, B. (2018) Host plant forensics and olfactory-based detection in Afro-tropical mosquito disease vectors. *PLOS* 12, e0006185. https://doi.org/10.1371/journal.pntd.0006185

Oliveira Sousa, J., Albuquerque, B.C., Coura, J.R. and Suárez-Mutis, M.C. (2019) Use and retention of long-lasting insecticidal nets (LLINs) in a malaria risk area in the Brazilian Amazon: A 5-year follow-up intervention. *Malaria Journal* 18, 100.

Qualls, W.A., Müller, G.C., Revay, E.E., Allan, S.A., Arheart, K.L. *et al.* (2014) Evaluation of attractive toxic sugar bait (ATSB®) – barrier for control of vector and nuisance mosquitoes and its effect on nontarget organisms in sub-tropical environments in Florida. *Acta Tropica* 131, 104–110.

Qualls, W.A., Müller, G.C., Khallaayoune, K., Revay, E.E., Zhioua, E. *et al.* (2015a) Control of sand flies with attractive toxic sugar baits (ATSB®) and potential impact on non-target organisms in Morocco. *Parasites and Vectors* 8, 87.

Qualls, W.A., Müller, G.C., Traore, S.F., Traore, M.M., Arheart, K.L. *et al.* (2015b) Indoor use of attractive toxic sugar bait (ATSB®) to effectively control malaria vectors in Mali, West Africa. *Malaria Journal* 14, 301.

Revay, E.E., Schlein, Y., Tsabari, O., Kravchenko, V., Qualls, W. *et al.* (2015) Formulation of attractive toxic sugar bait (ATSB®) with safe EPA-exempt substance significantly diminishes the *Anopheles sergentii* population in a desert oasis. *Acta Tropica* 150, 29–34.

Rochlin, I., Faraji, A., Ninivaggi, D.V., Barker, C.M. and Kilpatrick, A.M. (2016) Anthropogenic impacts on mosquito populations in North America over the past century. *Nature Communications*, 7(1–14), 13604. https://www.nature.com/articles/ncomms13604.

Ryan, S.J., Carlson, C.J., Mordecai, E.A. and Johnson, L.R. (2019) Global expansion and redistribution of *Aedes*-borne virus transmission risk with climate change. *PLOS Neglected Tropical Diseases*, 13(3), e0007213. https://journals.plos.org/plosntds/article?id=10.1371/journal.pntd.0007213

Saghafipour, A., Vatandoost, H., Zahraei-Ramazani, A.R., Yaghoobi-Ershadi, M.R., Rassi, Y. *et al.* (2011) Control of zoonotic cutaneous leishmaniasis vector, *Phlebotomus papatasi*, using attractive toxic sugar baits (ATSB). *PLOS ONE* 6(4), e0173558. https://journals.plos.org/plosone/article?id=10.1371/journal.pone.0173558

Schlein, Y. and Müller, G. (2006) Sugar questing mosquitoes in arid areas gather on scarce blossoms that can be used for control. *International Journal for Parasitology* 36, 10–11.

Schlein, Y. and Müller, G. (2008) An approach to mosquito control: Using the dominant attraction of flowering tamarix jordanis trees against culex pipiens. *Journal of Medical Entomology* 45(3), 384–390.

Stewart, Z.P., Oxborough, R.M., Tungu, P.K., Kirby, M.J., Rowland, M.W. and Irish, S.R. (2013) Indoor Application of Attractive Toxic Sugar Bait (ATSB) in Combination with Mosquito Nets for Control of Pyrethroid-Resistant Mosquitoes. *PLOS ONE* 8(12), e84168.

Tenywa, F.C., Kambagha, A., Saddler, A. and Maia, M.F. (2017) The development of an ivermectin-based attractive toxic sugar bait (ATSB®) to target *Anopheles arabiensis*. *Malaria Journal* 16, 338.

Traore, M.M., Junnila, A., Traore, S.F., Doumbia, S., Revay, E.E, *et al.* (2020) Large-scale field trial of attractive toxic sugar baits (ATSB®) for the control of malaria vector mosquitoes in Mali, West African. *Malaria Journal* 19, 72.

Yuval, B. (1992) The other habit: sugar-feeding by mosquitoes. *Bulletin of the Society for Vector Ecology* 17, 150–156.

11 The Use of Repellents for Mosquito Control

J.D. CHARLWOOD*

Global Health and Tropical Medicine, Instituto de Hygiene e Medicina Tropical, Lisbon, Portugal

Abstract

Strategies for preventing vector-borne diseases include the use of repellents to prevent biting. The effectiveness of repellents like DEET, Picaridin and plant-based alternatives are described, highlighting mixed results in reducing malaria transmission. Despite uncertainties, community willingness to use repellents suggests potential for their inclusion in disease prevention strategies. Challenges such as consistent use and ethical considerations are important. The development and efficacy of synthetic pyrethroid insecticides like metofluthrin and transfluthrin as spatial repellents and killing agents against mosquitoes, tailored to specific ecological settings is described. Contextual factors, such as sleeping habits, vector behaviour and environmental conditions, influence the choice between repellents, killing agents or a combination. Studies show their potential in reducing outdoor biting rates and malaria transmission, particularly in regions with varying ecological settings. The spread of the invasive malaria vector *Anopheles stephensi* in East Africa is highlighted, along with proposed solutions such as using emanators in vehicles to combat mosquito transmission. Overall, the importance of tailored approaches to disease prevention based on local conditions and community participation is emphasized.

> 'We suggest a combination of directed wind and volatile repellent as a mosquito deterrent strategy for the backyard setting'
>
> – Hoffman and Miller (2002)

*Corresponding author: jdcharlwood@gmail.com

DOI: 10.1079/9781789248845.0011

Introduction

One way of avoiding vector-borne disease is not to be bitten by the insect or tick that transmits the disease in the first place. The bites of many blood-sucking insects are painful and can cause irritation or infection. In some places, particularly in northern latitudes, their very number can be daunting. Linnaeus, with justification, complained of their number during his trip to Lapland. In Alaska, it has been calculated, by staff at the University of Alaska Natural History Museum, that 17 trillion mosquitoes (equivalent to 43 million kilos of mosquito) emerge every year. Tropical ulcers, aka jungle rot, are often initiated by scratching a mosquito bite, so it is not just disease like malaria or dengue that might be avoided. The use of repellents to achieve this has a long history. The indefatigable enquirer, Herodotus, describes in his *Histories*, the use by the Ancient Egyptians of a lamp that may have been used as a spatial repellent (Herodotus, 1966):

> The Egyptians who live in the marsh country use an oil extracted from the castor-oil plant. This plant, which grows wild in Greece, they call Kiki, and the Egyptian variety is very prolific and has a disagreeable smell. Their practise is to sow it along the banks of rivers and lakes, and when the fruit is gathered it is either bruised and pressed, or else boiled down, and the liquid thus obtained is of an oily nature and quite as good as olive oil for burning in lamps, although the smell is unpleasant.

If the vapour from the lamp was considered effective and alternative sources of oil for lamps were available, they probably put up with the smell because of its anti-mosquito effect. The marsh country mosquito would probably have been *Mansonia* sp. These attack throughout the night and are known for their painful bites. As pointed out by Charlwood (2003), if they used lamps at all it implies that, despite the mosquitoes, they stayed up. Without a repellent, they might not have done so.

Repellents fall into two categories: topical repellents and spatial repellents.

Topical Repellents

The most widely used, and still the most effective against both *Aedes* (Rodriguez *et al.*, 2015) and anophelines (Curtis *et al.*, 1987), is DEET (N,N-diethyl-m-toluamide) (Fradin and Day, 2002). This was developed in 1946 by the American armed forces and brought to market in 1956. It is available in many formulations, most of which differ in the concentration of the active ingredient. It remains the standard by which other repellents are judged. A more recently developed repellent is Picaridin. This is a colourless, nearly odourless active ingredient. Picaridin is as effective in repelling mosquitoes as products containing similar concentrations of DEET. There are also a number of other plant-based repellents available, including citronella (obtained from two different species of lemongrass: *Cymbopogon nardus* and *Cymbopogon winterianus*) and oil of eucalyptus (*Eucalyptus maculata citriodon*) the principal ingredient of which is para-menthane-diol (PMD). Plant-based repellents are produced naturally in response to, or as a way of avoiding, phytophagous insects. Maia and Moore (2011) provide an exhaustive survey of such potential repellents. They point out that many of these plants produce compounds that are highly toxic to mammals and/or irritating to the skin, and 'natural' does not mean 'safe'.

Given that malaria transmission may occur outdoors a number of trials have been undertaken to determine if widespread use of repellents reduces transmission. A repellent using oil of eucalyptus (PMD) reduced the incidence of *Plasmodium vivax* in a randomised control trial in Bolivia (Hill *et al.*, 2007), and in northern Ghana, the protective efficacy of using a repellent containing PMD and LLINs against malaria infection of both *Plasmodium falciparum*/*P. vivax* and *P. falciparum* was 34 and 47%, respectively (Dadzie *et al.*, 2013). After a 6 month trial in Afghanistan, a group of households using Mosbar, containing 20% DEET and 1% permethrin formulated in a bar of coconut substrate, reported less than half the number of *P. falciparum* malaria episodes compared to a group of households using a placebo (3.7% compared to 8.9%) (Rowland *et al.*, 2004). In Ethiopia, 3078 individuals in an intervention and 3004 in a control group were enrolled in a study in which an unspecified repellent (Buzz-Off® petroleum jelly), widely used in the country, in conjunction with LLINs was compared to a control group only using LLINs. The repellent significantly reduced prevalence in the intervention group (Deressa *et al.*, 2014). Nevertheless, Cochrane reviews (Maia *et al.*, 2018; Gabaldón Figueira *et al.*, 2023) concluded that there was insufficient evidence to conclude whether topical or spatial repellents prevent malaria. Both reviews suggested that there was a need for better designed trials to generate higher certainty of evidence before well-informed recommendations can be made, despite the evaluation and development of plant-based repellents becoming far more rigorous in recent years (Maia and Moore, 2011). In Cambodia, notwithstanding having a successful distribution system, the daily use of Picardin was suboptimal in a large, randomised control trial, and no post-intervention differences in PCR plasmodium prevalence were observed between study groups (Sluydts *et al.*, 2016). Daily compliance and appropriate use of the repellents remains the main obstacle in such trials. You can take a horse to water, but you can't make him drink. In Tanzania, where biting rates were high and where repellent and placebo lotion were provided free of charge to 940 households for a period of 14 months between July 2009 and August 2010, the majority of the participants were, however, willing to continue to use repellents (Sangoro *et al.*, 2014).

Repellents are relatively expensive. Formulations that can be produced locally should be encouraged. One such repellent is Mosbar. In Papua New Guinea, the protection provided during a 4 hour collecting period was 51.0% when the repellent was applied 4.5 and 2.5 hour before collection began, and

69.9% when the repellent was applied just before the commencement of collections (Charlwood and Dagaro, 1987).

As discovered by Gryseels *et al.* (2015) in Cambodia, adherence to daily compliance is a major limitation and indicates that malaria in low endemic settings can probably only be reduced substantially by introducing measures after researching and optimizing community involvement strategies. However, as pointed out by Hill *et al.* (2007), in areas where vectors feed in the early evening, effectiveness of treated nets can be significantly increased by using repellent between dusk and bedtime. Topical repellents are likely to be useful in areas invaded by *Anopheles stephensi*, especially since they appear to be more susceptible to their effects than other anophelines, such as *Anopheles gambiae* (Curtis *et al.*, 1987). Their subsidized sale may reduce malaria transmission by this early evening outdoor biting mosquito.

One kind of repellent that does not work, but remains on the market in many forms, is devices that emit sound, or ultrasound, that the manufactures claim is avoided by hungry mosquitoes. Such devices first appeared in 1976. They were tested by Gillies (personal communication), just as later devices have been tested by other entomologists of repute. No one has ever found an effect, and indeed the initial purveyors of the product were threatened with legal action for making false claims. The continuing availability of such devices says as much about the gullibility and desire of people to avoid mosquito bites as it does about the greed of the charlatans still selling such products in 2024.

The development of bracelets or anklets containing repellent for long duration of protection requires a slow release of the repellent. As this technology develops, they may become more useful but presently many do not appear to work (Rodriguez *et al.*, 2017). Incorporating the repellent into polymer systems (substances made up of macromolecules composed of many repeating subunits) holds the most promise for this (Mapossa *et al.*, 2021). Polymeric micro- and nanocapsules, micro/solid lipid nanoparticles, nanoemulsions/microemulsions, liposomes, nanostructured micellar hydrogels and cyclodextrins all provide slow release of mosquito repellent, improving the duration of repellence (Mapossa *et al.*, 2021).

Spatial Repellents

In many places, before modern alternatives were available, people in many countries, in Africa at least, burnt leaves and such, in the hopes that the smoke would drive the mosquitoes away. In some places they still do this. The smoke may force the mosquitoes who were resting inside to leave, but these mosquitoes are not the ones who would later attack the householders. These would be new arrivals of hungry, unfed females. Thus, without a long-lasting effect, such techniques are unlikely to be much use. In addition, the inhalation of residual smoke, especially by children and babies, who would be the first to go to bed, is a potential problem.

Mosquito coils and mats

Until a few years ago, the only other spatial repellents available have been mosquito coils. They were invented by Eiichiro Ueyama, a Japanese entrepreneur. Ueyama experimented with mixtures of pyrethrum powder and other ingredients and with various manufacturing techniques, but although he tried for more than 10 years, he failed to create a system that was long lasting. His wife Yuki suggested the spiral shape, and Ueyama then spent a further 7 years perfecting the product, finally marketing the first coils in 1902. They were handmade until the 1950s. They became popular once manufacturing was introduced (Fig. 11.1).

They also produce smoke over a 6–8 hour period, but when used indoors, they can generally be placed under a bed (a likely resting place for mosquitoes). Outdoors, the smoke is unlikely to be a problem. They are made of sawdust, which smoulders like a cigarette. The heat from the sawdust vaporises a repellent (generally some form of pyrethroid), and it is this, not the smoke, that repels the mosquitoes.

In the aptly titled 'The coil works (against mosquitoes in Papua New Guinea) (Charlwood and Jolley, 1984)', a crossover experiment found that lighting a coil in a room reduced entry rates of *Anopheles farauti* and *Anopheles koliensis* by about tenfold. At the time of writing, it was estimated to cost 2p

Fig. 11.1. Mosquito coil, made in China and sold in Mozambique. (Photo: J.D. Charlwood.)

for 8 hour of protection. Outdoors, restaurants often use them under tables in order to protect people's feet from marauding mosquitoes. In Dar es Salaam, the number of coils sold in local shops was correlated with the number of *Culex quinquefasciatus* collected in light traps (Chavasse *et al.*, 1996). Their efficacy depends on the mosquito not being resistant to the insecticide used. In the Kilombero valley of Tanzania, during a crossover experiment, coils with 0.03% transfluthrin as the active ingredient reduced landing rates by 80% among *Anopheles arabiensis* but did not affect the pyrethroid-resistant *Anopheles funestus* (Maia *et al.*, 2016). There also appeared to be diversion from users to non-users. This poses a number of ethical problems, a potential problem for all repellents or techniques that provide personal, rather than community, protection.

The ethics of personal protection

When a personal protection method that does not kill mosquitoes is used on an individual basis, mosquitoes are diverted to other hosts, thus enhancing their risk to contract a vector-borne disease (Moore *et al.*, 2007). This begs the question 'Is it ethical to promote such techniques?' even though when used on a wide scale, there may eventually be a 'community effect' in which non-users are protected. The required proportion of users to non-users for a community effect – in which exposure, even to non-users, is reduced – might be determined in semi-field situations.

Perhaps the answer depends on the affordability and availability of the intervention in question. Repellents are generally expensive (although mosquito coils are not) and are not widely available, but those that contain DEET, or Picardin, at least, work. Perhaps the solution would be to subsidize them so that the price barrier is no longer a problem. This may be particularly important in areas where *An. stephensi* is likely to occur, or where transmission occurs outdoors before people go to bed. The advantages of using the repellent might also be stressed in public health campaigns. This is not to suggest that they should be made available for free.

An alternative to mosquito coils is mosquito mats. These also require heat in order to work. Small electric heaters, designed for this purpose, are generally available. When these are used, the repellent effect should last for 12 hour. When the correct heater is unavailable, people tend to improvise (Figs 11.2 and 11.3), but this reduces the effect of the mat to less than an hour.

Passive Emanators

The recent development of synthetic pyrethroid insecticides with high vapour action at ambient temperatures has led to the development of devices that work without requiring the application of heat (Ujihara *et al.*, 2004; Ogoma *et al.*, 2012; IVCC, 2020). Metofluthrin is a repellent that shows promise as an active ingredient when vapourized indoors, although it is possible that

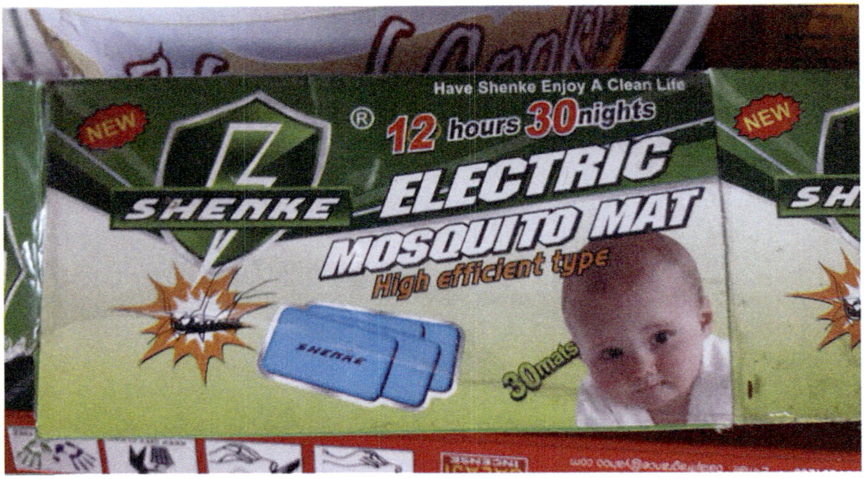

Fig. 11.2. An electronic 'mosquito mat' repellent sold in Tanzania. (Photo: J.D. Charlwood.)

Fig. 11.3. An alternative, and less efficient, manner of heating a mosquito mat. In this case, the effect lasts for minutes rather than hours. (Photo: J.D. Charlwood.)

its main effect is as a killing agent (Rapley *et al.*, 2009). When used in paper-based emanators, it reduced outdoor biting rates of *Aedes canadensis* and *Aedes vexans* by approximately 90% (Lucas *et al.*, 2007). These reductions are probably attributable to a disruption of orientation towards the host resulting from neural excitement, which occurs at an early stage of pyrethroid toxicity (Kawada *et al.*, 2005). The formulation used in these early studies was, however, effective for only a few days. A formulation of 5% metofluthrin on

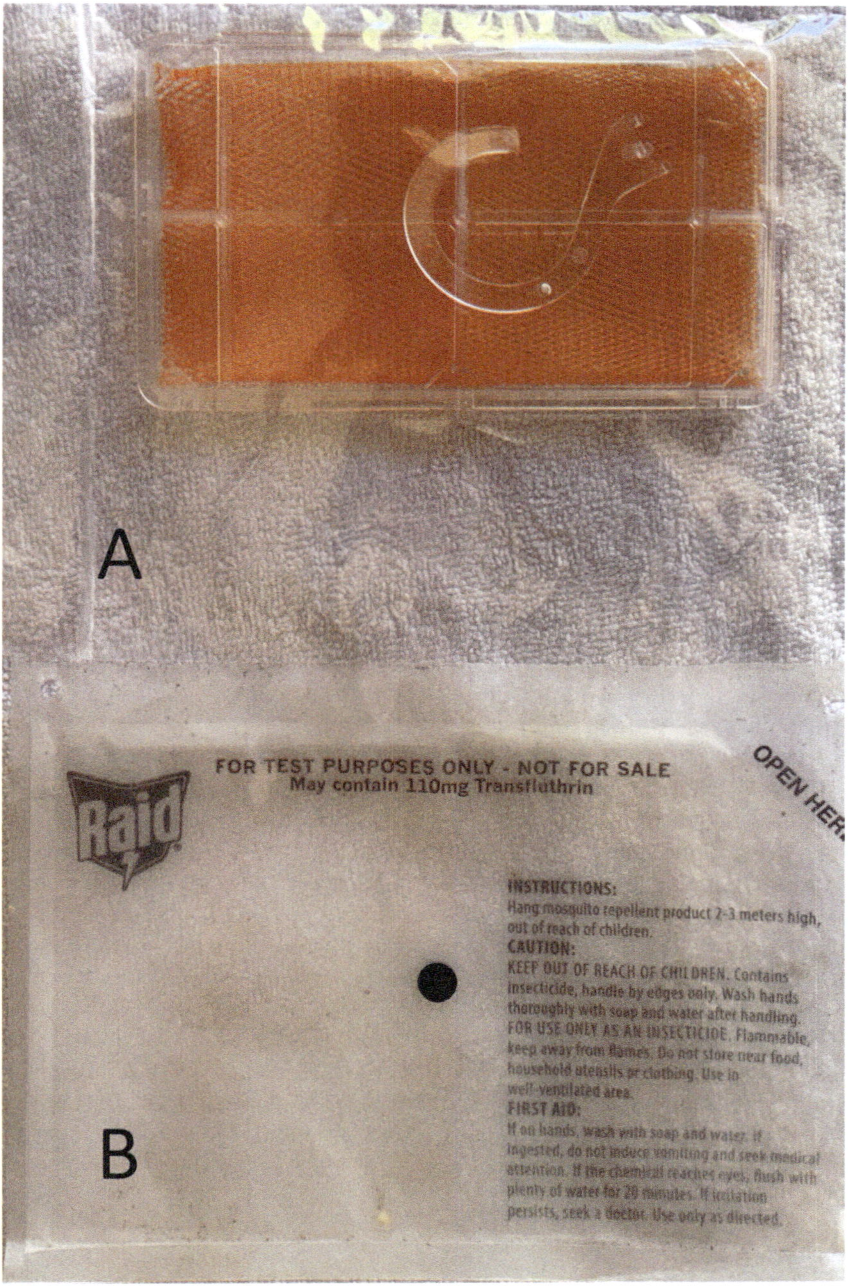

Fig. 11.4. (A) Metofluthrin emanator, as used in studies in Cambodia and (B) transfluthrin emanator that consists of two plastic sheets that can be pulled apart to expose the repellent. (Photo: J.D. Charlwood.)

a plastic lattice, designed to control evaporation rates, was found to reduce resting mosquito densities inside houses in Indonesia for 1 month (Kawada *et al.*, 2006). Although not designed as a malaria control tool (due to their high cost), slow-release emanators made of dual-layer polyethylene mesh, impregnated with 10% (w/w) metofluthrin (Fig. 11.4a), were tested to see if they reduced landing rates of outdoor biting of secondary vectors of malaria in Cambodia and subsequently to see if they reduced malaria incidence and prevalence in two villages (Ou Chra and Pu Cha in Mondolkiri Province) where the principal vector, *Anopheles dirus*, was actively transmitting malaria, albeit at a low intensity (Charlwood *et al.*, 2017).

For the outdoor biting study in Pailin, metofluthrin emanators were hung, either individually or in a group of four, in the middle of each side of a square 1m high and 1.5m on a side. Inside the square landing, collections were undertaken for the first 4 hour of the night (the period of maximum mosquito activity) (Charlwood *et al.*, 2016). Four treatment collections and four control collections were undertaken each night, and the location of treatment and control and the person doing the collection were alternated every second night. After the 4 hours of landing collection the collectors slept in Furvela tent-traps, with or without an emanator close to the trap intake. Similar collections were undertaken in three other locations of Cambodia, where the principal anopheline species collected differed. In Pailin, where *Anopheles minimus* was the most common mosquito, a single emanator reduced landing rates by 48% and four emanators by 67%. A similar result was obtained in another of the other study sites (Pursat) but not in another (Koh Kong).

In a ‘double’ crossover experiment, emanators in Mondolkiri were provided to all the houses in one village, with an equivalent village 5 km distant acting as the control. Emanators were replaced every month for 3 months, after which time treatment and control village were swapped. After a further 3 months, they were swapped again. At each changeover point, a prevalence survey of all inhabitants was undertaken. Perhaps due to the low levels of malaria transmission in Mondolkiri, no significant relationships were found in *Plasmodium* cases or seroconversion rates between villages, surveys or by intervention (Charlwood *et al.*, 2017).

Another pyrethroid that can be used as a spatial repellent is transfluthrin. This behaves in much the same way as metofluthrin although it is considerably less potent (Kim *et al.*, 2023). Hessian strips impregnated with transfluthrin when applied over the eaves of houses in Tanzania, reduced entry rates by malaria vectors (Mmbando *et al.*, 2018).

Possible Product Profiles?

In many ways, the questions about products now generally considered to be spatial repellents and in development are, ‘Where they are likely to be used?’ and ‘What are the characteristics of the ideal product?’ In addition to the usual and expected ones (long duration, no toxicity for humans, low price, etc..), what might be the ideal product profile? Should they act as a repellent only, with no killing effects? As a repel-and-kill agent, or, indeed, largely as a killing agent?'

I would suggest that there might be need for all three types, depending on their 'use ecosystem'. This 'use-ecosystem' has to take account of the actual ecological setting in which they might be used. There is definitely not a one-size-fits-all product.

In areas like Gash Barka, Eritrea, people sleep outside (underneath mosquito nets). I camped outside a compound for 2 weeks in 2016. It is very hot during the day, so it is nice to feel comfortable at night. People stayed up late outside of their nets. *An. arabiensis* is the only vector. At the time, it was present at low density. There were plenty of alternative hosts available. Under these circumstances, a killing emanator outdoors would be the best option. Mosquitoes that bite animals are left undisturbed. Under other circumstances, but with the same vector, a repellent, also used outdoors, is perhaps the best option. Such would be the case in Muleba, Tanzania. Following an intervention against the predominant *An. gambiae s.s.*, responsible for intense malaria transmission in the area the gambiae/arabiensis, ratio changed from 98% *An. gambiae* and 2% *An. arabiensis* to 84% *An. arabiensis* and 16% *An. gambiae* (LeClair *et al.*, 2017).

Densities of *An. arabiensis* increased compared to pre-intervention times, so more adults were produced than in periods prior to the intervention. In Ifakara, *An. gambiae* larvae survived better than *An. arabiensis* during periods of heavy of rainfall (Charlwood and Edoh, 1996). It rains a lot in Muleba. The implication in Muleba is that, previously, the *An. gambiae* were out competing the *An. arabiensis* larvae, but once control with IRS and LLINs had been enacted, the endophilic *An. gambiae* were controlled, and the less affected, primarily exophagic, An. *arabiensis* occupied the vacant larval niche. Should the effect of the intervention be reduced (as the insecticide lost its potency) and the low *An. gambiae* population start to reinvade, it would surely find it more difficult to occupy the larval niche if it was already occupied by *An. arabiensis*, which, considering the rainfall, was doing the best it could.

A similar situation may occur in places where *An. funestus* is replaced by *Anopheles rivulorum*. In such situations, rather than killing the mosquitoes that might bite outdoors (which might empty the larval niche), using a repellent that does not kill stops the mosquito from biting people without necessarily reducing the population. Adult *An. arabiensis* females are known to have catholic feeding habits. In Muleba, there are plenty of cattle to serve as alternative hosts for the mosquito. Mosquitoes that bite them are left alone. In the absence of alternate hosts, a push-pull system using attractive traps could be used.

Mosquitoes that enter houses should be killed, even in areas where no large-scale intervention is taking place. This includes, if possible, all genera. Prevention is better than cure. Reducing house entry rates by simple measures, such as screening openings between roof and walls, would seem the best solution. No repellent product is needed. Mosquitoes that bite animals outdoors are left undisturbed.

In Cambodia, many houses are built on 2m high stilts. Children play outside under these houses, especially during the morning and evening (Liverani *et al.*, 2017). Dengue is a major problem in Cambodia. The sight of a mother on the back of a motorbike cradling a sick baby in her lap and holding a pole to which a serum bottle is attached was a common site during the 2 years I spent there. Dengue vectors bite predominantly in the late afternoon when children are outside (Charlwood *et al.*, 2014). Repel and kill is a good option in this situation.

Houses often have large, wide verandas. People sit on these in the evening to watch television or videos. They are potentially exposed to *An. dirus*, which (perhaps because of its history as a monkey-biting mosquito) is basically the only mosquito able to enter a house built 2m off the ground. Again, a repel-and-kill emanator would seem to be the best option. *An. dirus* is an extremely anthropophagic mosquito, again because of its history. Its presence signals a nearby primary forest. The population density can go through rapid changes. The need for an emanator may be infrequent. Under those circumstances, the most useful system is still likely to be the humble mosquito coil. Each coil works for one night, is cheap and available. Because the repellent needs heat to vapourize, they do not lose their efficacy over time. It is easy and cheap to have access to at least two coils.

On the other hand, if the product is required for *Aedes* control, then a metofluthrin emanator would be recommended. Under these circumstances, given the ease with which it can be moved, it would obviously double for use against *An. dirus* at night.

Control of *Anopheles stephensi*

The history of the long-range dispersal of vectors by humans is a long and continuing one. It is now recognized that the invasive malaria vector *An. stephensi* is making its way down the highways of East Africa. It arrived by boat to Ethiopia from India (perhaps via Djibouti) (Faulde *et al.*, 2014). It is now travelling by road and has been recorded in Kenya from Ethiopia. This is an urban invader that breeds in water tanks, common in places where the municipal water supply is erratic (in Eritrea, it came on at an outside tap at two am on what seemed like a random series of days. Everyone in Asmara has a tank. It may be protected by its high altitude). The potential problem cannot be overstated. The mosquito bites outdoors in the early evening, just the time when an immunologically naive population is likely to be outdoors themselves. Even if the Kenyan outbreak is eliminated the, now established, *An. stephensi* from Ethiopia will always be a 'source' population and so reinvasion is always likely.

In many cars and lorries, in many countries, drivers buy emanators of air freshener (in the shape of trees or similar) (Fig. 11.5).

Emanators suitable for use inside lorry cabs (something to hang up underneath the mirror), desirable in their own right, but which repel or kill mosquitoes that might enter the cab, may reduce transmission (Charlwood, 2023).

Fig. 11.5. Sweet smelling emanator of the sort commonly used in motorcars. It is suggested that these might incorporate a repellent to ward off mosquitoes such as *Aedes albopictus* in Europe and America and *An. stephensi* in Africa. (Photo: J.D. Charlwood.)

They might also smell, or look, good. If coils can smell of lemon, why not an emanator? Emanators could be sold at recognized vendors for cash. The emanators come in plastic bags, which have a saturated atmosphere of the product. They can be stored for long periods in this way. I would suggest opening the emanator and leaving it in the cab the night before the start of a long drive. At the point of sale, the vendor might scan a barcode on a sheet of such barcodes. This is sent to the phone of the driver, and a central database, available to epidemiologists and the like. Should the lorry be stopped (at customs say), then the presence of an emanator gives the driver wellness tokens in the form of ADA (Chapter 14, this volume). The driver only gets tokens, however, if it can be confirmed by his barcode (sent to the phone of the person distributing tokens who can look it up) that the emanator is less than 3 weeks old. At the point of distribution, the number of stops without a Wellness Token being given is recorded. The information on rates of use, say, can be determined by a central database. It also means that it is more difficult to cheat the system with an old, ineffective, repellent.

How valuable the tokens are or become may depend on how often they can be earned – if it is only once per emanator (i.e. every 3 weeks) or once every stop until the emanator runs out. Establishment of a Wellness Token system in itself would seem to be a useful thing to do. The tokens then function in a system such as that described by Lietaer *et al.* (2012). It helps people ease out of a monetary monoculture towards a more stable monetary ecosystem. It is imperative however that the electronic setup is not carbon expensive. Drivers who do not have an up-to-date emanator just do not get tokens. They may be recorded as 0 and given a leaflet explaining why having such an emanator might be a good thing. If wellness tokens become more valuable, then the cash transaction to buy one using standard currency, say, for £3, may become less important than the loss of tokens (which might, to the person buying the emanator, be considered to be worth the equivalent of £10). If there is to be a subsidy of such things, it goes towards the value of wellness tokens, not the commercial price. This would help to make the tokens more valuable.

Making such an up-to-date emanator mandatory for cross-border passage, like a vaccine certificate, destroys the above system, because uptake is involuntary, and it works on punishment rather than reward. It may break down if the official demand is reduced. But reward at the border may be a no-brainer. While setting up such a system might be a priority, the use of such emanators inside cabs in numerous other sites (especially ports) could also become widespread.

Emanators used in this way may also help reduce the spread of *Aedes albopictus* in Europe. Were it possible to develop a repellent that was effective for 6 months, their use commercially would be assured. Failing that, determining periods of maximum dispersal might enable the targeted use (in time) of emanators that might need replacing every month.

References

Charlwood, J.D. (2003) Did Herodotus describe the first use of mosquito repellents?. *Trends in Parasitology* 19, 555–556.

Charlwood, J.D. (2023) A self-sustaining system to reduce migration of *Anopheles stephensi* in East Africa. *Acta Tropica* 245, 106981.

Charlwood, J.D. and Dagaro, H. (1987) Repellent soap for use against malaria vectors in Papua New Guinea. *Papua New Guinea Medical Journal* 30, 301–303.

Charlwood, J.D. and Edoh, D. (1996) Larval distribution of members of the *Anopheles gambiae* Giles complex (Diptera, Culicidae), identified by the PCR reaction, in the environs of Ifakara, Tanzania. *Journal of Medical Entomology* 33, 202–204.

Charlwood, J.D. and Jolley, D.J. (1984) The coil works (against mosquitoes in Papua New Guinea). *Transactions of the Royal Society of Tropical Medicine and Hygiene* 78, 678.

Charlwood, J.D., Tomás, E.V.E., Kelly-Hope, L. and Briët, O.J.T. (2014) Evidence of an 'invitation' effect in feeding sylvatic *Stegomyia albopicta* from Cambodia. *Parasites and Vectors* 7, 324.

Charlwood, J.D., Nenhep, S., Protopopoff, N., Sovannaroth, S., Morgan, J.C. *et al.* (2016) The effect of the spatial repellent metofluthrin on landing rates of outdoor biting anophelines in Cambodia, S.E. Asia. *Medical and Veterinary Entomology* 30(2), 229–234. DOI: 10.1111/mve.12168.

Charlwood, J.D., Hall, T., Nenhep, S., Rippon, E., Branca-Lopes, A. *et al.* (2017) Spatial repellents and malaria transmission in an endemic area of Cambodia with high mosquito net usage. *Malaria World Journal* 8, 11.

Chavasse, D.C., Lines, J.D. and Ichimori, K. (1996) The relationship between mosquito density and mosquito coil sales in Dar es Salaam. *Transactions of the Royal Society of Tropical Medicine and Hygiene* 90(5), 493. DOI: 10.1016/s0035-9203(96)90291-6.

Curtis, C.F., Lines, J.D., Ijumba, J., Callaghan, A., Hill, N. *et al.* (1987) The relative efficacy of repellents against mosquito vectors of disease. *Medical and Veterinary Entomology* 1, 109–119.

Dadzie, S., Boakye, D., Asoala, V., Koram, K., KiszewskI, A. *et al.* (2013) A community-wide study of malaria reduction: Evaluating efficacy and user-acceptance of a low-cost repellent in Northern Ghana. *American Journal of Tropical Medicine and Hygiene* 88, 309–314. DOI: 10.4269/ajtmh.2012.12-0370.

Deressa, W., Yihdego, Y.Y., Kebede, Z., Batisso, E., Tekalegne, A. *et al.* (2014) Effect of combining mosquito repellent and insecticide treated net on malaria prevalence in Southern Ethiopia: A cluster-randomised trial. *Parasites and Vectors* 7, 132. DOI: 10.1186/1756-3305-7-132.

Faulde, M.K., Rueda, L.M. and Khaireh, B.A. (2014) First record of the Asian malaria vector *Anopheles stephensi* and its possible role in the resurgence of malaria in Djibouti, Horn of Africa. *Acta Tropica* 139, 39–43.

Fradin, M.S. and Day, J.F. (2002) Comparative efficacy of insect repellents against mosquito bites. *New England Journal of Medicine* 347, 13–18. DOI: 10.1056/NEJMoa011699.

Gabaldón Figueira, J.C., Wagah, M.G., Adipo, L.B., Wanjiku, C. and Maia, M.F. (2023) Topical repellents for malaria prevention. *Cochrane Database Systematic Reviews* 8(8), CD015422. DOI: 10.1002/14651858.CD015422.pub2.

Gryseels, C., Sambunny, U.K., Sluydts, V., Durnez, L., Phoeuk, P. *et al.* (2015) Factors influencing the use of topical repellents: Implications for the effectiveness of malaria elimination strategies. *Science Reports* 5, 16847. DOI: 10.1038/srep16847.

Herodotus (1966). Translated by A. de Sélincourt. *The Histories*. Penguin Books, London.

Hill, N., Lenglet, A., Arnez, A.M. and Carneiro, I. (2007) Plant based insect repellent and insecticide treated bed nets to protect against malaria in areas of early evening biting vectors: Double blind randomised placebo controlled clinical trial in the Bolivian Amazon. *British Medical Journal* 335, 1023. DOI: 10.1136/bmj.39356.574641.55.

Hoffman, E.J. and Miller, J.R. (2002) Reduction of mosquito (Diptera: Culicidae) attacks on a human subject by combination of wind and vapour phase DEET repellent. *Journal of Medical Entomology* 39, 935–938.

IVCC (2020) *An Expert Review of Spatial Repellents for Mosquito Control*. Available at: https://www.ivcc.com/wp-content/uploads/2020/08/An-Expert-Review-of-Spatial-Repellents-for-Mosquito-Control.pdf.(accessed (accessed 23 September 2024).

Kawada, H., Maekawa, Y. and Takagi, M. (2005) Field trial on the spatial repellency of metofluthrin-impregnated plastic strips for mosquitoes in shelters without walls (Beruga) in Lombok, Indonesia. *Journal of Vector Ecology* 30, 181–185.

Kawada, H., Iwassaki, T., Loan, L.L.E., Tien, T.K., Mai, N.T.N. *et al.* (2006) Field evaluation of spatial repellency of metofluthrin-impregnated latticework plastic strips against *Aedes aegypti* (L.) and analysis of environmental factors affecting its efficacy in My Tho City, Tien Giang, Vietnam. *American Journal of Tropical Medicine and Hygiene* 75, 1153–1157.

Kim, D.Y., Hii, J. and Chareonviriyaphap, T. (2023) Transfluthrin and metofluthrin as effective repellents against pyrethroid-susceptible and pyrethroid-resistant *Aedes aegypti* (L.) (Diptera: Culicidae). *Insects* 14(9), 767. DOI: 10.3390/insects14090767.

LeClair, C., Cronery, J., Kessy, E., Tomás, E.V.E., Rowland, M. *et al.* (2017) 'Repel all biters': Combination mosquito nets enhance collections of endophilic *Anopheles gambiae* and *An. arabiensis* in CDC light-traps. *Malaria Journal* 16, 336.

Lietaer, B., Arnsperger, C., Goerner, S. and Brunnhuber, S. (2012) *Money and sustainability: The missing link*. Report to the Club of Rome. Triarcy Press, Axminster, UK.

Liverani, M., Charlwood, J.D., Lawford, H. and Yeung, S. (2017) Field assessment of a novel spatial repellent for malaria control: A feasibility and acceptability study in Mondulkiri, Cambodia. *Malaria Journal* 16, 412.

Lucas, J.R., Shono, Y., Iwasaki, T., Ishiwatari, T., Spero, N. *et al.* (2007) U.S. laboratory and field trials of metofluthrin (SumiOne) emanators for reducing mosquito biting outdoors. *Journal of the American Mosquito Control Association* 23, 47–54.

Maia, M.F. and Moore, S.J. (2011) Plant-based insect repellents: A review of their efficacy, development and testing. *Malaria Journal* 10(Suppl. 1), S11.

Maia, M.F., Kreppel, K., Mbeyela, E., Roman, D., Mayagaya, V. *et al.* (2016) A crossover study to evaluate the diversion of malaria vectors in a community with incomplete coverage of spatial repellents in the Kilombero Valley, Tanzania. *Parasites and Vectors* 9, 451.

Maia, M.F., Kliner, M., Richardson, M., Lengeler, C. and Moore, S.J. (2018) Mosquito repellents for malaria prevention. *Cochrane Database Systematic Reviews* 2(2), CD011595. DOI: 10.1002/14651858.CD011595.pub2.

Mapossa, A.B., Focke, W.W., Tewo, R.K., Androsch, R. and Kruger, A. (2021) Mosquito-repellent controlled-release formulations for fighting infectious diseases. *Malaria Journal* 20, 165. DOI: 10.1186/s12936-021-03681-7.

Mmbando, A.S., Ngowo, H., Limwagu, A., Kilalangongono, M., Kifungo, K. *et al.* (2018) Eave ribbons treated with the spatial repellent, transfluthrin, can effectively protect against indoor-biting and outdoor-biting malaria mosquitoes. *Malaria Journal* 17, 368.

Moore, S.J., Davies, C.R., Hill, N. and Cameron, M.M. (2007) Are mosquitoes diverted from repellent-using individuals to non-users? Results of a field study in Bolivia. *Tropical Medicine and International Health* 12, 532–539. DOI: 10.1111/j.1365-3156.2006.01811.x.

Ogoma, S.B., Ngonyani, H., Simfukwe, E.T., Mseka, A., Moore, S.J. *et al.* (2012) Spatial repellency of transfluthrin-treated hessian strips against laboratory reared *Anopheles arabiensis* mosquitoes in a semi-field tunnel cage. *Parasites and Vectors* 5, 54.

Rapley, L.P., Russell, R.C., Montgomery, B.L. and Ritchie, S.A. (2009) The effects of sustained release metofluthrin on the biting, movement, and mortality of *Aedes aegypti* in a domestic setting. *American Journal of Tropical Medicine and Hygiene* 81, 94–99.

Rodriguez, S.D., Drake, L.L., Price, D.P., Hammond, J.I. and Hansen, I.A. (2015) The efficacy of some commercially available insect repellents for *Aedes aegypti* (Diptera: Culicidae) and *Aedes albopictus* (Diptera: Culicidae). *Journal of Insect Science* 15, 140. DOI: 10.1093/jisesa/iev125.

Rodriguez, S.D., Chung, H.N., Gonzales, K.K., Vulcan, J., Li, Y. *et al.* (2017) Efficacy of some wearable devices compared with spray-on insect repellents for the yellow fever mosquito, *Aedes aegypti* (L.) (Diptera: Culicidae). *Journal of Insect Science (Online)* 17, 1. DOI: 10.1093/jisesa/iew117.

Rowland, M., Downey, G., Rab, A., Freeman, T., Mohammad, N. *et al.* (2004) DEET mosquito repellent provides personal protection against malaria: A household randomized trial in an Afghan refugee camp in Pakistan. *Tropical Medicine and International Health* 9, 335–342. DOI: 10.1111/j.1365-3156.2004.01198.x.

Sangoro, O., Kelly, A.H., Mtali, S. and Moore, S.J. (2014) Feasibility of repellent use in a context of increasing outdoor transmission: A qualitative study in rural Tanzania. *Malaria Journal* 13, 347. DOI: 10.1186/1475-2875-13-347.

Sluydts, V., Durnez, L., Heng, S., Gryseels, C., Canier, L. *et al.* (2016) Efficacy of topical mosquito repellent (picaridin) plus long-lasting insecticidal nets versus long-lasting insecticidal nets alone for control of malaria: A cluster randomised controlled trial. *Lancet Infectious Diseases* 16(10), 1169–1177. DOI: 10.1016/S1473-3099(16)30148-7.

Ujihara, K., Mori, T., Iwasaki, T., Sugano, M., Shono, Y. *et al.* (2004) Metofluthrin: A potent new synthetic pyrethroid with high vapor activity against mosquitoes. *Bioscience and Biotechnological Biochemistry* 68, 170–174.

12

Underused Alternative Techniques for Mosquito Control

ROBERT T. JONES[1]* AND LOUISA A. MESSENGER[2]*

[1]*Department of Disease Control, London School of Hygiene & Tropical Medicine, London, United Kingdom*
[2]*Parasitology and Vector Biology (PARAVEC) Laboratory, School of Public Health, University of Nevada, Las Vegas, Nevada; Department of Environmental and Occupational Health, School of Public Health, University of Nevada, Las Vegas, Nevada*

Introduction

The emergence of widespread insecticide resistance among major malaria vector species (World Health Organization, 2020), and shifts in mosquito behaviour that enable some populations to evade traditional chemical interventions (Lefèvre *et al.*, 2009; Russell *et al.*, 2011; Gatton *et al.*, 2013; Kreppel *et al.*, 2020), has renewed interest in expanding the arsenal of mosquito control tools to include complementary environmental management and biological strategies (Gari and Lindtjørn, 2018). This search for new tools includes a look back at those used in the past, which may be underused because of a lack of evidence for their efficacy. Here, we discuss two underused techniques and consider the evidence available to support their use in the future of mosquito control.

Larvivorous Fish

Attempts to control mosquitoes through reducing the size of the immature population can include the manipulation or modification of available larval

*Corresponding authors: Robert.jones@lshtm.ac.uk; louisa.messenger@unlv.edu

DOI: 10.1079/9781789248845.0012

habitats, or the addition of microbial or chemical substances that kill or inhibit the development of immature stages. A third approach employs biological predators of immature mosquitoes. These can be frogs or invertebrate predators, but most frequently the strategy uses fish.

The use of fish as a method of vector control is founded on the requirement for mosquitoes to lay their eggs in, on or near water sources. The eggs develop into larvae and then pupae in a range of aquatic habitats, including streams and rivers, rice fields and mangrove swamps, as well as bodies of water in domestic environments. If fish can be used to reduce the size of the vector population, it is expected that the transmission of pathogens to humans will be reduced. Concordant with this, larval predators have been used to control populations of mosquitoes that transmit malaria parasites as well as those that transmit arboviruses and filaria (Kim *et al.*, 2002; Howard *et al.*, 2007; Morales-Pérez *et al.*, 2017). However, reducing larval densities may not have a sufficient or direct impact on adult mosquito numbers and may have undesirable effects on mosquito populations, such as larger adults borne from a reduction in larval competition for resources (Steinwascher, 2018). It has, therefore, been necessary to conduct controlled trials to determine the efficacy of this form of intervention.

The use of larvivorous fish has been promoted in the past by the World Health Organization (WHO) as an environmentally-friendly alternative to insecticide-based interventions for the control of malaria in specific situations where mosquito breeding habitats are well defined and water conditions are suitable (World Health Organization, 2003). However, while the current WHO response framework for vector control lists larvicides and environmental management among the proven and cost-effective interventions suitable for public health use, larvivorous fish are not included in the recommendations (World Health Organization, 2017).

The use of larvivorous fish is simple and may appear harmless, but there can be challenges in finding and treating all mosquito larval habitats in an area, and the release of non-native larvivorous fish has the potential to adversely affect local ecosystems through competition with indigenous species or the predation of non-target organisms. These risks have to be considered and weighed against the evidence for the efficacy of the intervention.

Larvivorous Species

Many fish species are natural predators of larvae, but it is the western mosquitofish *Gambusia affinis* (Fig. 12.1) and the eastern mosquitofish *Gambusia holbrooki* that have been most widely used for mosquito control. These mosquitofish were once considered to be subspecies of *G. affinis* but are now recognized as two separate species, with *G. holbrooki* being native to drainages east of Mobile Bay, Alabama, USA, and *G. affinis* having a larger natural range throughout the Mississippi River drainage, into Texas and eastern

Fig. 12.1. Larvivorous fish used in the control of mosquitoes. The western mosquitofish *G. affinis* (top), guppy *P. reticulata* (middle) and Nile tilapia *O. niloticus* (bottom). (Source: https://commons.wikimedia.org/wiki/File:Gambusia_affinis.jpg; https://commons.wikimedia.org/wiki/File:Guppy_-_Poecilia_reticulata.jpg; https://commons.wikimedia.org/wiki/File:Til%C3%A1pia_ou_Sarotherodon_niloticus_2.jpg.)

Mexico (Wooten *et al.*, 1988). These fish have now been introduced across many regions of the world, including parts of South America, Australia, Asia, Europe and Northern Africa (Sitaraman *et al.*, 1975; Gamal, 2012; Cabrera *et al.*, 2017). The guppy *Poecilia reticulata*, native to parts of South America, has also been widely introduced throughout temperate and tropical regions for mosquito control (Juliano *et al.*, 1989).

Larvivorous fish for mosquito control should be small enough to move about in shallow waters, and be prolific breeders, even in confined spaces. They must be hardy enough to survive handling and transportation and have a preference for predating mosquito larvae even in the presence of alternative prey (Chandra *et al.*, 2008). They should also be compatible with existing aquatic life in the chosen environment, and the use of indigenous fish, rather than introduced species, is favoured where possible.

Outside of fish, other larvivorous species investigated as control agents include various copepods, such as *Mesocyclops aspericornis* (Lardeux *et al.*, 2002). Indeed, treatment of large water storage tanks, with local *Mesocyclops*, combined with source reduction, was associated with the elimination of *Aedes* mosquitoes from several communes in northern Vietnam (Kay *et al.*, 2002). Hemipteran predators have also been studied (Nam *et al.*, 2000; Valbon *et al.*, 2019), and even some plants, such as the carnivorous bladderwort, whose larvicidal efficiency for the mosquito vectors *Aedes aegypti* and *Aedes albopictus* has been investigated in no-choice, laboratory experiments (Couret *et al.*, 2020).

Evidence for Efficacy

The use of larvivorous fish to control *Anopheles* mosquitoes to prevent malaria transmission

Studies investigating impacts of interventions on mosquito populations are of limited value if they do not collect corresponding data on disease epidemiology, since a reduction in immature or adult mosquitoes does not directly result in a reduction in disease cases. A Cochrane review on the use of larvivorous fish to prevent malaria transmission identified no randomized controlled trials or quasi-experimental studies that reported confirmed episodes of malaria among community members or entomological inoculation rates (Walshe *et al.*, 2017). Indeed, even density of adult vector mosquitoes was not reported from such study designs. As a result, the review concluded that there was no direct evidence that indicates this intervention impacts malaria transmission in endemic communities.

Whether or not the introduction of larvivorous fish impacts immature anopheline mosquito presence or density has been more fully investigated, using larval habitats as diverse as domestic water containers, wells and rice fields (Yu and Lee, 1986; Haq and Srivastava, 2013), and provides some indirect evidence of the potential effectiveness of this intervention (Walshe *et al.*, 2017).

In the first 120 days following the introduction of the indigenous fish *Aphanius dispar* to domestic tanks in Gujarat, India, reductions in *Anopheles stephensi* larval and pupal densities were found to range between 84 and 100% (Haq and Srivastava, 2013). For the remainder of the year-long controlled time series study, reductions varied between 71 and 100% without the need for restocking. The impact of the fish on the larval density was highly significant relative to untreated control tanks. The use of fish in domestic tanks was considered appropriate for *An. stephensi* because it is an urban vector of malaria in India that selects tanks, wells and temporary water storage containers for breeding (Thomas *et al.*, 2016). *Anopheles stephensi* was also the target vector species in earlier studies conducted in India using the non-indigenous fish *G. affinis* and *P. reticulata* (Sitaraman *et al.*, 1975, 1976). In each study, the use of the introduced fish led to reductions in larval densities, but *P. reticulata* were found to be less efficient predators of anopheline larvae, and the authors concluded that they would have to be introduced and maintained in large numbers for effective larval control (Sitaraman *et al.*, 1976).

Both wells and domestic water containers were treated with *A. dispar* in a randomised controlled trial in Ethiopia (Fletcher *et al.*, 1992). The fish were found to be equally effective in controlling mosquito larvae in all the types of breeding site investigated, with the overall proportion of sites with fish that harboured mosquito larvae being 2%, as compared to 34% in sites left unstocked. These results are comparable to those reported from a controlled interrupted time-series study in India, where the number of wells containing anopheline larvae was reduced from 33% before the introduction of fish to less than 1% one month later. The non-indigenous *G. affinis* was found to have a relatively low tolerance for high salinity and pH and was replaced in some wells with the indigenous *Aplocheilus blochii* (Menon and Rajagopala, 1978).

In the Republic of Comoros, where *Anopheles gambiae* s.s. breeds in man-made water reservoirs due to the high permeability of the soil, all larval breeding places existing in the village of Hantsambou were provided with imported *P. reticulata* or a temephos treatment. The village of Bandamadji, 3 km farther, provided a control arm to the study. In those containers treated with fish, the percentage positive for *An. gambiae* decreased from 41% at baseline to 6% after 1 year (Sabatinelli *et al.*, 1991). The parasite index for *Plasmodium falciparum* and spleen rate in 5–9 year-old school children was also found to be reduced significantly over the course of the study between the treatment and control villages, but because the intervention was coupled with an insecticide treatment, it is not possible to disentangle the effects of the fish alone.

There is limited evidence for reductions in anopheline larval densities when fish have been introduced to larger bodies of water, such as fishponds, pools and canals (Mahmoud, 1985; Howard *et al.*, 2007; Imbahale *et al.*, 2011). *Oreochromis niloticus* (formerly *Tilapia nilotica*) were introduced into abandoned fishponds in a controlled study in Kenya. Mosquito densities immediately dropped in the treated ponds, and after correction for a natural increase in the control pond, it was determined that the introduction of fish caused a highly significant reduction of *An. gambiae* s.l. (Howard *et al.*, 2007). Although these results are encouraging, particularly given that the fish are native to southern Africa and did not need to be restocked, there was a large seasonal effect (Walshe *et al.*, 2017). It should also be noted that only two treated ponds and one control pond were used in the study, and ecological changes in the control pond may have contributed to the observed effects. In Kenyan man-made canals, the population of *An. gambiae* reduced by 50% following the introduction of *G. affinis*, but greater reductions were seen when *Bacillus thuringiensis israelensis* (*Bti*) was used alone or in combination with the fish (Imbahale *et al.*, 2011).

In rice fields, the combined release of indigenous *Aplocheilus latipes* and the African cichlid fish *Tilapia mossambicus niloticus* in Korea resulted in a maximum suppression of *Anopheles sinensis* larvae of approximately 80%, recorded 5 weeks after introduction (Yu and Lee, 1986). The two fish species were used at densities of 2 fish per m^2 and 2 fish per 10 m^2, respectively, in this controlled interrupted time series study. In a subsequent study of the same design, *Aplocheilus* found naturally breeding at a density of 0.8 fish per m^2 were responsible for a reduction in *An. sinensis* larvae by approximately 50%. These fish were later supplemented with *Tilapia* to improve mosquito control (Kim *et al.*, 2002). Finally, in Tajikistan, *P. reticulata* introduced to rice fields reduced the larval density to similar levels as achieved by *G. affinis*, but the effects were less sustained (Zvantsov *et al.*, 2008; Walshe *et al.*, 2017). The authors noted that *P. reticulata* had to be restocked because they had poor survival rates compared to *G. affinis*, which bred successfully.

The use of larvivorous fish to control other mosquito genera and the pathogens they transmit

A variety of fish species have been used to eliminate *Aedes* mosquitoes from the large containers, open freshwater wells, concrete irrigation ditches and industrial tanks in which they mature. The viviparous species *P. reticulata* adapts well to these types of confined water bodies and has been most commonly used (World Health Organization, 2009), but other species have been investigated, including inland silverside (*Menidia beryllina*), killifish (*Rivulus marmoratus*), *Astronotus ocellatus* and *Macropodus opercularis* (Middaugh *et al.*, 1985; Consoli *et al.*, 1991; Taylor *et al.*, 1992).

As with *Anopheles*, there is evidence that some fish species are effective predators of *Aedes* larvae (de Góes Cavalcanti *et al.*, 2007), but evidence on the efficacy of larvivorous fish in the control of *Aedes* populations in the field is limited. The study of *A. dispar* in Gujarat, described above, monitored the impact of fish in domestic tanks on *Ae. aegypti* mosquitoes as well as on *An. stephensi*. The controlled time series study found similar rates of reductions of both species and concluded that the use of *Aphanius* fish has the potential to limit populations of these mosquitoes throughout the year without reapplication (Haq and Srivastava, 2013). However, as noted above, only larval densities were reported and not adult densities.

Further studies have indicated that the presence of introduced larvivorous fish has been associated with reductions in *Aedes* larval indices in Taiwan (Wang *et al.*, 1990), China (Wu *et al.*, 1987), Mexico (Martínez-Ibarra *et al.*, 2002) and Brazil (De Oliveira Lima *et al.*, 2010), but these studies have not demonstrated a direct impact on disease. For example, in Karnataka, India, *G. affinis* and *P. reticulata* were introduced in indoor cement tanks in an intervention that also included information, education and communication campaigns. Dramatic reductions in chikungunya cases were reported after the combined intervention, but no control villages were used in the trial (Ghosh *et al.*, 2011). In this particular study, where fish were used alone, no sustained improvement in the Breteau Index was reported.

In a controlled intervention study in Cambodia, community volunteers distributed *P. reticulata* to the water jars and storage tanks of approximately 1000 households. After 1 year, the container index of eligible containers in the intervention area was approximately 80% lower than that in the control area (Seng *et al.*, 2008). In addition, small or discarded containers that were considered unsuitable for guppy distribution in the intervention area also had 51% less infestation than those in the control area, suggesting an overall reduction in mosquito numbers and a community-wide protective effect (Seng *et al.*, 2008).

In a study reporting human bait collections as well as mosquito larval densities, the introduction of *G. affinis* and *P. reticulata* to open wells and ponds in French Polynesia failed to reduce mean adult *Aedes* catches post-treatment beyond a downward trend observed throughout the pretreatment period (Lardeux, 1992). Larvae disappeared from all open wells and ponds following the introduction of the fish but remained in smaller bodies of water more suitable for *Aedes* maturation. Copepods introduced to these sites had limited impacts.

Morales-Pérez attempted to identify associations between household fish use in water containers and dengue risk following a cluster-randomised controlled trial in Guerrero, Mexico (Morales-Pérez *et al.*, 2017). A total of 1730 out of 10,111 (17%) households had fish in their water containers. The presence of fish was associated with lower levels of recent dengue virus infection in children aged 3–9 years (odds ratio 0.64). Nonetheless, there is a lack of appropriate studies from which evidence can be drawn for the efficacy and community effectiveness of larvivorous fish in the control of *Aedes*-borne diseases (Han *et al.*, 2015).

To control *Culex* mosquitoes, the native tiger barb *Puntius tetrazona* and rosy tetra *Hyphessobrycon rosaceus* were added to ditches in West Bengal, India, and removed after 30 days. All larvae were eliminated in the ditches containing *H. rosaceus*, and there was a 78% reduction in larval density in those containing *P. tetrazona*. Both reductions were significant compared to control ditches, which maintained approximately 250 larvae/dip (Barik *et al.*, 2018). In Australia, two indigenous species, *Melanotaenia duboulayi* and *Hypseleotris galii*, were introduced to small plots for field evaluations, but failed to adequately reduce *Culex* larval populations, despite being used at high densities (>30 kg/ha). However, when supplemented with a *Bacillus sphaericus* treatment, the fish were able to prevent recolonization by *Culex annulirostris*, which was not possible using a treatment of *B. sphaericus* alone (Hurst *et al.*, 2006). In the controlled interrupted time series in Kenya, described above, a 75% reduction in culicine mosquitoes was reported in fishponds treated with *O. niloticus* (*T. nilotica*; Howard *et al.*, 2007). This compared to a reduction of approximately 95% for *Anopheles* species. The reductions in *Culex pipiens* and *Culex tritaeniorhynchus* following the introduction of both native and non-indigenous fish into rice fields in Korea were similar to those of *An. sinensis*, although larvae of these species were much less abundant (Yu and Lee, 1986).

In small Korean rice plots, single mud loaches (*Misgurnus mizolepi*) consumed within 60 hour all 1200 3rd instar larvae of *Cx. pipiens pallens* or *Culex inatomii* provided to them (Lee, 2000), and the same species consumed similar quantities of *Cx. pipiens molestus* in aerobic septic tanks over 24 hour (Chae *et al.*, 2014). Over 3 years, fathead minnows (*Pimephales promelas*) were able to establish self-sustaining populations and suppressed levels of *Culex tarsalis* mosquito larvae in reservoirs in Wyoming, USA, where adults transmit West Nile virus (Watchorn *et al.*, 2018). These studies provide further evidence that larvivorous fish have the potential for mosquito control in these environments, but again, there is a need for appropriate controlled studies, ideally with epidemiological outcomes to determine the impact they have, if any, on the transmission of filaria and arboviruses.

Limitations as an Intervention

Biodiversity and ecosystem risks

The primary concern with the introduction of non-native fish for mosquito control is the risk of adverse environmental impact. It has been argued that *Gambusia* are among the most voracious and destructive of fish and, where introduced for purposes of mosquito control, endanger native aquatic organisms (Rupp, 1996; Hamer *et al.*, 2002). Given that these fish are omnivores, they are expected to have far broader tastes than are revealed in univariate laboratory experiments, so aside from their predation of mosquito larvae, they may also eat the eggs and young of other fish and amphibians (Komak and Crossland, 2000; Kumar *et al.*, 2015).

In Australia, *G. holbrooki* was introduced to waterways in the 1920s and has been directly linked with significant reductions in native fish populations (Arthington and Marshall, 1999). Open-air tank experiments have shown that this species profoundly affected the breeding of the Pacific blue-eye *Pseudomugil signifer*. In the presence of *G. holbrooki*, *P. signifer* did not gain weight or grow in total length, ovaries remained morphologically undeveloped and fecundity was greatly reduced (Howe *et al.*, 1997). Further, native fish species have been found in the gut of wild-caught *G. holbrooki* (Ivantsoff and and Aarn, 1999). *Pseudomugil signifer* is itself a larvivorous fish, having consumption rates in laboratory experiments that are comparable to those of *G. holbrooki* when fed *Aedes vigilax* and *Cx. annulirostris* larvae (Willems *et al.*, 2005; Griffin, 2014), making a strong argument for using such native species rather than exotic pest species for the purposes of mosquito control. In any case, careful assessments under natural habitat conditions should be made before any fish are released.

Community acceptance and engagement

The willingness of local communities to accept the introduction of fish into water containers is essential and should be established prior to any intervention (World Health Organization, 2009). Indeed, community members can assist with the distribution of fish, as well as monitoring and restocking, and are expected to reduce the costs of this control method (Kusumawathie *et al.*, 2008).

The acceptance of fish as an intervention has been qualitatively assessed in communities in Cambodia, where *P. reticulata* were used in combination with pyriproxyfen (Shafique *et al.*, 2019). The majority of participants in intervention villages preferred the guppy fish over other vector control methods due to ease of use and rearing, quick reproduction and propensity to eat larvae. The use of fish was also well accepted in a randomised controlled trial in Ethiopia, where study participants were aware of the role of the fish in malaria prevention and found the fish useful in keeping their water free of other aquatic organisms (Fletcher *et al.*, 1992). Others have noted that there should be no problem with acceptance of *T. nilotica* for malaria control in Kenya, since the local communities already farm this fish species as a source of protein and income (Howard *et al.*, 2007).

Other considerations

Other challenges associated with the use of fish for mosquito control include the expense and task of rearing the fish on a large scale, and their limited utility in some aquatic sites where temperature, pH and organic pollution may exceed the narrow requirements of the species (World Health Organization, 2009). Many sites may be too transitory, too anaerobic or have less obvious deterrents, such as piscivorous fish. Even the presence of floating debris can inhibit the efficacy of fish (Phan-Urai *et al.*, 1976). Early studies in Asia reported that, although guppies were of value in mosquito control, some sites were vulnerable to unauthorized fish harvesting, as residents netted the fish for use as duck food (Bay and Self, 1972).

A particular requirement for fish used in the control of *Aedes* mosquitoes is that the species may need to be able to survive in the chlorinated water supplied by public water systems, and studies have been conducted to identify which fish are most tolerant to chlorine (Cavalcanti *et al.*, 2009). Where temporary pools of water in remote areas dry up seasonally, the annual killifish *Nothobranchius guentheri* has been proposed as a solution; these fish can undergo diapause during the embryonic stages to survive periodic drought and so maintain permanent populations when other fish would disappear. The introduction of juveniles or embryos of *N. guentheri* to ponds with high densities of *Culex quinquefasciatus* larvae has been shown to result in total eradication of the mosquito population in semifield studies (Matias and Adrias, 2010).

There is the possibility of mosquitoes responding to the presence of fish in their oviposition sites. Studies in the flood plain of The Gambia River revealed that fewer culicines oviposited in sites with fish, suggesting that ovipositing culicine females may use fish chemical cues to avoid water that contains fish (Louca *et al.*, 2009). These findings are supported by laboratory studies – binary choice experiments to investigate whether fish exudates from *G. affinis* deter oviposition by mosquitoes found that *Cx. tarsalis* was deterred significantly from egg laying by the presence of fish exudates in oviposition cups. By contrast, ovipositing by *Cx. quinquefasciatus* was slightly reduced, and the container-breeding species *Ae. aegypti* was not deterred from egg laying by the presence of the fish exudates (Van Dam and Walton, 2008). These observations correlate with each species' risk of predation by fish in natural habitats; from an evolutionary perspective, fish would not be a strong selective agent for a species that prefers containers and polluted water, where fish are not frequently encountered.

These studies have important implications for the use of larvivorous fish in the control of mosquito vectors, as they reveal that fish not only have a role in eating larvae but also in deterring egg laying. If mosquitoes have a strong response to the presence of fish in their breeding sites, they may be diverted to nearby sites that are not suitable for treatment with fish, such as small water containers (Angelon and Petranka, 2002). It will be important in these situations that the alternative sites are the focus of education campaigns and are regularly cleaned or emptied. Further, the responses of target mosquitoes to selected larvivorous fish must be determined experimentally, as fish-associated kairomones do not elicit consistent responses across mosquito species (Van Dam and Walton, 2008; Hurst *et al.*, 2010).

Use of fish with other mosquito control interventions

Integration of larvivorous fish with biolarvicides may result in a safer and more effective tool. However, the usefulness of this integrated approach depends upon the survival and ecological fitness of the larvivorous fish employed, since any direct effects of larvicides on the fish may diminish their potential in mosquito control. A study investigating possible toxicological effects of sublethal concentrations of pyriproxifen on *Xiphophorus maculatus* fish indicated that application of the larvicide at a concentration reported to control the emergence of *Ae. aegypti* larvae may decrease the swimming performance of larvivorous fish and their ability to ingest *Ae. aegypti* larvae (Caixeta *et al.*, 2016). Similarly, pirimiphos-methyl used at the effective field concentration was found to significantly reduce the critical swimming speed of crimson-spotted rainbowfish, *M. duboulayi*. However, exposure to *Bti*, *B. sphaericus* and the insect growth regulator s-methoprene larvicides at 10 times their effective field concentration had no effect on the critical swimming speed of *M. duboulayi*, suggesting these are more suitable options for integrated pest management programmes (Hurst *et al.*, 2007). It is, therefore, important that such interactions between interventions are suitably characterized before they are implemented in communities.

Conclusions

The use of fish for purposes of controlling mosquitoes has a history of locally dramatic, but typically temporary, success in decreasing larval numbers (Gachelin *et al.*, 2018). More recent and rigorous investigations continue to report a high rate of larval predation in laboratory experiments, and large reductions in mosquito densities in some semifield studies, but the evidence for the effectiveness of fish in preventing disease transmission remains insufficient (Han *et al.*, 2015; Walshe *et al.*, 2017). While cluster-randomised controlled trials are typically recommended for the generation of robust evidence, demonstrating the effect of larvivorous fish as a scalable intervention comes with many challenges, and our desire for this formal testing may need to be revised.

The limitations identified in early studies, including that *Gambusia* do not always breed well and have to be restocked several times a year (Gachelin *et al.*, 2018), have been exacerbated by a recognition of the potential harmful effects these fish can have on ecosystems (Rupp, 1996). The chances of accidental introduction of fish into local aquatic ecosystems, and the damage they may cause once established, must be weighed against the benefits of using this approach.

The future of the introduction of larvivorous fish for mosquito control may lie in their complementary use with other interventions and the employment of new technologies. For example, a global positioning system technology to facilitate rapid rechecking of release sites for the continuing presence of fish may improve sustainability (Ghosh and Dash, 2007), and where appropriate, native fish could be used in some water containers to reduce the reliance on chemical treatment. This, combined with public awareness and engagement, could form a value component of a broader integrated vector management programme (Das *et al.*, 2018).

Zooprophylaxis

Livestock breeding is a major economic activity in rural malaria-endemic areas, potentially providing ample hosts that can serve as alternate blood meal sources for mosquitoes, especially as humans become inaccessible due to the deployment of routine insecticidal interventions, including insecticide-treated nets (ITNs)/long-lasting insecticidal nets (LLINs) or indoor residual spraying (IRS).

Zooprophylaxis was first recommended by WHO in 1982 as a potential component of integrated vector management (IVM) schemes, using wild or domestic animals (usually cattle or other livestock) as decoy or dead-end hosts to divert vectors from biting humans (World Health Organization, 1982). By reducing human blood feeding, zooprophylaxis can impact disease transmission through modification of different parameters related to vectorial capacity, including mosquito density, longevity, fecundity, biting frequency and forage-related mortality. IVM has been promoted as a sustainable approach to malaria control involving coordinated deployment of chemical and non-chemical strategies, targeted to specific ecological settings to maximize efficiency while minimizing cost and adverse environmental impacts (World Health Organization, 2004; World Health Organization, 2020). In this regard, zooprophylactic effects can be passive, active or combined with insecticidal public health interventions (World Health Organization, 1982)

Passive zooprophylaxis can reduce disease prevalence as a serendipitous result of the normal presence of animals in a community and is measured as the association between domestic animal ownership and parasitaemia (Bulterys *et al.*, 2009; Yamamoto *et al.*, 2009), vector blood meal index (Adugna, 1996; Hadis *et al.*, 1997; Seyoum *et al.*, 2002; Massebo *et al.*, 2013), sporozoite rate (Iwashita *et al.*, 2014) or density (Mahande *et al.*, 2007a; Muriu *et al.*, 2008; Kaburi *et al.*, 2009). Active zooprophylaxis involves strategic positioning of livestock and/or livestock shelters between mosquito breeding sites and human settlements (World Health Organization, 1982; Hewitt *et al.*, 1994; Tirados *et al.*, 2011) and is evaluated using paired cohorts of individuals living with or without cattle in close proximity or with cattle kept at known distances (Bøgh *et al.*, 2001, 2002). Combination zooprophylaxis may describe the interactions between ITN/LLIN or IRS coverage, livestock ownership and risk of malaria (Kaburi *et al.*, 2009; Mahande et al. 2007b; Iwashita *et al.*, 2014; Hasyim *et al.*, 2018), or the impact of insecticidal treatment of cattle by sponging or dipping on vector mortality, density and malaria incidence in households with treated or untreated animals (Hewitt and Rowland, 1999; Rowland *et al.*, 2001; Habtewold *et al.*, 2004).

Passive zooprophylaxis was described anecdotally as early as 1903, where domestic animals in northern Italy indirectly protected humans from mosquito bites and has been credited with playing a major role in the elimination of malaria from Europe and much of the USA following an increase in livestock keeping (Bruce-Chwatt, 1982; Service, 1991). Correspondingly, in some areas where livestock densities were reduced, an increase in malaria prevalence was also reported (Service, 1991). In Guyana, after DDT IRS successfully eradicated *Anopheles darlingi* in the 1950s, a shift in agricultural practices towards crop production led to a dramatic decline in livestock numbers, accompanied by several malaria outbreaks 16 years later, attributed to a switch in the behaviour of formally zoophilic *Anopheles aquasalis* to feed on humans (Giglioli, 1963). A fivefold upsurge in malaria, transmitted by the highly zoophilic vector *Anopheles aconitus*, was observed in Indonesia, when a poor rice crop forced farmers to sell their water buffaloes, reducing the cattle to human ratio from 1:25 to 1:50. Some of the earliest examples of active zooprophylaxis come from the USSR, where health administrators advised that livestock sheds should be arranged in a continuous line along the periphery of human settlements, with houses built 250–300m away from them, to control biting mosquitoes (Brumpt, 1945).

In general, the effectiveness of zooprophylaxis is predicated on vector species behaving opportunistically, with some proclivity to feed on domestic animals instead of humans (Iwashita *et al.*, 2014). Mosquito blood meal choice is generally influenced by several factors, including host availability, nutritional requirements, intrinsic host preference and vector density (Takken and Verhulst, 2013). In practice, there is contradicting evidence for the efficacy of zooprophylaxis (Kirnowordoyo and Supalin, 1986; Service, 1991; Bourna and Rowland, 1995; Ghebreyesus *et al.*, 2000; Bøgh *et al.*, 2001, 2002; Charlwood, 2001; Saul, 2003; Donnelly *et al.*, 2015; Asale *et al.*, 2017). It has been argued that the use of domestic animals results in the polar effect of zooprophylaxis – zoopotentiation, where livestock attract vectors towards humans, or alter vector density and lifespan, thereby increasing the likelihood of disease transmission (Bourna and Rowland, 1995; Saul, 2003).

Key Determinants for Zooprophylaxis or Zoopotentiation

Table 12.1 presents key studies summarizing the evidence for zooprophylaxis or zoopotentiation among major anopheline mosquito species. The majority of research has focused on *Anopheles arabiensis*, a primary vector of malaria across eastern and south-east Africa (Sinka *et al.*, 2012). Compared to its more strictly anthropophilic and endophilic counterpart *An. gambiae* sensu stricto, *An. arabiensis* can both rest and feed indoors and outdoors depending largely on host location and coverage of insecticidal interventions. This species also exhibits opportunistic anthropophagic and zoophagic behaviour (Tirados *et al.*, 2006; Fornadel *et al.*, 2010; Massebo *et al.*, 2015). In areas where hosts of *An. arabiensis* principally reside outdoors, and/or there has been recent deployment of LLINs or IRS, this species may shift from endophilic to exophilic behaviour in response to host availability and deterrence/contact irritancy

Table 12.1. Summary of key studies investigating the evidence for zooprophylaxis in major malaria vector species.

Study site, Country	Study design	Major malaria vector species	Outcome(s)	Study findings	References
Ziway, Ethiopia	Mosquito collections via CDC light traps (CDC-LTs) and human-landing catches (HLCs) at three sites: Mixed human-cattle dwelling Separate human-cattle dwelling (cattle shed) Human-only dwelling	*An. arabiensis* *An. pharoensis*	Malaria incidence in children < 10 years	Zoopotentiation in mixed dwellings: higher *An. arabiensis* human biting rate (HBR) in the presence of cattle or goats vs without livestock (3.50 [95% CI: 1.34–5.67], 3.38 [95% CI: 1.47–5.30] vs 1.43 [95% CI: 0.52–2.33]). No clear *An. pharoensis* HBR pattern due to lower biting density. Zooprophylaxis in separate human-cattle dwellings: mean parasite rates of 26.67% [95% CI: 17.9–37.0], 15.05% [95% CI: 10.2–21.0] and 23.85% [95% CI: 18.4–30.1] in children living in mixed dwellings, separate human-cattle dwellings or without livestock, respectively.	(Seyoum *et al.*, 2002)
Kilombero valley, South-East Tanzania	Reared *An. arabiensis* released among fungus-treated or untreated cattle in semifield system and small-scale field trials.	*An. arabiensis*	Proportion of mosquitoes recaptured and classified as alive, fed, fertile and infected (with fungus)	90% of *An. arabiensis* fed on fungus-treated calves and became infected immediately; 70% of *An. arabiensis* fungus infections occurred after 3 days.	(Lyimo *et al.*, 2012)

Continued

Table 12.1. Continued

Study site, Country	Study design	Major malaria vector species	Outcome(s)	Study findings	References
Mwea, Central Kenya	Mosquito collections using pyrethrum spray catch (PSC), from houses with or without LLINs, stratified by number of cattle.	*An. gambiae* s.l.; *An. funestus* s.l.	HBR, blood meal analysis and sporozoite rate	Zooprophylaxis: significant protection from mosquito bites and decrease in human blood meal index (HBI) in houses with LLINs and four or less cattle. No significant protection with more than five animals.	(Kaburi *et al.*, 2009)
Macha, Zambia	Nested case-control malaria survey, including documented animal ownership and other malaria risk factors.	*An. arabiensis*	Malaria incidence	Zooprophylaxis: cattle ownership associated with reduced *P. falciparum* infection (aOR = 0.19 [95% CI: 0.05–0.69]). Risk of *P. falciparum* infection (aOR = 0.13 [95% CI: 0.03–0.56]) was reduced with household domestic animals (cats and dogs) or livestock (cattle and goat ownership). Effect observed with ownership of ≥1 cow, ≥1 goat/person, >0.3 dogs/person or ≥0.13 cats/person.	(Bulterys *et al.*, 2009)
Mwea, Central Kenya	Mosquito collections from houses by CDC-LTs and PSC.	*An. arabiensis; An. coustani; An. pharoensis; An. funestus* s.l.	Mosquito density and blood meal analysis	Zooprophylaxis: proportions of *An. arabiensis* fed on bovine significantly higher indoors (71.8%) than outdoors (41.3%). For *An. funestus* s.l., significantly less were fed on bovine indoors (51.4%) than outdoors (78%).	(Muriu *et al.*, 2008)

Lower Moshi, North Tanzania	Mosquito collections using odour-baited entry traps (OBETs), CDC-LTs and indoor PSCs and hand catches in pit traps outdoors. Mosquito collections from experimental huts to observe resting behaviour.	*An. arabiensis; An. gambiae* s.s.	Mosquito density and blood meal analysis	Zooprophylaxis: 90.3% of mosquitoes were trapped by cattle odour and 9.7% were trapped by human odour (p=0.0005). Low attraction to odours from sheep (9.7%), goat (7.2%) or pig (7.3%) compared to cattle. Lower mosquito densities collected from houses with cattle (mean = 22.5) compared to those without livestock (mean = 9.0). A lower HBI was recorded in both outdoor (0.1–0.3) and indoor (0.4–0.9) collected mosquitoes.	(Mahande *et al.*, 2007a,b)
Lower Moshi, North Tanzania	Mosquitoes exposed to deltamethrin sprayed grazing or non-grazing cattle.	*An. arabiensis*	Mosquito knock-down/mortality and blood meal analysis	50% of mosquitoes which fed on grazing cattle died for 21 days after deltamethrin treatment. Treated cows caused higher mosquito mortality (mean = 2) compared to untreated cows (mean = 0.3) or humans (mean = 0.8).	(Mahande *et al.*, 2007a,b)
Kisumu, Kenya	Mosquitoes exposed to ivermectin-treated cattle or untreated cattle.	*An. gambiae* s.s.; *An. arabiensis*	Mosquito survivorship and fecundity	90% of *An. gambiae* s.s. that fed on ivermectin-treated cattle within 2 weeks of treatment failed to survive more than 10 days post blood meal. No eggs were deposited by *An. gambiae* s.s. that fed on ivermectin-treated cattle within 10 days of treatment.	(Fritz *et al.*, 2009)

Continued

Table 12.1. Continued

Study site, Country	Study design	Major malaria vector species	Outcome(s)	Study findings	References
Mbita, Kenya	Mosquito collections from houses with high ITN coverage (>80%), using PSC, and livestock census.	*An. arabiensis; An. gambiae* s.s.; *An. funestus* s.s.	Mosquito density, blood meal analysis and sporozoite rate	Zooprophylaxis: proportion of *An. gambiae* s.s., *An. funestus* s.s. and *An. arabiensis* fed on cattle were 1.2% [95% CI: 0.6–2.3], 2.0% [95% CI: 0.6–5.1] and 40.5% [95% CI: 36.9–44.2], respectively.	(Iwashita *et al.*, 2014)
				Proportion of *An. gambiae* s.s., *An. funestus* s.s. and *An. arabiensis* fed on humans were 49.5% [95% CI: 42.3–56.7], 63.9% [95%CI: 60.2–67.4] and 12.0% [95% CI: 9.7–14.6], respectively.	
				An. arabiensis abundance halved by ITN use.	
				An. arabiensis HBI significantly increased 1.01 times for each goat/sheep tethered within 500m from a household and decreased with the number of cattle tethered in the same radius.	
				Sporozoite rate reduced by 9.2% as proportion of *Anopheles* feeding on cattle increased by 10%.	

Arba Minch, Ethiopia	Mosquitoes collected from three different sites: A: Separate human-cattle dwelling (cattle shed) B: Mixed human-cattle dwelling C: Separate human-cattle dwelling (humans on elevated platforms)	*An. arabiensis; An. quadriannulatus*	Blood meal analysis and sporozoite rate	Zooprophylaxis in separate human-cattle dwellings; proportion of mosquitoes that fed on livestock significantly higher than on humans in site C (93.7% vs 3.1%). No effect: proportion of mosquitoes that fed on humans and livestock did not differ significantly in sites A or B. Higher proportions of cattle feeding in site C (93.7%) vs sites A (42.7%) and B (54.7%).	(Habtewold *et al.*, 2001)
North-West Frontier province, Pakistan	Community randomised trial: three Afghan refugee settlements received deltamethrin treatment of livestock; three were untreated controls.	*An. stephensi; An. culicifacies*	Mosquito mortality and parity, parasite prevalence	Zooprophylaxis: 56% [95% CI: 14–78%] reduction in *P. falciparum* malaria incidence; 31% [95% CI: 5–50%] reduction in *P. vivax* malaria incidence. In treated villages, *An. stephensi* and *An. culicifacies* densities reduced by 47 and 46%, respectively. In treated villages, *An. stephensi* and *An. culicifacies* parity reduced by 27 and 22%, respectively.	(Rowland *et al.*, 2001)
North-West Frontier province, Pakistan	Mosquito collections using HLCs in the presence of a cow, a goat or human only.	*An. stephensi*	Mosquito density	Zoopotentiation: man-biting by *An. stephensi* increased by 38% [95% CI: 8–68%] in the presence of a cow and by 50% [95% CI: 16–84%] in the presence of two goats.	(Hewitt *et al.*, 1994)

Continued

Table 12.1. Continued

Study site, Country	Study design	Major malaria vector species	Outcome(s)	Study findings	References
North-West Frontier province, Pakistan	Cross-sectional malaria prevalence surveys, including questionnaires to identify individual and household malaria risk factors.	*An. stephensi; An. culicifacies; An. subpictus*	Malaria incidence in children 4–15 years	Zoopotentiation: parasite prevalence (15.2%) significantly greater among children of families which kept cattle compared to those who did not (9.5%) ('compound effect'). Villages with a greater proportion of cattle-owning families recorded higher malaria prevalence than villages with fewer cattle owners ('village effect').	(Bourna and Rowland, 1995)
Zambezia, Mozambique	Cross-sectional malaria prevalence surveys, including questionnaires to identify individual and household malaria risk factors.	*An. gambiae* s.l.; *An. funestus* s.l.	Malaria incidence in children 1–15 years	Zoopotentiation: increased risk of malaria incidence (OR = 3.2 [95% CI: 2.1–4.9]) associated with pig keeping.	(Temu *et al.*, 2012)

Arba Minch, Ethiopia	Experiment 1: HLC outdoors with a ring of 20 cattle (10m radius) Experiment 2: human-baited trap (HBT) and cattle-baited trap (CBT) with or without a ring of cattle Experiment 3: HLC indoors or outdoors Experiment 4: HLC indoors with or without ring of cattle outdoors	*An. arabiensis; An. pharoensis*	Mosquito density / trap	Zooprophylaxis: indoor catch of *An. arabiensis* significantly reduced by 49% with cattle ring (exp. 4). No significant effect on *An. arabiensis* density when humans are outdoors (exp. 1; mean = 24.8 and 37.2 mosquitoes/night, $p>0.22$, caught outdoors by HLC with or without a ring of cattle, respectively). Zooprophylaxis: significantly less *An. pharoensis* when humans are outdoors (exp. 1; mean = 4.9 vs 8.7 mosquitoes/night, $p<0.05$, caught outdoors by HLC with or without a ring of cattle, respectively). *An. arabiensis* density in HBTs was 25 times greater than in CBT (exp. 2; 34.0 vs 1.3, $p<0.001$), whereas for *An. pharoensis* there was no significant difference.	(Tirados *et al.*, 2011)
Nouna, Burkina Faso	Case-control study, including questionnaires to identify individual and household malaria risk factors. Household mosquito collections using CDC-LTs and PSCs.	*An. gambiae* s.l.; *An. funestus* s.l.	Mosquito density and parasite prevalence	Zoopotentiation: positive correlation between donkeys and *An. gambiae* abundance indoors (Pearson's R = 0.21; p=0.0002). No difference between malaria cases and controls associated with animal ownership after adjusting for bed net use and education level.	(Yamamoto *et al.*, 2009)

Continued

Table 12.1. Continued

Study site, Country	Study design	Major malaria vector species	Outcome(s)	Study findings	References
Nyamarambe, Kenya	Case-control study, including questionnaires to identify individual and household malaria risk factors.	*An. gambiae* s.l	Parasite prevalence	Zoopotentiation: cattle in long grass outside houses significantly increased malaria risk (OR = 1.81 [95% CI: 1.19–2.73], p=0.005).	(Githinji *et al.*, 2009)
Konso District, Ethiopia	Mosquito collections in a village (cattle:human ratio ~0.6:1) and a temporary cattle camp (cattle:human ratio ~17:1) using PSCs, pit shelter traps (PITs), HLCs, CBTs, CDC-LTs and OBETs.	*An. arabiensis*	Mosquito density / trap, blood meal analysis and sporozoite rate	Zoopotentiation: HBI for outdoor and indoor mosquitoes were 51 and 66%, respectively. Significantly more mosquitoes caught using HLCs vs CBTs (mean 163.8 vs 26.0 mosquitoes/trap/ night, respectively, p<0.001). 59–91% of resting mosquitoes took blood meals from cattle.	(Tirados *et al.*, 2006)
Tigray, Ethiopia	Cross-sectional malaria prevalence surveys, including questionnaires to identify individual and household malaria risk factors.	ND	Malaria incidence in children < 10 years	Zoopotentiation in mixed dwellings: animals sleeping inside houses was significantly associated with malaria incidence (aRR = 1.92 [95% CI: 1.29–2.85]). Cattle or sheep/goat ownership not associated with malaria infection (1–2 cows: aRR = 0.75 [95% CI: 0.39–1.45]; 3–4 cows: aRR = 1.18 [95% CI: 0.65–2.14]; ≥five cows: aRR = 1.18 [95% CI: 0.64–2.17]; 1–4 sheep/goats: aRR = 0.93 [95% CI: 0.58–1.50]; ≥five sheep/goats: aRR = 0.81 [95% CI: 0.54–1.22]).	(Ghebreyesus *et al.*, 2000)

Farafenni town, the Gambia	Paired cohort study of children living within 20m or >50m away from a cow. Household mosquito collections using PSCs.	*An. gambiae* s.l	Mosquito density, blood meal analysis and sporozoite rate	Zooprophylaxis: presence of cattle reduced HBI of *An. arabiensis* from 82 to 52% (p=0.039) but not for *An. gambiae* s.s. or *An. melas*. No significant difference between *An. gambiae* s.l. sporozoite rate in compounds with (0.97%) or without cattle (1.28%).	(Bøgh *et al.*, 2001)
Farafenni town, the Gambia	Paired cohort study of children living within 20m or >50m away from a cow.	*An. gambiae* s.l	Malaria prevalence in children < 7 years	No significant difference in parasite prevalence (OR = 1.69 [95%CI: 0.67–4.24], p=0.26) or high-density parasitaemia (OR = 0.73 [95% CI: 0.27–1.97], p=0.54) between children in compounds with or without cows, after adjusting for wealth.	(Bøgh *et al.*, 2002)
Sile, Ethiopia	Mosquitoes collected from HLCs next to: Deltamethrin-treated cattle Untreated cattle	*An. arabiensis* *An. pharoensis* *An. tenebrosus*	Mosquito density / trap	No effect of untreated ox on HLC for *An. arabiensis* but ox odour increased mosquito density (mean catch/person/night = 22 without cattle odour vs 32 with, p<0.05). For *An. pharoensis*, HLC was significantly reduced in the presence of untreated ox (mean catch/person/night = 50 without vs 26 with, p<0.01) but increased in presence of cattle odour (mean catch/person/night = 6 without vs 18 with, p<0.001).	(Habtewold *et al.*, 2004)

Continued

Table 12.1. Continued

Study site, Country	Study design	Major malaria vector species	Outcome(s)	Study findings	References
Ledi, Aliberti, Sille, Erbore, Itang and Jawe, Ethiopia	Mosquitoes collected from: Human dwellings Cattle sheds Mixed dwellings	*An. arabiensis*	Blood meal analysis	Zooprophylaxis: proportion of human blood meals identified was highest from *An. arabiensis* caught in human dwellings (91.5%), followed by mixed dwellings (20.2%) and cattle sheds (3.5%) (p<0.0001).	(Hadis *et al.*, 1997)
Khammouane Province, Lao PDR	Mosquitoes collected using CDC-LTs.	*An. philippinensis*	Mosquito density	Anophelines were more common in homes with cattle compared with those without (IRR = 2.32 [95% CI: 1.29–4.17], p=0.005). No significant effect of keeping large animals (pigs, cows or buffalo) underneath the house.	(Hiscox *et al.*, 2013)
Mataral, Bolivia	Mosquitoes collected using CDC-LTs, HLCs (indoors and outdoors with animal baits) and ORCs.	*An. pseudopunctipennis*	Blood meal analysis	*An. pseudopunctipennis* preferred small ruminants (goat/sheep; forage ratio 1.99 [95% CI: 1.80–2.19]) to equids (horse/donkey; 1.95 [95% CI: 1.38–2.52]) to humans (1.47 [95% CI: 1.25–1.69]) to cows (1.15 [95% CI: 0.65–1.66]). *An. pseudopunctipennis* avoided pigs (forage ratio 0.34 [95% CI: 0.20–0.48]) and chickens (0.03 [95% CI: 0–0.75]).	(Lardeux *et al.*, 2007)

Kumasi, Ghana	Mosquitoes collected using HLCs within and 20m away from: Pen A – with cattle and no net Pen B – with cattle and untreated net Pen C – with cattle and deltamethrin-treated net Pen D – no cattle and no net	*An. gambiae* s.s.; *An. ziemanni*	Mosquito density and sporozoite rate	Zooprophylaxis: presence of cattle reduced number of *An. gambiae* s.s. in HLCs by 66% ($p<0.0001$); no significant effect on *An. ziemanni*. Cattle presence did not influence HLCs performed 20m away.	(Maia *et al.*, 2012)
Baringo, Kenya	Mosquitoes collected using outdoor CDC-LTs and PSCs.	*An. arabiensis; An. funestus* s.l; *An. pharoensis; An. coustani*	Mosquito density, blood meal analysis and sporozoite rate	*An. arabiensis* density decreased in presence of animals (OR = 0.40, $p=0.031$) and increasing distance to animal shelters (OR = 0.88, $p<0.0001$).	(Mala *et al.*, 2011)
Mwea, Kenya	Cross-sectional malaria prevalence surveys, including questionnaires to identify individual and household malaria risk factors in two rice irrigated and two non-irrigated villages. Mosquitoes collected using PSCs.	*An. arabiensis*	Malaria prevalence in children 0–9 years, mosquito density and blood meal analysis	Zooprophylaxis: irrigated villages had higher *An. arabiensis* density (average number of mosquitoes collected per house = 33 and 17 vs 3 and 1, respectively) and significantly less malaria (prevalence of 0 and 9% vs 17% and 54%, respectively), than those without irrigation. Greater proportions of *An. arabiensis* feeding on cattle in irrigated (49 and 48%) vs non-irrigated villages (21 and 6%).	(Mutero *et al.*, 2004)

Continued

Table 12.1. Continued

Study site, Country	Study design	Major malaria vector species	Outcome(s)	Study findings	References
Antula, Guinea Bissau	Mosquitoes collected using indoor resting catches (IRCs) and outdoor UV LTs.	*An. gambiae* s.s; *An. melas*	Mosquito density	Zoopotentiation: presence of pigs indoors associated with increased mosquito abundance ($\chi^2 = 17.63$, $p<0.001$) but the presence of goats was not ($\chi^2 = 1.08$, $p<0.30$).	(Pålsson *et al.*, 2004)
North-West Frontier province, Pakistan	Mosquitoes collected using HLCs next to cattle treated with deltamethrin, permethrin or lambda-cyhalothrin.	*An. stephensi; An. culicifacies*	Mosquito mortality	No evidence of increased diversion of host-seeking mosquitoes from treated livestock to humans. Proportion of blood-fed survivors reduced by >50% for the first 2 weeks after livestock treatment with deltamethrin.	(Hewitt and Rowland, 1999)
Adami Tulu, Ethiopia	Cross-sectional malaria prevalence surveys, including questionnaires to identify individual and household malaria risk factors.	*An. arabiensis; An. pharoensis*	History of febrile illness in last 2 weeks in children < 5 years	Zoopotentiation: sharing a house with livestock significantly increased malaria risk (OR = 1.3 [95% CI: 1.1–1.6]).	(Deressa *et al.*, 2007)
North-West Frontier province, Pakistan	Cross-sectional malaria prevalence surveys, including questionnaires to identify individual and household malaria risk factors.	ND	Malaria prevalence in children 5–18 years	Higher malaria burden recorded among children of families which kept cattle (11.2%) compared to those who did not (7.1%).	(Idrees and Jan, 2001)

Kilombero Valley, Tanzania	Mosquitoes collected using ORCs, IRCs, resting catches in animal sheds and indoor CDC-LTs and livestock census.	*An. gambiae* s.l. (mainly *An. arabiensis*); *An. funestus* s.l.	Mosquito density, blood meal analysis and sporozoite rate	No significant difference in mean mosquito density in households with and without livestock. At houses with cattle, significantly more mosquitoes found resting inside cattle sheds than inside houses or in outdoor resting boxes. Zooprophylaxis: HBI of indoor *An. arabiensis* and *An. funestus* s.l. was significantly lower at households with livestock than without ($\chi^2 = 42.93$, $p<0.001$ and $\chi^2 = 28.44$, $p<0.001$, respectively). Zoopropohylaxis: significantly higher sporozoite rate of *An. gambiae* s.l. at houses without livestock than with ($\chi^2 = 4.62$, $p=0.03$).	(Mayagaya *et al.*, 2015)
Muheza, Tanzania	Mosquitoes exposed to L-lactic acid (Abate) – treated or untreated goat in semifield system.	*An. gambiae* s.s.	Mosquito recapture locations	Goat treated with L-lactic acid attracted *An. gambiae* s.s. at higher rates (70.24% [95% CI: 67.48–73.0%]) than untreated ones (29.76% [95% CI: 25.52–34.0%]). *An. gambiae* s.s. attracted to treated goat at equivalent degree (47.3%) to humans (52.7%).	(Kemibala *et al.*, 2020)

Continued

Table 12.1. Continued

Study site, Country	Study design	Major malaria vector species	Outcome(s)	Study findings	References
Awassa, Metahara, Ziway, Ethiopia	Mosquitoes collected from mixed dwellings and animal shelters.	*An. arabiensis; An. pharoensis; An. coustani; An. demeilloni; An. christyi; An. tenebrosus*	Blood meal analysis	Higher HBI in anophelines collected from mixed dwellings (range: 26–92%) compared to animal shelters (range: 9–48%). Zoopotentiation: higher positivity for human blood in *An. arabiensis* and *An. pharoensis* collected from mixed dwellings (88 and 84%) compared to animal shelters (43 and 9%), respectively.	(Adugna, 1996)
Mwea, Kenya	Mosquitoes collected indoors using PSCs in rice irrigated and non-irrigated villages.	*An. arabiensis; An. funestus s.l.; An. pharoensis; An. maculipalpis; An. pretoriensis; An. coustani; An. rufipes*	Mosquito density, blood meal analysis and sporozoite rate	Zooprophylaxis: HBI for *An. arabiensis* and *An. funestus* s.l. was significantly higher in the non-irrigated village compared with irrigated villages (12:1 and 5:1 human:cattle ratio, respectively). EIR for *An. funestus* s.l. 26-fold higher in non-irrigated compared to irrigated villages; attributed to *An. parensis* dominating in irrigated villages vs *An. funestus* s.s. in non-irrigated village. No significant difference in EIR for *An. arabiensis* between villages.	(Muturi *et al.*, 2008)
The Kombos, the Gambia	Case-control study of malaria prevalence, including questionnaires to identify household malaria risk factors. Mosquitoes collected indoors using CDC-LTs and PSCs.	*An. gambiae* s.l.; *An. melas*	Malaria prevalence, mosquito density, blood meal analysis and sporozoite rate	Zoopotentiation: compounds of children with severe malaria had significantly more animals than compounds of mild cases (OR = 1.86; χ^2 = 4.8, p=0.03).	(Adiamah *et al.*, 1993)

Kanungu district, Uganda	Cross-sectional malaria prevalence surveys, including questionnaires to identify individual and household malaria risk factors.	*An. funestus* s.l.; *An. gambiae* s.l.	Malaria prevalence	Zooprophylaxis: keeping animals inside the house at night decreased malaria prevalence (aOR = 0.29 [95% CI: 0.09–0.94]), after controlling for ethnicity, wealth and bed net ownership.	(Donnelly *et al.*, 2016)
Gilgel-Gibe dam, Ethiopia	Mosquito collections indoors using IRCs and CDC-LTs and outdoors using ORCs for all potential natural resting sites.	*An. arabiensis; An. coustani; An. demeiloni*	Blood meal analysis and sporozoite rate	Zooprophylaxis: *An. arabiensis* took significantly higher proportions of bovine (26.2%) and ovine blood (25.7%) than human (6%).	(Yewhalaw *et al.*, 2014)
Chano, Ethiopia	Mosquito collections indoors using PSCs and CDC-LTs and outdoors using ORCs for pit shelters.	*An. arabiensis; An. marshalli; An. garnhami; An funestus* s.l.; *An. pharoensis; An. tenebrosus*	Blood meal analysis	Zooprophylaxis: higher bovine blood index (BBI) (69%) than HBI (44%) in *An. arabiensis*. Significantly higher HBI in indoor resting *An. arabiensis* (38%) compared to outdoor resting (13%; $\chi^2 = 58.8$, $p<0.0001$). Proportion of bovine blood meals in *An. arabiensis* similar for indoor resting (65%), outdoor resting (68%) and CDC-LTs (72%) samples.	(Massebo *et al.*, 2013)

Continued

Table 12.1. Continued

Study site, Country	Study design	Major malaria vector species	Outcome(s)	Study findings	References
Chano, Ethiopia	Mosquito collections indoors using PSCs and CDC-LTs and outdoors using ORCs for pit shelters.	*An. arabiensis; An. marshalli; An. garnhami; An funestus* s.l.; *An. pharoensis; An. tenebrosus*	Blood meal analysis	Zooprophylaxis: *An. arabiensis* and *An. marshalli* showed strong preferences for bovine meal over human (despite threefold higher prevalence of humans); relative feeding preferences of *An. arabiensis* and *An. marshalli* to feed on cattle were 4.7 and 6 times higher than to feed on humans, respectively.	(Massebo *et al.*, 2015)
15 provinces, Indonesia	Nationwide cross-sectional malaria prevalence surveys, including questionnaires to identify individual and household malaria risk factors.	20 *Anopheles* species; most abundant was *An. vagus*	Malaria prevalence	Zoopotentiation: raising medium-size animals in the house was a significant predictor of malaria prevalence (aOR = 2.809 [95% CI: 2.207–3.575], $p<0.0001$), after adjusting for gender, age, access to community health facilities, sewage canal condition and use of mosquito nets and ITNs.	(Hasyim *et al.*, 2018)

aOR: adjusted odds ratio; aRR: adjusted rate ratio; BBI: bovine blood index; CBT: cattle-baited trap; CDC-LT: Centers for Disease Control and Prevention light trap; CI: confidence interval; EIR: entomological inoculation rate; HBI: human blood meal index; HBR: human biting rate; HBT: human-baited trap; HLC: human-landing catch; IRC: indoor resting collection; IRR: incidence rate ratio; ITN: insecticide-treated net; LLIN: long-lasting insecticidal net; LT: light trap; ND: not described; OBET: odour-baited entry trap; OR: odds ratio; ORC: outdoor resting collection; PDR: People's Democratic Republic; PIT: pit shelter trap; PSC: pyrethrum spray catch.

induced by indoor vector control interventions (Padonou *et al.*, 2012; Kreppel *et al.*, 2020).

In general, the outcomes of using zooprophylaxis as a malaria control strategy vary across endemic regions, requiring site-specific epidemiological evidence of its effectiveness prior to implementation. Fig. 12.2 summarizes key interacting environmental factors that influence zooprophylaxis.

There are three major factors for zooprophylaxis to be effective. First, the principal vector species must have some zoophilic and exophagic tendencies. Where the predominant mosquito species is highly anthropophilic and human hosts are available, keeping livestock nearby is unlikely to elicit zooprophylaxis. In general, there is evidence to support zooprophylactic control of *An. arabiensis* and *Anopheles pharoensis* in Africa (Habtewold *et al.*, 2001; Seyoum *et al.*, 2002; Mahande *et al.*, 2007; Bulterys *et al.*, 2009; Kaburi *et al.*, 2009; Tirados *et al.*, 2011) and *An. stephensi*, *Anopheles culicifacies*, *An. sinensis* and *Anopheles farauti* in Asia (Hewitt *et al.*, 1994; Rowland *et al.*, 2001; Hassanali *et al.*, 2008), with some caveats. In areas with sympatric malaria vector species, zooprophylactic effects may be variable across mosquito populations. For example, in irrigated rice fields in Kenya, where *An. arabiensis* and *An. funestus* s.l. co-occur, significantly higher proportions of *An. arabiensis* that had fed on bovine hosts were found indoors vs outdoors, while the opposite trend was observed for *An. funestus* s.l. (Muriu *et al.*, 2008). Similarly, in areas of *An. arabiensis* and *An. pharoensis* in Ethiopia, contrasting results have been reported with one species being more readily diverted from humans to feeding on livestock than the other, depending on study area (Seyoum *et al.*, 2002; Tirados *et al.*, 2011). Vector feeding behaviours can also be heterogeneous within a species; *An. arabiensis* populations in Ethiopia have both been identified with high proportions fed on cattle (>50%) and found to bypass a ring of

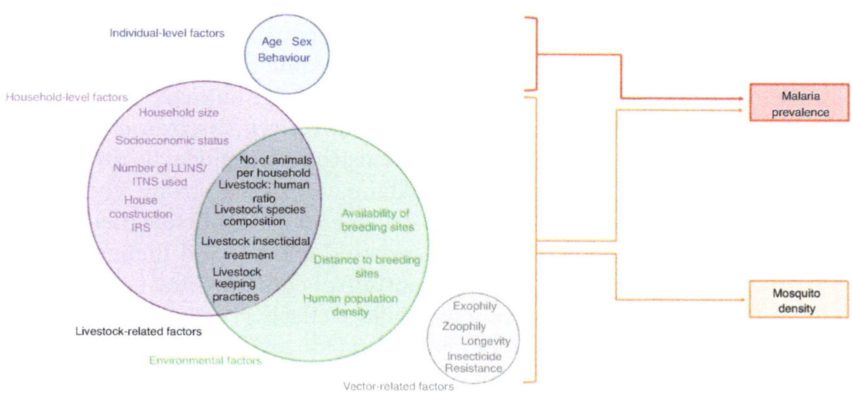

Fig. 12.2. Key interacting environmental factors that influence zooprohylaxis/ zoopotentiation.

more than 20 cattle to obtain a human blood meal outdoors (Tirados *et al.*, 2006; Tirados *et al.*, 2011). Mathematical modelling studies support that for vectors with a low biting index, an increase in animal host density can significantly decrease disease transmission, while the same does not hold true for weakly zoophilic species (Saul, 2003; summarized in Table 12.2).

Second, there must be a degree of habitat separation between human and animal quarters, particularly during peak vector feeding time. In general, when animals are housed inside at night or nearby, malaria risk is usually increased, even when local mosquito species have a preference for animal hosts (Hewitt *et al.*, 1994; Bourna and Rowland, 1995; Hadis *et al.*, 1997; Ghebreyesus *et al.*, 2000; Kawaguchi *et al.*, 2004; Pålsson *et al.*, 2004; Iwashita *et al.*, 2014), and when livestock are housed in separate shelters some distance away, malaria risk is decreased (Tirados *et al.*, 2006; Tirados *et al.*, 2011; Mala *et al.*, 2011). For example, in Ethiopia, higher *An. arabiensis* human biting rates (HBR) and parasite rates were observed in shared human-cattle dwellings and human-only dwellings (mean HBR of 8.45 and 5.97, with mean parasite rates of 26.67 and 23.85%, respectively), compared to separate human dwellings and cattle sheds (mean HBR of 4.64 and mean parasite rate of 15.05%) (Seyoum *et al.*, 2002). In Kenya, householders were at significantly increased risk of malaria with cattle herding directly outside in long grass (OR = 1.81 [95% CI: 1.19–2.73], p=0.005) (Githinji *et al.*, 2009). However, not all studies have reported a clear effect of livestock location on vector density or malaria prevalence (Bøgh *et al.*, 2002; Habtewold *et al.*, 2004; Mala *et al.*, 2011; Hiscox *et al.*, 2013). In the Gambia, no significant difference in parasite prevalence (OR = 1.69 [95% CI: 0.67–4.24], p=0.26) or high-density parasitaemia (OR = 0.73 [95% CI: 0.27–1.97], p=0.54) was apparent between children sleeping in compounds within 20m of or more than 50m away from the nearest cow, after adjusting for wealth (Bøgh *et al.*, 2002); although other livestock, including goats, donkeys and horses were commonly found among compounds but were unaccounted for in the analysis. In Laos People's Democratic Republic (PDR), cattle ownership increased the risk of anopheline house entry (IRR = 2.32 [95% CI: 1.29–4.17], p=0.005), but keeping large animals (pigs, cows or buffalo) underneath the house had no significant effect (Hiscox *et al.*, 2013).

In Pakistan, two distinct, mutually exclusive zooprophylactic area effects have been described (Bourna and Rowland, 1995): a strong correlation between malaria prevalence in families which kept cattle outside their houses compared to families who did not ('compound effect'), and higher malaria prevalence in villages with a greater proportion of cattle-owning families compared to those with fewer cattle owners ('village effect'). The 'village effect' has been attributed to increased vector density and longevity resulting from more readily available blood meals from cattle (Sota and Mogi, 1989; Saul, 2003) and potential creation of breeding sites (in puddles, hoof prints, watering sites) and refuges for adult mosquitoes in animal enclosures (Bulterys *et al.*, 2009; Githinji *et al.*, 2009; Yamamoto *et al.*, 2009). The 'compound effect' has been proposed to raise HBRs and malaria prevalence when cattle attract too large

Table 12.2. Summary of key modelling/computer simulation studies investigating the evidence for zooprophylaxis in major malaria vector species.

Data source	Major malaria vector species	Study findings	References
Pakistan; Ethiopia	*An. stephensi; An. arabiensis*	Zooprophylaxis is predicted in two situations: if vector population density does not increase with livestock introduction, or if livestock numbers and availability to vectors are sufficiently high such that the increase in vector density is counteracted by the diversion of bites from humans to animals.	(Franco *et al.*, 2014)
		Insecticide treatment of livestock is predicted to be more beneficial in settings with highly zoophilic vectors, like South Asia; this intervention could also interrupt malaria transmission in regions with anthropophilic vectors, like *An. arabiensis* in Africa under specific circumstances: high treatment coverage of the livestock population using a product with long-lasting insecticidal efficacy and a small or negligible repellency effect, or by increasing the attractiveness of treated livestock to malaria vectors.	
South Korea	*An. sinensis*	Risk of *P. vivax* outbreaks was reduced when the animal population exceeded a threshold population size.	(Nah *et al.*, 2010)
		In seasonal and non-seasonal environments, even an unintentional reduction in animal population may increase R_0 over the outbreak threshold.	
		When the relative animal population size is larger than a certain threshold, the effect of increasing animals is more apparent than the effect of reducing mosquitoes to control malaria outbreaks, suggesting that even passive zooprophylaxis may be effective.	

Continued

Table 12.2. Continued

Data source	Major malaria vector species	Study findings	References
ND	ND	As the distance between humans and animals increases, initially there is a decrease in the number of bites followed by an increase before it plateaus. An increase in number of bites per person was observed with increasing distance between hosts, implying as the animal 'zone of influence' moves away from that of humans, the net diversion of mosquitoes from humans to animals occurs with decreasing frequency. This implies there is an optimum distance between humans and animal groups, such that zoophilic vectors can be diverted. An increase in the number of animals with a fixed number of persons led to a decrease in the number of bites per person; with a fixed number of animals, the number of bites per person depends on the number of persons. When the number of animals is large, there are two extremes; initially an increase in the number of persons leads to a decrease in the number of bites per person, then there is an increase, but further increase in the number of persons leads to an asymptotic decrease in the frequency of bites. This non-linear effect does not occur when the distance between host groups and mosquito habitats is small; in such situations the number of bites per person is directly associated with the size of the human group.	(Hassanali *et al.*, 2008)
Tanzania; the Gambia	*An. gambiae* s.s; *An. arabiensis*	ITNs reduce transmission for both users and non-users by lowering mosquito survival and extending the length of the mosquito feeding cycle, except for *An. arabiensis* in the presence of cattle; in this case failure to increase feeding cycle length results in excito-repellent diversion of mosquitoes to cattle, reducing HBI and transmission intensity. Together significant impacts on EIR of users and, to a lesser extent, non-users, are predicted at high coverage levels.	(Killeen and Smith, 2007)

Tanzania; the Gambia; Pakistan	*An. gambiae* s.s; *An. arabiensis; An. funestus* s.l.; *An. stephensi; An. culicifacies; An. gambiae* s.l.	Mixed-habitat model predicted that placing livestock among humans leads only to a higher malaria prevalence, indicating habitat separation is essential for effective zooprophylaxis. If blood-host density is far below the satiation level of blood feeding then zooprophylaxis will unconditionally fail. In such areas where blood feeding is not satiated, malaria can be controlled by placing livestock 'sufficiently' far away from human dwellings (where a 'sufficient' distance is approximately given by the mosquito's daily blood-searching distance).	(Kawaguchi *et al.*, 2004)
ND	ND	Introduction of domestic animals increases mosquito density (due to more available blood meals) and in some situations this leads to increased human biting and concomitant malaria infection. Even if the average number of bites per human is reduced owing to the introduction of domestic animals, mosquito density will still be kept at high levels, due to the alternative animal blood source and potential creation of new habitats, suitable for mosquito breeding. Malaria control was only predicted when the mosquito population had attained its maximum density prior to introduction of domestic animals, or if an extremely large number of easily accessible (relative to humans) domestic animals were introduced for which the mosquitoes have high biting efficiency.	(Sota and Mogi, 1989)

Continued

Table 12.2. Continued

Data source	Major malaria vector species	Study findings	References
ND	ND	For endemic and epidemic malaria with significant foraging-associated vector mortality, changing animal numbers and accessibility had little impact; altering human accessibility had a much greater effect.	(Saul, 2003)
		Decreased malaria transmission by feeding on animals is likely to be offset by the increased survival of vectors, resulting from greater feeding success.	
		Reducing vector breeding sites (thus reducing the number of mosquitoes) and separating animals and humans as far as possible from breeding sites (thus maximizing mortality while search from blood meals and oviposition sites) are important, practical control measures.	
		Separation of humans from animals can be augmented using ITNs; use of animals to attract vectors to insecticide may lead to substantial reductions in vectorial capacity.	
Tanzania; the Gambia	*An. gambiae* s.s; *An. arabiensis; An. funestus* s.l.; *An. gambiae* s.l.	Increased cattle populations would not affect malaria transmission in Tanzania but could drastically reduce transmission in The Gambia or where *An. arabiensis* was the dominant vector.	(Killeen Ly *et al.*, 2001)

EIR: entomological inoculation rate; HBI: human blood index; ND: not described; R_0: basic reproductive number a number of mosquitoes, a proportion of which become diverted to feed on nearby individuals (Burkot *et al.*, 1989; Schultz, 1989; Sota and Mogi, 1989; Hewitt *et al.*, 1994).

These scenarios have been reproduced in mathematical models, corroborating that in the presence of moderately zoophilic vectors, such as *An. arabiensis*, the introduction of livestock would increase malaria transmission except in two cases: (i) if vector population density does not increase with livestock introduction; or (ii) if livestock numbers and availability are so great as to counteract the effect of increased vector density associated with such livestock introduction (Franco *et al.*, 2014). These phenomena explain the inverse relationship often observed between the relative numbers of livestock required for zooprophylaxis and highlight the fragility of this effect, with increasing livestock above a certain density resulting in zoopotentiation (Hassanali *et al.*, 2008; Kaburi *et al.*, 2009). A threshold of livestock and optimal positioning of domestic animals have yet to be defined; however, results from Kenya indicate that a distance of ≤25m between cattle sheds and human dwellings may be sufficient to avoid a zoopotentiation effect with *An. arabiensis* (Mayagaya *et al.*, 2015), and human-landing catches in Ghana suggest that cattle presence did not influence *An. gambiae* sensu stricto host-seeking behaviour 20m away (Maia *et al.*, 2012). This distance is likely to differ between regions and major vector species. For example, incidence of *Plasmodium vivax* transmitted by *An. culicifacies* was reduced in households in Sri Lanka that were located 70m away from cattle sheds (RR = 0.70 [95% CI: 0.47–1.03]) (Van Der Hoekl *et al.*, 1998). Computer simulations of mixed-habitat models predict that the separation required between humans and animal hosts for effective zooprophylaxis is approximate to the vector's daily foraging distance (Kawaguchi *et al.*, 2004).

Finally, the extent of zooprophylaxis can be augmented by insecticide treatment of animals (Rowland *et al.*, 2001; Fritz *et al.*, 2009; Belay *et al.*, 2019; Imbahale *et al.*, 2019; Khaligh *et al.*, 2021), addition of chemical lures (Kemibala *et al.*, 2020), deployment of IRS (Kawaguchi *et al.*, 2004) or ITNs/LLINs (Kaburi *et al.*, 2009; Iwashita *et al.*, 2014) or use of other bioinsecticides, such as entomopathogenic fungus (Lyimo *et al.*, 2012). While passive or active zooprophylaxis can impact disease transmission by diverting mosquitoes to different hosts, this usually does not affect vector abundance. The advantage of insecticide zooprophylaxis is the ability to reduce the survival and fecundity of vectors (Lyimo *et al.*, 2012). In particular, when ITNs/LLINs have been evaluated in combination with zooprophylactic strategies, additive protection from mosquito biting has been observed (Kaburi *et al.*, 2009), attributable to a repellent effect, diverting vectors to feed on cattle, if animal hosts are available in sufficient numbers. Several studies from Kenya, in areas of high ITN/LLIN coverage, have described this 'push-pull' system (Cook *et al.*, 2007), whereby *An. arabiensis* are 'pushed' out of human contact by ITNs and are further 'pulled' by cattle, resulting in a decrease in both vector density and sporozoite rate (Saul, 2003; Iwashita *et al.*, 2014). Modelling data further indicate that, while availability of cattle as alternate hosts for *An. arabiensis* reduces the impact of ITNs/LLINs on mosquito survival and entomological inoculation rate (EIR), at high ITN/LLIN coverage, the supplementary reduction in human blood index and transmission intensity afforded by mosquito diversion to feed on cattle will altogether contribute to a reduction in malaria (Killeen and Smith, 2007).

Cattle sponging with deltamethrin can also induce significant mortality among susceptible vector populations in semifield trials (Mahande *et al.*, 2007) as well as affect malaria transmission at the community level, as demonstrated by a 56% [95% CI: 14–78%] and 31% [95% CI: 5–50%] reduction in *P. falciparum* and *P. vivax* malaria incidence in refugee settlements in Pakistan (Rowland *et al.*, 2001). In this same cluster-randomised controlled trial, pyrethroid cattle sponging had the collateral economic benefits of improving livestock health (increased weight and elimination of tick infestations) and productivity (milk yield), thereby increasing social desirability of the intervention and sustaining participant acceptability throughout the study. However, topical insecticide treatment of livestock is limited by the propensity for some vector species to feed on animal legs, where insecticide can easily wash off and by the short residual activity of pyrethroid insecticides, which require reapplication every 6 weeks (Rowland *et al.*, 2001; Habtewold *et al.*, 2004). Given these challenges and the subsequent rise in global pyrethroid resistance (World Health Organization, 2020), a more appropriate longer-lasting alternative may be the use of systemic endectocides to treat livestock, including ivermectin, which has recently demonstrated promising efficacy against mosquito vectors across a number of semifield trials (Fritz *et al.*, 2009; Poché *et al.*, 2015; Pooda *et al.*, 2015; Lyimo *et al.*, 2017; Chaccour *et al.*, 2018; Belay *et al.*, 2019; Dreyer *et al.*, 2019; Pasay *et al.*, 2019).

Other Modifying Factors and Study Design Limitations

Zooprophylactic effects may vary between animal host species. In the Lake Victoria Basin, Kenya, *An. arabiensis* density increased by 10% with each additional goat/sheep tethered within 20m of a household and increased its human blood feeding with the overall number of goats/sheep within a 500m radius of the house; the opposite effect was observed with cattle abundance across the same area (Iwashita *et al.*, 2014). Similar observations were reported in Zambia where animal ownership was protective of malaria (OR = 0.13 [95% CI: 0.03–0.56]) when the ratio of cattle:humans > 0, dogs:humans > 0.3, pigs:humans > 0 and goats:humans was ≥1, but the latter was a significant risk factor for malaria when goats:humans was <1 (Bulterys *et al.*, 2009). Pig keeping was also found to significantly increase the risk of malaria incidence among communities in Mozambique (Temu *et al.*, 2012) and Guinea Bissau (Pålsson *et al.*, 2004), but in Papua New Guinea, sporozoite rates were lower in *An. farauti* among villages with high pig populations compared to other villages (Birley and Charlwood, 1987). A positive correlation between donkeys and

An. gambiae s.l. abundance indoors was reported in Burkina Faso (Yamamoto *et al.*, 2009) and other equines (horses) have been shown to exert a stronger zooprophylactic effect on anopheline and culicine mosquitoes in the Gambia (Bøgh *et al.*, 2001; Kirby *et al.*, 2008). These variable effects per species may reflect differences in animal behaviour and/or size (related to relative production of odour/attractants); goats, pigs and sheep often spend time directly outside house walls under extended eaves and may release semiochemical plumes that could potentially attract vectors closer to the house. Furthermore, goats, sheep and pigs are generally smaller than cattle and, when present at lower abundance relative to humans, may attract vectors to the household but may not be abundant enough to sufficiently deviate vectors from feeding on humans.

Dissecting the relative contributions of different ecological and entomological factors to zooprophylaxis can be complicated by a range of experimental limitations. Some studies have failed to determine the accurate geolocation of overnight tethering points for livestock preventing accurate assessments of any zooprophylactic 'area effect'; householder interviews may provide imprecise information about host availability, as cattle belonging to a household are not necessarily the one tethered closest to it (Bourna and Rowland, 1995; Iwashita *et al.*, 2014). Similarly, other studies have focused exclusively on cattle, without considering other animals in the area that may influence zooprophylaxis, such as goats, donkeys, horses, dogs, etc. (Bøgh *et al.*, 2002); in some environments, dogs have even been shown to play an important role as alternate blood sources for *Anopheles coluzzii* (a normally anthropophilic member of the *An. gambiae* complex) (Sousa *et al.*, 2001). A number of reports are based on indirect inferences from blood meal analysis alone (White *et al.*, 1980; Adugna, 1996; Hadis *et al.*, 1997); such data from indoor or outdoor resting collections fail to capture accurately where mosquitoes might have acquired a blood meal (Tirados *et al.*, 2006). Furthermore, many entomological studies are reliant on easier, more feasible indoor collections of resting mosquitoes, compared to more challenging outdoor traps, which may bias samples towards endophilic and endophagic species that also tend to be more anthropophilic (Hewitt *et al.*, 1994). Moreover, within indoor and outdoor sampling methods, relative vector species composition and blood meal indices can also vary (Massebo *et al.*, 2015). Another potential confounder is that a number of zooprophylaxis studies do not present findings disaggregated by confirmed species complex members (e.g. from *An. gambiae* s.l. or *An. funestus* s.l. groups), usually because of restrictions to study design or resources, or if data were collected prior to routine adoption of PCR-based species identification techniques (Bøgh *et al.*, 2001, 2002; Muriu *et al.*, 2008; Yamamoto *et al.*, 2009). In particular, this limitation is important to distinguish between differential anthropophilic vs zoophilic behaviours of *An. gambiae* s.s and *An. arabiensis*, which are morphologically indistinguishable. Finally, wealth index is a very important effect modifier of zooprophylaxis, which is often not accounted for (Donnelly *et al.*, 2016). Increased socio-economic status (reflected by cattle ownership) may be correlated with greater access to healthcare and diagnosis, anti-malaria drugs, vector control interventions, improved house construction and better overall health and nutritional status (Bøgh *et al.*, 2002; Randolph *et al.*, 2007; Bulterys *et al.*, 2009; Temu *et al.*, 2012; Degarege *et al.*, 2019). While ITN/LLINs have been shown to enhance protective zooprophylaxis in some circumstances, other studies indicate that these effects diminish when intervention use is controlled for (Bøgh *et al.*, 2002; Yamamoto *et al.*, 2009). Measurements of socio-economic status should be included in future zooprophylaxis studies; however, financial metrics, such as maternal education level (Yamamoto *et al.*, 2009) or commodity ownership (Ghebreyesus *et al.*, 2000) are preferable to those that include livestock ownership to avoid issues of collinearity (Donnelly *et al.*, 2015).

Conclusions

There remains considerable debate regarding the efficacy of zooprophylaxis, which has manifested as reluctance to employ this strategy as part of IVM programmes. Diversion of feeding to alternative hosts needs to outweigh the potential increase in mosquito population size, longevity and attraction to humans by zooprophylactic livestock. In most scenarios, animals are only likely to provide a prophylactic effect when the major malaria vector species is sufficiently zoophilic and opportunistic, and animals are deployed to form a barrier between vectors and humans; highly anthropophilic species are generally unaffected by presence/absence of livestock. In areas where strategic cattle positioning is not feasible, livestock should be located as far away from humans as possible and excluded from inside human dwellings at night, even if mosquito species are predominantly zoophilic. Where this is not culturally acceptable due to concerns of livestock security, ITN/LLINs or other insecticidal treatment of animals can be used to protect householders. In general, deployment of zooprophylaxis first requires site-specific epidemiological evidence of its effectiveness. However, in practice, the scalability of this technique may be difficult to determine using conventional experimental designs and may even vary considerably within a malaria-endemic region with differing environmental and social factors. Further investigation is warranted in three priority areas to maximize the likelihood of zooprophylaxis and minimize zoopotentiation: (i) estimation of the distance threshold and conditions whereby zoopotentiation transitions into zooprophylaxis for specific livestock, host and vector species combinations; (ii) more accurate assessments of host feeding preferences using a wider range of indoor and outdoor collection methods combined with PCR confirmation of individual vectors to species level; and (iii) inclusion of socio-economic factors and other vector control interventions as key covariates in empirical studies. The development of guidelines for the use of zooprophylaxis in malaria IVM schemes will require strong interdisciplinary collaborations between medical and veterinary entomologists and the agricultural sector. While zooprophylaxis still has potential as a supplementary method in rural endemic areas where gains in malaria control have stalled in recent years, the landscape of rapid urbanization across Africa does call into question the longer-term widespread utility of this strategy.

References

Adiamah, J.H., Koram, K.A., Thomson, M.C., Lindsay, S.W., Todd, J. *et al.* (1993) Entomological risk factors for severe malaria in a peri-urban area of The Gambia. *Annals of Tropical Medicine and Parasitology* 87(5), 491–500.

Adugna, N.P.B. (1996) Determination of the human blood index of some anopheline mosquitoes using ELISA. *Ethiopian Medical Journal* 34, 1–10.

Angelon, K.A. and Petranka, J.W. (2002) Chemicals of predatory mosquitofish (*Gambusia affinis*) influence selection of oviposition site by *Culex* mosquitoes. *Journal of Chemical Ecology* 28(4), 797–806.

Arthington, A. and Marshall, C. (1999) Diet of the exotic mosquitofish, *Gambusia holbrooki*, in an Australian lake and potential for competition with indigenous fish species. *Asian Fisheries Science* 12(1), 1–16.

Asale, A., Duchateau, L., Devleesschauwer, B., Huisman, G. and Yewhalaw, D. (2017) Zooprophylaxis as a control strategy for malaria caused by the vector *Anopheles arabiensis* (Diptera: Culicidae): A systematic review. *Infectious Diseases of Poverty* 6(1), 160.

Barik, M., Bhattacharjee, I., Ghosh, A. and Chandra, G. (2018) Larvivorous potentiality of *Puntius tetrazona* and *Hyphessobrycon rosaceus* against *Culex vishnui* subgroup in laboratory and field based bioassay. *BMC Research Notes* 11, 804.

Bay, E.C. and Self, L.S. (1972) Observations of the guppy, *Poecilia reticulata* Peters, in *Culex pipiens fatigans* breeding sites in Bangkok, Rangoon, and Taipei. *Bulletin of the World Health Organization* 46(3), 407–416.

Belay, A., Petros, B., Gebre-Michael, T. and Balkew, M. (2019) Effect of LongRange™ eprinomectin on *Anopheles arabiensis* by feeding on calves treated with the drug. *Malaria Journal* 18(1), 332.

Birley, M.H. and Charlwood, J.D. (1987) Sporozoite rate and malaria prevalence. *Parasitology Today* 3, 231–232.

Bøgh, C., Clarke, S.E., Pinder, M., Sanyang, F. and Lindsay, S.W. (2001) Effect of passive zooprophylaxis on malaria transmission in The Gambia. *Journal of Medical Entomology* 38(6), 822–828.

Bøgh, C., Clarke, S.E., Walraven, G.E.L. and Lindsay, S.W. (2002) Zooprophylaxis, artefact or reality? A paired-cohort study of the effect of passive zooprophylaxis on malaria in The Gambia. *Transactions of the Royal Society of Tropical Medicine and Hygiene* 96(6), 593–596.

Bourna, M. and Rowland, M. (1995) Failure of passive zooprophylaxis: Cattle ownership in Pakistan is associated with a higher prevalence of malaria. *Transactions of the Royal Society of Tropical Medicine and Hygiene* 89(4), 351–353.

Bruce-Chwatt, L. (1982) *Essential Malariology*, 2nd edn. William Heinemann Medical Books, London.

Brumpt, E. (1945) Zooprophylaxie du paludisme. *Annaux de Parasitologie* 20, 191–206.

Bulterys, P.L., Mharakurwa, S. and Thuma, P.E. (2009) Cattle, other domestic animal ownership, and distance between dwelling structures are associated with reduced risk of recurrent *Plasmodium falciparum* infection in Southern Zambia. *Tropical Medicine and International Health* 14(5), 522–528.

Burkot, T.R., Dye, C. and Graves, P.M. (1989) An analysis of some factors determining the sporozoite rates, human blood indexes, and biting rates of members of the *Anopheles punctulatus* complex in Papua New Guinea. *American Journal of Tropical Medicine and Hygiene* 40(3), 229–234.

Cabrera, M.B., Bogan, S., Posadas, P., Somoza, G.M., Montoya-Burgos, J.I, *et al.* (2017) Risks associated with introduction of poeciliids for control of mosquito larvae: First record of the non-native *Gambusia holbrooki* in Argentina. *Journal of Fish Biology* 91(2), 704–710.

Caixeta, E.S., Silva, C.F., Santos, V.S.V., Olegário de Campos Júnior, E. and Pereira, B.B. (2016) Ecotoxicological assessment of pyriproxyfen under environmentally realistic exposure conditions of integrated vector management for *Aedes aegypti* control in Brazil. *Journal of Toxicology and Environmental Health – Part A: Current Issues* 79(18), 799–803.

Cavalcanti, L., Paula, F., Pontes, R., Heukelbach, J. and Lima, J. (2009) Survival of larvivorous fish used for biological control of *Aedes aegypti* larvae in domestic containers with different chlorine concentrations. *Journal of Medical Entomology* 46(4), 841–844.

Chaccour, C.J., Ngha'Bi, K., Abizanda, G., Irigoyen Barrio, A., Aldaz, A. *et al.* (2018) Targeting cattle for malaria elimination: Marked reduction of *Anopheles arabiensis* survival for over six months using a slow-release ivermectin implant formulation. *Parasites and Vectors* 11(1), 287.

Chae, S., Kwon, Y., Min, K., Kim, H., Kim, N. *et al.* (2014) Predation and control efficacies of *Misgurnus mizolepis* (Cypriniformes: Cobitidae) toward *Culex pipiens molestus* (Diptera: Culicidae) and fish toxicity of temephos in laboratory and septic tank conditions. *Journal of Medical Entomology* 51(4), 817–823.

Chandra, G., Bhattacharjee, I., Chatterjee, S.N. and Ghosh, A. (2008) Mosquito control by larvivorous fish. *Indian Journal of Medical Research* 127(1), 13–27.

Charlwood, D. (2001) Zooprophylaxis: Are we in Plato's cave?. *Trends in Parasitology* 17, 517.

Consoli, R.A., Guimarães, C.T., do Carmo, J.A., Soares, D.M. and dos Santos, J.S. (1991) *Astronotus ocellatus* (Cichlidae: Pisces) and *Macropodus opercularis* (Anabatidae: Pisces) as predators of immature *Aedes fluviatilis* (Diptera: Culicidae) and *Biomphalaria glabrata* (Mollusca: Planorbidae). *Memórias Do Instituto Oswaldo Cruz* 86(4), 419–424.

Cook, S.M., Khan, Z.R. and Pickett, J.A. (2007) The use of push-pull strategies in integrated pest management. *Annual Review of Entomology* 52, 375–400.

Couret, J., Notarangelo, M., Veera, S., Leclaire-Conway, N., Ginsberg, H.S. *et al.* (2020) Biological control of *Aedes* mosquito larvae with carnivorous aquatic plant, *Utricularia macrorhiza*. *Parasites and Vectors* 13(1), 208.

Das, M.K., Rajesh Kumar Rao, M. and Kulsreshtha, A.K. (2018) Native larvivorous fish diversity as a biological control agent against mosquito larvae in an endemic malarious region of Ranchi district in Jharkhand, India. *Journal of Vector Borne Diseases* 55(1), 34–41.

de Góes Cavalcanti, L.P., Soares Pontes, R.J., Ferreira Regazzi, A.C., Paula, F.J., Frutuoso, R.L. *et al.* (2007) Efficacy of fish as predators of *Aedes aegypti* larvae, under laboratory conditions. *Revista de Saude Publica* 41(4), 638–644.

de Oliveira Lima, J.W., De Góes Cavalcanti, L.P., Pontes, R.J.S. and Heukelbach, J. (2010) Survival of *Betta splendens* fish (Regan, 1910) in domestic water containers and its effectiveness in controlling *Aedes aegypti* larvae (Linnaeus, 1762)) in Northeast Brazil. *Tropical Medicine and International Health* 15(12), 1525–1532.

Degarege, A., Fennie, K., Degarege, D., Chennupati, S. and Madhivanan, P. (2019) Improving socioeconomic status may reduce the burden of malaria in sub Saharan Africa: A systematic review and meta-analysis. *PLoS ONE* 14(1), e0211205.

Deressa, W., Ali, A. and Berhane, Y. (2007) Household and socioeconomic factors associated with childhood febrile illnesses and treatment seeking behaviour in an area of epidemic malaria in rural Ethiopia. *Transactions of the Royal Society of Tropical Medicine and Hygiene* 101(9), 939–947.

Donnelly, B., Berrang-Ford, L., Ross, N.A. and Michel, P. (2015) A systematic, realist review of zooprophylaxis for malaria control. *Malaria Journal* 14(1), 313.

Donnelly, B., Berrang-Ford, L., Labbé, J., Twesigomwe, S., Lwasa, S. *et al.* (2016) *Plasmodium falciparum* malaria parasitaemia among indigenous Batwa and non-indigenous communities of Kanungu district, Uganda. *Malaria Journal* 15(1), 254.

Dreyer, S.M., Leiva, D., Magaña, M., Pott, M., Kay, J. *et al.* (2019) Fipronil and ivermectin treatment of cattle reduced the survival and ovarian development of field-collected *Anopheles albimanus* in a pilot trial conducted in Northern Belize. *Malaria Journal* 18(1), 296.

Fletcher, M., Teklehaimanot, A. and Yemane, G. (1992) Control of mosquito larvae in the port city of Assab by an indigenous larvivorous fish, *Aphanius dispar*. *Acta Tropica* 52(2–3), 155–166.

Fornadel, C.M., Norris, L.C., Glass, G.E. and Norris, D.E. (2010) Analysis of *Anopheles arabiensis* blood feeding behavior in southern Zambia during the two years after introduction of insecticide-treated bed nets. *American Journal of Tropical Medicine and Hygiene* 83(4), 848–853.

Franco, A.O., Gomes, M.G.M., Rowland, M., Coleman, P.G. and Davies, C.R. (2014) Controlling malaria using livestock-based interventions: A One Health approach. *PLoS ONE* 9(7), e101699.

Fritz, M.L., Siegert, P.Y., Walker, E.D., Bayoh, M.N., Vulule, J.R. *et al.* (2009) Toxicity of bloodmeals from ivermectin-treated cattle to *Anopheles gambiae* s.l. *Annals of Tropical Medicine and Parasitology* 103(6), 539–547.

Gachelin, G., Garner, P., Ferroni, E., Verhave, J.P. and Opinel, A. (2018) Evidence and strategies for malaria prevention and control: A historical analysis. *Malaria Journal* 17(1), 96.

Gamal, Z.A. (2012) Effectiveness of *Gambusia holbrooki* fish in domestic water containers and controlling *Aedes aegypti* larvae (Linnaeus, 1762) in Southwest Saudi Arabia (Jeddah). *Journal of the Egyptian Society of Parasitology* 42(1), 1–10.

Gari, T. and Lindtjørn, B. (2018) Reshaping the vector control strategy for malaria elimination in Ethiopia in the context of current evidence and new tools: Opportunities and challenges. *Malaria Journal* 17(1), 454.

Gatton, M.L., Chitnis, N., Churcher, T., Donnelly, M.J., Ghani, A.C. *et al.* (2013) The importance of mosquito behavioural adaptations to malaria control in Africa. *Evolution* 67(4), 1218–1230.

Ghebreyesus, T.A., Haile, M. and Witten, K.H. (2000) Household risk factors for malaria among children in the Ethiopian highlands. *Transactions of the Royal Society of Tropical Medicine and Hygiene* 94(1), 17–21.

Ghosh, S.K. and Dash, A.P. (2007) Larvivorous fish against malaria vectors: a new outlook. *Transactions of the Royal Society of Tropical Medicine and Hygiene* 101(11), 1063–1064.

Ghosh, S.K., Chakaravarthy, P., Panch, S.R., Krishnappa, P., Tiwari, S. *et al.* (2011) Comparative efficacy of two poeciliid fish in indoor cement tanks against Chikungunya vector *Aedes aegypti* in villages in Karnataka, India. *BMC Public Health* 11(1), 599.

Giglioli, G. (1963) Ecological change as a factor in renewed malaria transmission in an eradicated area: A localized outbreak of *A. aquasalis* – transmitted malaria on the Demerara river estuary, British Guinea, in the fifteenth year of *A. darlingi* and malaria eradication. *Bulletin of the World Health Organization* 29, 131–145.

Githinji, S., Herbst, S. and Kistemann, T. (2009) The human ecology of malaria in a highland region of South-West Kenya. *Methods of Information in Medicine* 48(5), 451–453.

Griffin, L. (2014) Laboratory evaluation of predation on mosquito larvae by Australian mangrove fish. *Journal of Vector Ecology* 39(1), 197–203.

Habtewold, T., Walker, A.R., Curtis, C.F., Osir, E.O. and Thapa, N. (2001) The feeding behaviour and plasmodium infection of *Anopheles* mosquitoes in Southern Ethiopia in relation to use of insecticide-treated livestock for malaria control. *Transactions of the Royal Society of Tropical Medicine and Hygiene* 95(6), 584–586.

Habtewold, T., Prior, A., Torr, S.J. and Gibson, G. (2004) Could insecticide-treated cattle reduce Afrotropical malaria transmission? Effects of deltamethrin-treated Zebu on *Anopheles arabiensis* behaviour and survival in Ethiopia. *Medical and Veterinary Entomology* 18(4), 408–417.

Hadis, M., Lulu, M., Makonnen, Y. and Asfaw, T. (1997) Host choice by indoor-resting *Anopheles arabiensis* in Ethiopia. *Transactions of the Royal Society of Tropical Medicine and Hygiene* 91(4), 376–378.

Hamer, A.J., Lane, S.J. and Mahony, M.J. (2002) The role of introduced mosquitofish (*Gambusia holbrooki*) in excluding the native green and golden bell frog (*Litoria aurea*) from original habitats in South-eastern Australia. *Oecologia* 132(3), 445–452.

Han, W.W., Lazaro, A., McCall, P.J., George, L., Runge-Ranzinger, S. *et al.* (2015) Efficacy and community effectiveness of larvivorous fish for dengue vector control. *Tropical Medicine and International Health* 20(9), 1239–1256.

Haq, S. and Srivastava, H.C. (2013) Efficacy of *Aphanius dispar* (Rüppell) an indigenous larvivorous fish for vector control in domestic tanks under the Sardar Sarovar Narmada project command area in district Kheda, Gujarat. *Journal of Vector Borne Diseases* 50, 137–140.

Hassanali, A., Nedorezov, L.V. and Sadykov, A.M. (2008) Zooprophylactic diversion of mosquitoes from human to alternative hosts: A static simulation model. *Ecological Modelling* 212(1–2), 155–161.

Hasyim, H., Dhimal, M., Bauer, J., Montag, D., Groneberg, D.A. *et al.* (2018) Does livestock protect from malaria or facilitate malaria prevalence? A cross-sectional study in endemic rural areas of Indonesia. *Malaria Journal* 17(1), 302.

Hewitt, S. and Rowland, M. (1999) Control of zoophilic malaria vectors by applying pyrethroid insecticides to cattle. *Tropical Medicine and International Health* 4(7), 481–486.

Hewitt, S., Kamal, M., Muhammad, N. and Rowland, M. (1994) An entomological investigation of the likely impact of cattle ownership on malaria in an Afghan refugee camp in the North West Frontier Province of Pakistan. *Medical and Veterinary Entomology* 8, 160–164.

Hiscox, A., Khammanithong, P., Kaul, S., Sananikhom, P., Luthi, R. *et al.* (2013) Risk factors for mosquito house entry in the Lao PDR. *PLoS ONE* 8(5), e62769.

Howard, A.F.V., Zhou, G. and Omlin, F.X. (2007) Malaria mosquito control using edible fish in western Kenya: Preliminary findings of a controlled study. *BMC Public Health* 7(1), 199.

Howe, E., Howe, C., Lim, R. and Burchett, M. (1997) Impact of the introduced poeciliid *Gambusia holbrooki* (Girard, 1859) on the growth and reproduction of *Pseudomugil signifer* (Kner, 1865) in Australia. *Marine and Freshwater Research* 48(5), 425–434.

Hurst, T.P., Brown, M.D., Kay, B.H. and Ryan, P.A. (2006) Evaluation of *Melanotaenia duboulayi* (Atheriniformes: Melanotaeniidae), *Hypseleotris galii* (Perciformes: Eleotridae), and larvicide Vectolex® WG (*Bacillus sphaericus*) for integrated control of *Culex annulirostris*. *Journal of the American Mosquito Control Association* 22(3), 418–425.

Hurst, T.P., Kay, B.H., Ryan, P.A. and Brown, M.D. (2007) Sublethal effects of mosquito larvicides on swimming performance of larvivorous fish *Melanotaenia duboulayi* (Atheriniformes: Melanotaeniidae). *Journal of Economic Entomology* 100(1), 61–65.

Hurst, T.P., Kay, B.H., Brown, M.D. and Ryan, P.A. (2010) *Melanotaenia duboulayi* influence oviposition site selection by *Culex annulirostris* (Diptera: Culicidae) and *Aedes notoscriptus* (Diptera: Culicidae) but not *Culex quinquefasciatus* (Diptera: Culicidae). *Environmental Entomology* 39(2), 545–551.

Idrees, M. and Jan, A.H. (2001) Failure of zooprophylaxis: cattle ownership increase rather than reduce the prevalence of malaria in district dir N.W.F.P. of Pakistan. *The Sciences* 1(2), 52–54.

Imbahale, S.S., Mweresa, C.K., Takken, W. and Mukabana, W.R. (2011) Development of environmental tools for anopheline larval control. *Parasites and Vectors* 4(1), 130.

Imbahale, S.S., Montaña Lopez, J., Brew, J., Paaijmans, K., Rist, C. *et al.* (2019) Mapping the potential use of endectocide-treated cattle to reduce malaria transmission. *Scientific Reports* 9(1), 5826.

Ivantsoff, W. and Aarn (1999) Detection of predation on Australian native fishes by *Gambusia holbrooki: Marine and Freshwater Research* 50(5), 467–468.

Iwashita, H., Dida, G.O., Sonye, G.O., Sunahara, T., Futami, K. *et al.* (2014) Push by a net, pull by a cow: can zooprophylaxis enhance the impact of insecticide treated bed nets on malaria control. *Parasites and Vectors* 7(1), 52.

Juliano, R., Guerrero, R. and Ronquillo, I. (1989) The introduction of exotic aquatic species in the Philippines. In: de Silva, S.S. (ed.) *Exotic Aquatic Organisms in Asia: Proceedings of a Workshop on Introduction of Exotic Aquatic Organisms in Asia.* Asian Fisheries Society, pp. 83–90.

Kaburi, J.C., Githuto, J.N., Muthami, L., Ngure, P.K., Mueke, J.M. *et al.* (2009) Effects of long-lasting insecticidal nets and zooprophylaxis on mosquito feeding behaviour and density in Mwea, central Kenya. *Journal of Vector Borne Diseases* 46(3), 184–190.

Kawaguchi, I., Sasaki, A. and Mogi, M. (2004) Combining zooprophylaxis and insecticide spraying: a malaria-control strategy limiting the development of insecticide resistance in vector mosquitoes. *Proceedings of the Royal Society B: Biological Sciences* 271(1536), 301–309.

Kay, B.H., Sinh Nam, V., Tien, T., Thi Yen, N., Vu Phong, T. *et al.* (2002) Control of *Aedes* vectors of dengue in three provinces of Vietnam by use of *Mesocyclops* (Copepoda) and community-based methods validated by entomologic, clinical, and serological surveillance. *American Journal of Tropical Medicine and Hygiene* 66(1), 40–48.

Kemibala, E.E., Kemibala, E.E., Mafra-Neto, A., Dekker, T., Saroli, J, *et al.* (2020) A zooprophylaxis strategy using l-lactic acid (abate) to divert host-seeking malaria vectors from human host to treated non-host animals. *Malaria Journal* 19(1), 52.

Khaligh, F.G., Jafari, A., Silivanova, E., Levchenko, M., Rahimi, B. *et al.* (2021) Endectocides as a complementary intervention in the malaria control program: a systematic review. *Systematic Reviews* 10(1), 30.

Killeen, G.F. and Smith, T.A. (2007) Exploring the contributions of bed nets, cattle, insecticides and excitorepellency to malaria control: a deterministic model of mosquito host-seeking behaviour and mortality. *Transactions of the Royal Society of Tropical Medicine and Hygiene* 101(9), 867–880.

Killeen Ly, G.F., Mckenzie, F.E., Foy, B.D., Bsghs, C. and Beier, J.C. (2001) The availability of potential hosts as a determinant of feeding behaviours and malaria transmission by african mosquito populations. *Transactions of the Royal Society of Tropical Medicine and Hygiene* 95(5), 469–476.

Kim, H.C., Lee, J.H., Yang, K.H. and Yu, H.S. (2002) Biological control of *Anopheles sinensis* with native fish predators (*Aplocheilus* and *Aphyocypris*) and herbivorous fish, tilapia in natural rice fields in Korea. *Entomological Research* 32(4), 247–250.

Kirby, M.J., West, P., Green, C., Jasseh, M. and Lindsay, S.W. (2008) Risk factors for house-entry by culicine mosquitoes in a rural town and satellite villages in The Gambia. *Parasites and Vectors* 1(1), 41.

Kirnowordoyo, S. and Supalin, S. (1986) Zooprophylaxis as a useful tool for control of *A. aconitus* transmitted malaria in Central Java, Indonesia. *Journal of Communicable Diseases* 18(2), 90–94.

Komak, S. and Crossland, M.R. (2000) An assessment of the introduced mosquitofish (*Gambusia affinis holbrooki*) as a predator of eggs, hatchlings and tadpoles of native and non-native anurans. *Wildlife Research* 27(2), 185–189.

Kreppel, K.S., Viana, M., Main, B.J., Johnson, P.C.D., Govella, N.J, *et al.* (2020) Emergence of behavioural avoidance strategies of malaria vectors in areas of high LLIN coverage in Tanzania. *Scientific Reports* 10(1), 14527.

Kumar, R., Muhid, P., Dahms, H.U., Sharma, J. and Hwang, J.S. (2015) Biological mosquito control is affected by alternative prey. *Zoological Studies* 54, 1–18.

Kusumawathie, P.H.D., Wickremasinghe, A.R., Karunaweera, N.D. and Wijeyaratne, M.J.S. (2008) Costs and effectiveness of application of *Poecilia reticulata* (guppy) and temephos in anopheline mosquito control in river basins below the major dams of Sri Lanka. *Transactions of the Royal Society of Tropical Medicine and Hygiene* 102(7), 705–711.

Lardeux, F.J.R. (1992) Biological control of Culicidae with the copepod *Mesocyclops aspericornis* and larvivorous fish (poeciliidae) in a village of French Polynesia. *Medical and Veterinary Entomology* 6(1), 9–15.

Lardeux, F., Rivière, F., Séchan, Y. and Loncke, S. (2002) Control of the *Aedes* vectors of the dengue viruses and *Wuchereria bancrofti:* the French Polynesian experience. *Annals of Tropical Medicine and Parasitology* 96(Suppl. 2), S105–S116.

Lardeux, F., Loayza, P., Bouchité, B. and Chavez, T. (2007) Host choice and human blood index of *Anopheles pseudopunctipennis* in a village of the Andean valleys of Bolivia. *Malaria Journal* 6(1), 8.

Lee, D.K. (2000) Predation efficacy of the fish muddy loach, *Misgurnus mizolepis*, against *Aedes* and *Culex* mosquitoes in laboratory and small rice plots. *Journal of the American Mosquito Control Association* 16(3), 258–261.

Lefèvre, T., Gouagna, L.C., Dabiré, K.R., Elguero, E., Fontenille, D, *et al.* (2009) Beyond nature and nurture: phenotypic plasticity in blood-feeding behavior of *Anopheles gambiae* s.s. when humans are not readily accessible. *American Journal of Tropical Medicine and Hygiene* 81(6), 1023–1029.

Louca, V., Lucas, M.C., Green, C., Majambere, S., Fillinger, U, *et al.* (2009) Role of fish as predators of mosquito larvae on the floodplain of the Gambia river. *Journal of Medical Entomology* 46(3), 546–556.

Lyimo, I.N., Ng'Habi, K.R., Mpingwa, M.W., Daraja, A.A., Mwasheshe, D.D. *et al.* (2012) Does cattle milieu provide a potential point to target wild exophilic *Anopheles arabiensis* (Diptera: Culicidae) with entomopathogenic fungus? A bioinsecticide zooprophylaxis strategy for vector control. *Journal of Parasitology Research* 2012, 280583.

Lyimo, I.N., Kessy, S.T., Mbina, K.F., Daraja, A.A. and Mnyone, L.L. (2017) Ivermectin-treated cattle reduces blood digestion, egg production and survival of a free-living population of *Anopheles arabiensis* under semi-field condition in south-eastern Tanzania. *Malaria Journal* 16(1), 239.

Mahande, A.M., Mosha, F.W., Mahande, J.M. and Kweka, E.J. (2007a) Feeding and resting behaviour of malaria vector, *Anopheles arabiensis* with reference to zoo-prophylaxis. *Malaria Journal* 6(1), 100.

Mahande, A.M., Mosha, F.W., Mahande, J.M. and Kweka, E.J. (2007b) Role of cattle treated with deltamethrin in areas with a high population of *Anopheles arabiensis* in Moshi, northern Tanzania. *Malaria Journal* 6(1), 109.

Mahmoud, A.A. (1985) Mosquitofish *Gambusia affinis holbrooki* as a malaria vector control agent in Gezira irrigation canals of the Sudan. *Journal of the American Mosquito Control Association* 1(4), 524.

Maia, M.F., Abonuusum, A., Lorenz, L.M., Clausen, P.H., Bauer, B. *et al.* (2012) The effect of deltamethrin-treated net fencing around cattle enclosures on outdoor-biting mosquitoes in Kumasi, Ghana. *PLoS ONE* 7(9), e45794.

Mala, A.O., Irungu, L.W., Shililu, J.I., Muturi, E.J., Mbogo, C.M. *et al.* (2011) *Plasmodium falciparum* transmission and aridity: A Kenyan experience from the dry lands of Baringo and its implications for *Anopheles arabiensis* control. *Malaria Journal* 10(1), 121.

Martínez-Ibarra, J.A., Guillén, Y.G., Arredondo-Jiménez, J.I. and Rodríguez-López, M.H. (2002) Indigenous fish species for the control of *Aedes aegypti* in water storage tanks in southern México. *BioControl* 47(4), 481–486.

Massebo, F., Balkew, M., Gebre-Michael, T. and Lindtjørn, B. (2013) Blood meal origins and insecticide susceptibility of *Anopheles arabiensis* from Chano in south-west Ethiopia. *Parasites and Vectors* 6(1), 44.

Massebo, F., Balkew, M., Gebre-Michael, T. and Lindtjørn, B. (2015) Zoophagic behaviour of anopheline mosquitoes in southwest Ethiopia: opportunity for malaria vector control. *Parasites and Vectors* 8(1), 645.

Matias, J.R. and Adrias, A.O. (2010) The use of annual killifish in the biocontrol of the aquatic stages of mosquitoes in temporary bodies of fresh water; a potential new tool in vector control. *Parasites and Vectors* 3, 46.

Mayagaya, V.S., Nkwengulila, G., Lyimo, I.N., Kihonda, J., Mtambala, H, *et al.* (2015) The impact of livestock on the abundance, resting behaviour and sporozoite rate of malaria vectors in southern Tanzania. *Malaria Journal* 14(1), 17.

Menon, P.K. and Rajagopala, P.K. (1978) Control of mosquito breeding in wells by using *Gambusia affinis* and *Aplocheilus blochii* in Pondicherry town. *Indian Journal of Medical Research* 68, 927–933.

Middaugh, D.P., Hester, P.G., Meisch, M.V. and Stark, P.M. (1985) Preliminary data on use of the inland silverside, *Menidia beryllina*, to control mosquito larvae. *Journal of the American Mosquito Control Association* 1(4), 435–441.

Morales-Pérez, A., Nava-Aguilera, E., Legorreta-Soberanis, J., Cortés-Guzmán, A.J., Balanzar-Martínez, A. *et al.* (2017) "Where we put little fish in the water there are no mosquitoes:" a cross-sectional study on biological control of the *Aedes aegypti* vector in 90 coastal-region communities of Guerrero, Mexico. *BMC Public Health* 17(Suppl. 1), 433.

Muriu, S.M., Muturi, E.J., Shililu, J.I., Mbogo, C.M., Mwangangi, J.M, *et al.* (2008) Host choice and multiple blood feeding behaviour of malaria vectors and other anophelines in mwea rice scheme, Kenya. *Malaria Journal* 7(1), 43.

Mutero, C.M., Kabutha, C., Kimani, V., Kabuage, L., Gitau, G, *et al.* (2004) A transdisciplinary perspective on the links between malaria and agroecosystems in Kenya. *Acta Tropica* 89(2), 171–186.

Muturi, E.J., Muriu, S., Shililu, J., Mwangangi, J., Jacob, B.G, *et al.* (2008) Effect of rice cultivation on malaria transmission in central Kenya. *American Journal of Tropical Medicine and Hygiene* 78(2), 270–275.

Nah, K., Kim, Y. and Lee, J.M. (2010) The dilution effect of the domestic animal population on the transmission of *P. vivax* malaria. *Journal of Theoretical Biology* 266(2), 299–306.

Nam, V.S., Yen, N.T., Holynska, M., Reid, J.W. and Kay, B.H. (2000) National progress in dengue vector control in Vietnam: Survey for *Mesocyclops* (Copepoda), *Micronecta* (Corixidae), and fish as biological control agents. *American Journal of Tropical Medicine and Hygiene* 62(1), 5–10.

Padonou, G.G., Gbedjissi, G., Yadouleton, A., Azondekon, R., Razack, O, *et al.* (2012) Decreased proportions of indoor feeding and endophily in *Anopheles gambiae* s.l. populations following the indoor residual spraying and insecticide-treated net interventions in Benin (west Africa). *Parasites and Vectors* 5(1), 262.

Pålsson, K., Jaenson, T.G., Dias, F., Laugen, A.T. and Bjo, A. (2004) Endophilic anopheles mosquitoes in Guinea Bissau, west Africa, in relation to human housing conditions. *Journal of Medical Entomology* 41(4), 746–752.

Pasay, C.J., Yakob, L., Meredith, H.R., Stewart, R., Mills, P.C, *et al.* (2019) Treatment of pigs with endectocides as a complementary tool for combating malaria transmission by *Anopheles farauti* (s.s.) in Papua New Guinea. *Parasites and Vectors* 12(1), 124.

Phan-Urai, P., Nelson, M.J. and Phanthumachinda, B. (1976) Control of *Culex pipiens fatigans* (W) by the larvivorous fish *Poecillia reticulata* and by removal of debris from the breeding habitat. *The Southeast Asian Journal of Tropical Medicine and Public Health* 1, 56–60.

Poché, R.M., Burruss, D., Polyakova, L., Poché, D.M. and Garlapati, R.B. (2015) Treatment of livestock with systemic insecticides for control of *Anopheles arabiensis* in western Kenya . *Malaria Journal* 14(1), 351.

Pooda, H.S., Rayaisse, J.B., Hien, D.F.D.S., Lefèvre, T., Yerbanga, S.R. *et al.* (2015) Administration of ivermectin to peridomestic cattle: A promising approach to target the residual transmission of human malaria. *Malaria Journal* 14(1), 496.

Randolph, T.F., Schelling, E., Grace, D., Nicholson, C.F., Leroy, J.L, *et al.* (2007) Invited review: role of livestock in human nutrition and health for poverty reduction in developing countries. *Journal of Animal Science* 85(11), 2788–2800.

Rowland, M., Durrani, N., Kenward, M., Mohammed, N., Urahman, H, *et al.* (2001) Control of malaria in Pakistan by applying deltamethrin insecticide to cattle: a community-randomised trial. *Lancet* 357, 1837–1841.

Rupp, H.R. (1996) Adverse assessments of *Gambusia affinis*: an alternate view for mosquito control practitioners. *Journal of the American Mosquito Control Association* 12(2), 155–159.

Russell, T.L., Govella, N.J., Azizi, S., Drakeley, C.J., Kachur, S.P, *et al.* (2011) Increased proportions of outdoor feeding among residual malaria vector populations following increased use of insecticide-treated nets in rural Tanzania. *Malaria Journal* 10(1), 80.

Sabatinelli, G., Blanchy, S., Majori, G. and Papakay, M. (1991) Impact of the use of larvivorous fish *Poecilia reticulata* on the transmission of malaria in FIR of Comoros. *Annales de Parasitologie Humaine et Comparée* 66(2), 84–88.

Saul, A. (2003) Zooprophylaxis or zoopotentiation: the outcome of introducing animals on vector transmission is highly dependent on the mosquito mortality while searching. *Malaria Journal* 2(1), 32.

Schultz, G.W. (1989) Animal influence on man-biting rates at a malarious site in Palawan, Philippines. *Southeast Asian J Trop Med Public Health* 20(1), 49–53.

Seng, C.M., Setha, T., Nealon, J., Socheat, D., Chantha, N, *et al.* (2008) Community-based use of the larvivorous fish *Poecilia reticulata* to control the dengue vector *Aedes aegypti* in domestic water storage containers in rural Cambodia. *Journal of Vector Ecology* 33(1), 139–144.

Service, M.W. (1991) Agricultural development and arthropod-borne diseases: a review [Desenvolvimento agrícola e doenças veiculadas por artrópodes: Revisão.] *Revista de Saúde Pública, São Paulo* 25, 165–178.

Seyoum, A., Balcha, F., Balkew, M., Ali, A. and Gebre-Michael, T. (2002) Impact of cattle keeping on human biting rate of anopheline mosquitoes and malaria transmission around Ziway, Ethiopia. *East African Medical Journal* 79(9), 485–490.

Shafique, M., Lopes, S., Doum, D., Keo, V., Sokha, L, *et al.* (2019) Implementation of guppy fish (*Poecilia reticulata*), and a novel larvicide (pyriproxyfen) product (Sumilarv 2MR) for dengue control in Cambodia: a qualitative study of acceptability, sustainability and community engagement. *PLoS Neglected Tropical Diseases* 13(11), e0007907.

Sinka, M.E., Bangs, M.J., Manguin, S., Rubio-Palis, Y., Chareonviriyaphap, T. *et al.* (2012) A global map of dominant malaria vectors. *Parasites and Vectors* 5(1), 69.

Sitaraman, N.L., Karim, M.A. and Reddy, G.V. (1975) Observations on the use of *Gambusia affinis holbrooki* to control *A. stephensi* breeding in wells. results of two years' study in greater Hyderabad city - India. *Indian Journal of Medical Research* 63(10), 1509–1516.

Sitaraman, N.L., Mahadevan, S. and Swamidas, S. (1976) Biological control of *A. stephensi* larvae in wells by *Poecilia reticulatus* in greater Hyderabad city, India. *Journal of Communicable Diseases* 8(4), 315–319.

Sota, T. and Mogi, M. (1989) Effectiveness of zooprophylaxis in malaria control: a theoretical inquiry, with a model for mosquito populations with two bloodmeal hosts. *Medical and Veterinary Entomology* 3(4), 337–345.

Sousa, C.A., Joã, J., Pinto, J., Paulo, A., Almeida, G, *et al.* (2001) Dogs as a favored host choice of *Anopheles gambiae* sensu stricto (Diptera: Culicidae) of São Tomé, West Africa. *Journal of Medical Entomology* 38(1), 122–125.

Steinwascher, K. (2018) Competition among *Aedes aegypti* larvae. *PLoS ONE* 13(11), e0202455.

Takken, W. and Verhulst, N.O. (2013) Host preferences of blood-feeding mosquitoes. *Annual Review of Entomology* 58, 433–453.

Taylor, D.S., Ritchie, S.A. and Johnson, E. (1992) The killifish rivulus marmoratus: a potential biocontrol agent for *Aedes taeniorhynchus* and brackish water *Culex*. *Journal of the American Mosquito Control Association* 8(1), 80–83.

Temu, E.A., Coleman, M., Abilio, A.P. and Kleinschmidt, I. (2012) High prevalence of malaria in Zambezia, Mozambique: the protective effect of IRS versus increased risks due to pig-keeping and house construction. *PLoS ONE* 7(2), e31409.

Thomas, S., Ravishankaran, S., Justin, J.A., Asokan, A., Mathai, M.T, *et al.* (2016) Overhead tank is the potential breeding habitat of *Anopheles stephensi* in an urban transmission setting of Chennai, India. *Malaria Journal* 15(1), 274.

Tirados, I., Costantini, C., Gibson, G. and Torr, S. (2006) Blood-feeding behaviour of the malarial mosquito *Anopheles arabiensis*: implications for vector control. *Medical and Veterinary Entomology* 20, 425–437.

Tirados, I., Gibson, G., Young, S. and Torr, S.J. (2011) Are herders protected by their herds? An experimental analysis of zooprophylaxis against the malaria vector *Anopheles arabiensis*. *Malaria Journal* 10(1), 68.

Valbon, W.R., Haddi, K., Gutiérrez, Y., Cruz, F.M., Azevedo, K.E.X, *et al.* (2019) Life history traits and predatory performance of *Belostoma anurum* (Hemiptera: Belostomatidae), a biological control agent of disease vector mosquitoes. *Neotropical Entomology* 48(6), 899–908.

Van Dam, A.R. and Walton, W.E. (2008) The effect of predatory fish exudates on the ovipostional behaviour of three mosquito species: *Culex quinquefasciatus*, *Aedes aegypti* and *Culex tarsalis*. *Medical and Veterinary Entomology* 22(4), 399–404.

Van Der Hoekl, W., Konradsenl, F., Dijkstra, D.S., Amerasinghe, H. and Amerasinghe, F.I. (1998) Risk factors for malaria: a microepidemiological study in a village in Sri Lanka. *Transactions of the Royal Society of Tropical Medicine and Hygiene* 92(3), 265–269.

Walshe, D.P., Garner, P., Adeel, A.A., Pyke, G.H. and Burkot, T.R. (2017) Larvivorous fish for preventing malaria transmission. *Cochrane Database of Systematic Reviews* 2017(12), CD008090.

Wang, C.H., Hwang, J.S. and Lay, J.R. (1990) Preliminary study on the biological control of dengue vectors by fish in Liouchyou Prefecture, Pingtung county, Taiwan. *Gaoxiong Yi Xue Ke Xue Za Zhi* 6(7), 382–388.

Watchorn, R.T., Maechtle, T. and Fedy, B.C. (2018) Assessing the efficacy of fathead minnows (*Pimephales promelas*) for mosquito control. *PLoS ONE* 13(4), e0194304.

White, G.B., Tesfaye, F., Boreham, P.F.L. and Lemma, G. (1980) Malaria vector capacity of *Anopheles arabiensis* and *anopheles quadriannulatus* in Ethiopia: Chromosomal interpretation after 6 years storage of field preparations. *Transactions of the Royal Society of Tropical Medicine and Hygiene* 74, 683–684.

Willems, K.J., Webb, C.E. and Russell, R.C. (2005) A comparison of mosquito predation by the fish *Pseudomugil signifier* Kner and *Gambusia holbrooki* (Girard) in laboratory trials. *Journal of Vector Ecology* 30(1), 87–90.

Wooten, M.C., Scribner, K.T. and Smith, M.H. (1988) Genetic variability and systematics of Gambusia in the southeastern United States. *Copeia* 1988(2), 283–289.

World Health Organization (1982) *Manual on Environmental Management for Mosquito Control with Special Emphasis on Malaria Vectors.* WHO, Geneva, Switzerland.

World Health Organization (2003) *Use of Fish for Mosquito Control. Regional Office for the Eastern Mediterranean.* WHO, Geneva, Switzerland.

World Health Organization (2004) *Global Strategic Framework for Integrated Vector Management.* WHO, Geneva, Switzerland.

World Health Organization (2009) *Dengue Guidelines for Diagnosis, Treatment, Prevention and Control.* WHO, Geneva, Switzerland.

World Health Organization (2017) *Global Vector Control Response.* WHO, Geneva, Switzerland.

World Health Organization (2020) *World Malaria Report 2020.* WHO, Geneva, Switzerland.

Wu, N., Wang, S.S., Han, G.X., Xu, R.M., Tang, G.K. *et al.* (1987) Control of *Aedes aegypti* larvae in household water containers by Chinese cat fish. *Bulletin of the World Health Organization* 65(4), 503–506.

Yamamoto, S.S., Louis, V.R., Sié, A. and Sauerborn, R. (2009) The effects of zooprophylaxis and other mosquito control measures against malaria in Nouna, Burkina Faso. *Malaria Journal* 8(1), 283.

Yewhalaw, D., Kelel, M., Degaga, E.G. and Temam Beyene, S. (2014) Blood meal sources and sporozoiterates of Anophelines in Gilgel-Gibe dam area, Southwestern Ethiopia. *African Journal of Vector Biology*. Available at: https://www.researchgate.net/publication/347314384_Blood_Meal_Sources_and_Feeding_Behavior_of_Anophelines_Mosquitoes_in_Bure_District_Northwestern_Ethiopia (accessed 13 September 2024).

Yu, H.S. and Lee, J.H. (1986) Biological control of malaria vector (*Anopheles sinensis* Wied) by conbined use of larvivorous fish (*Aplacheilus latipes*) and herbivorous hybrid (*Tilapia mossombicus niloticus*) in rice paddies of Korea. *The Korean Journal of Entomology* 28(4), 229–236.

Zvantsov, A., Kadamov, D. and Fozilov, H. (2008) *Experiences and Prospects of Use of Larvivorous Fishes for Control/Prevention of Malaria in Tajikistan.* World Health Organization, Geneva, Switzerland.

13

Culture and Vector-Borne Disease

J. D. Charlwood*

Global Health and Tropical Medicine, Instituto de Hygiene e Medicina Tropical, Lisbon, Portugal

Abstract

The influence of culture on health beliefs and practices, tracing historical attitudes from romanticizing primitive cultures to the rise of germ theory, is described. The challenge of reconciling individual actions with societal well-being, urging a shift in social attitudes to address global health issues, is stressed. Emphasizing the persistence of diseases of poverty requires a revolution in social attitudes and humility in decision making to address global challenges and environmental concerns. Community involvement in health initiatives is crucial, but obstacles like bureaucratic hurdles can hinder progress. Specific examples, such as the ineffectiveness of insecticide-treated mosquito nets against certain species, highlight the need for accurate information and local knowledge in disease control efforts. Advocating for community-managed larval control projects, especially in combating malaria transmission, underscores the importance of engaging communities in health interventions, particularly in addressing challenges posed by day-biting mosquitoes like *Aedes* and urban vectors like *Anopheles stephensi*.

> The systematic neglect of culture in health is the single biggest barrier to advancing the highest attainable standard of health worldwide.
>
> (Napier *et al.*, 2014)

> For culture, the ideal of permanency and durability, toward which a description of the physical world inevitably strives, is unattainable.
>
> (Radin, 1933)

> Babies, including the greatest future scientists, are humanized, as well as socialized, by being talked to and encouraged to talk. Human communication claims

*Corresponding author: jdcharlwood@gmail.com

DOI: 10.1079/9781789248845.0013

> them and equips them, from birth to function as members of a specific communicating human group. In doing so, it provides them with its own ways of thinking and talking about experience. It introduces them to a conceptual world already culturally shaped, which is their human heritage.
>
> (Vickers, 1968)

> Refusal to accept responsibility for looking ahead and for planning in matters national and international is based upon refusal to employ in social affairs, in the field of human relations, the methods of observation interpretation and test, that are matters of course in dealing with physical things and to which we owe the conquest of physical nature. The net result is a state of imbalance, of profoundly disturbed equilibrium between our physical knowledge and our social-moral knowledge. This lack of harmony is a powerful factor in producing the present crisis with all its tragic features.
>
> (Dewey, 1946)

Introduction

Culture may be considered to be the sum of the values and beliefs, language, communication and practices shared by a group of people. It can be considered to consist of our knowledge, common sense, assumptions and expectations, as well as the rules, norms, laws and morals that govern society. As pointed out by Lauwereyns (2010), 'The key challenge in the establishment of an open society is easily identified as the submission of individual actions to review by others, with the well-being of the many in society as the final goal and the ultimate criterion for what may be deemed acceptable'. How this can be done harmoniously among people with different, bias-influenced, world views remains problematical. He points out, 'The mechanisms of bias are so ingrained in human thought that we encounter them in many different guises in a wide variety of contexts and situations'. This makes the development of an equitable society even more difficult. As Orwell (1945) said, 'Society has always to demand a little more from human beings than it will get in practice'. Within a particular community, individuals need to feel, or be, empowered to make appropriate decisions regarding their health. 'Health and sickness are to be determined by the traditions to which the healthy or sick person belongs and within this tradition again by the particular ideal of life a person has formed for himself' (Feyerabend, 1991). The value systems of local populations are, however, often ignored by health professionals.

People in the West have often believed in an ideal state of health and happiness, either, in a Golden Age, in the past or, among optimists, in a disease-free future in times to come. The early reports of the happy state among natives of the South Sea islands (Tahiti, in particular) led to the romantic idea, of an Arcadia. These tales inspired philosophers like Rousseau for there to be a 'return to nature' in which people lived in harmony with the natural world. This idea of health resulting from living in harmony with nature was actually an old one. Hippocrates, the father of medicine, stated, 'It is more important to know what sort of person has a disease than to know what sort of disease a person has'.

Few people actually followed this recommendation of Rousseau, and, after the white man had destroyed the valuable attributes of primitive cultures, as pointed out by Dubos (1959), his attitude became again one of contempt for the savage, and he recovered assurance in the superiority of his own civilization.

Despite some opposition, the approach to health and disease changed following the demonstration of the germ theory by Pasteur and Koch. Specific causes for a patient's illness rather than their overall living conditions were sought. Thus, the reduction in diseases such as tuberculosis was attributed to specific interventions, even though, following an improvement in living conditions (inspired in part by the hygienists of the 18th century), the trend was already in steep decline.

The discovery by Laveran of the malaria parasite in human blood also led to the 'germ theory' approach exemplified in other diseases. One of the main problems with this, and many other health-related problems in the tropics, however, is that they remain diseases of poverty, and it is this that is the root cause of the disease.

Hygiene (derived from the Greek Goddess Hygeia, the guardian of health) involves a social philosophy that must take into consideration the human and economic aspects of the cultural pattern for which it is designed. White (2004), in response to a campaign by the *British Medical Journal* to revitalise academic medicine suggested expanding medicine's biomedical world-view to include the impacts of social, emotional, environmental, familial and occupational factors on health and disease, replacing the emphasis on single 'causes' for most diseases with profiles of predisposing, precipitating and perpetuating factors. These may affect biological outcomes (e.g. it may be a cultural norm to sit outside in the evening and to talk and so expose people to outdoor malaria transmission) but are often dismissed as sociological problems rather than medical ones. The WHO enrolled a number of prominent economists to participate in the Commission on Macroeconomics and Health. As pointed out by Banerji (2002), the emphatic recommendations for perpetuating vertical programs against major communicable diseases (tuberculosis, AIDS, and malaria), on the grounds that such programs have proved convenient to 'donors', reveals the real motivations for an almost openly ideology-driven agenda. The blame for continuance of transmission can then be cast on to the people themselves rather than the health professionals. This may have led Djukanovic and Mach (1975) to state, 'The de-professionalization of the health services is the single most important step in raising health standards in the third world'.

As pointed out by Wijkman and Rockström (2011) in their report to the Club of Rome, 'What is needed is nothing less than a revolution, both in attitudes and in social and economic organization'. Unfortunately, it is now almost jejune to quote Jameson (1991) and say that 'it is easier to imagine the end of the World than the end of Capitalism'. As described by Vickers (1968), 'During the past two centuries, men gained knowledge and power, which vastly increased their ability to predict and control; and they used these powers to make a world increasingly unpredictable and uncontrollable'. Similarly, Darlington (1969) wrote, 'All man's progress has been made at the expense of damage to the environment which he cannot repair and could not foresee'. That was 50 years ago, and since then, the situation has not changed. Governments should reconsider existing relations between decision makers, experts and citizens in the management of technology (which, in our case, equals the means by which vector control is attempted). As Latour (1998) points out there is now a third power player involved in decisions, and that is the planet. The planet will, more and more, dictate policy. Never before experienced climatic events and their impact on all living things are becoming more frequent. We now live in a postmodern world in which history is meaningless. Policy makers, and authors, need a 'technology of humility' for assessing unknown and uncertain outcomes from their decisions (Jasanoff, 2003, 2007).

Care should always be taken when attempting a control measure that either affects, or involves, community participation. 'Fools rush in where angels fear to tread'. In particular, it is important that people know what an intervention will not do. Thus, information associated with any intervention needs to be accurate, informative and to the point. For example, mosquito nets impregnated with a pyrethroid insecticide have a limited effect against *Culex quinquefasciatus*, and people should know this. For this, of course, having a population that knows the difference between genera of mosquitoes would be useful. What people might do if they see anophelines is described in the chapter on sampling. Knowing the difference between an anopheline and a culicine or an aedes is not rocket science. Yet posters, websites, news items and such more often than not show one of the latter species when discussing malaria. The value of education cannot be overstated in empowering people. For example, community knowledge about malaria enhanced insecticide-treated net use in Ghana (Adongo *et al.*, 2005).

In Ifakara, Tanzania, in the 1990s, before the days of LLINS, bed nets impregnated with a pyrethroid insecticide that required retreatment every 6 months were sold on the basis that they would control mosquitoes, not just anophelines (Schellenberg *et al.*, 1999). However, they failed to have a noticeable effect on the mosquitoes in town (although they worked against malaria vectors in the surrounding villages), because those were largely *Cx. quinquefasciatus*, which is tolerant to these insecticides. I was told that many people living in towns had relatives in the villages. At the time, retreatment could be undertaken in town but not in the villages. Few people took advantage of this, perhaps because, given the manifest continued presence of the *Cx. quinquefasciatus*, townspeople told their rural relatives that it didn't work. The proportion of nets that were retreated was, consequently, extremely low. This was perhaps because the public health messages were that the nets would kill mosquitoes (not reduce malaria transmission). Had 'this will reduce malaria but not all mosquitoes' been the message, then perhaps retreatment rates would have improved, because transmission in towns would anyway have been low, so there would be no indication that nets were not working (since rates among users might have gone down). Indeed, the effect of control in areas where transmission was intense would probably have been noticeable even among individual households. For example, before control (using impregnated nets treated with PBO and a pyrethroid and with IRS with pirimiphos-methyl, commercially known as Actellic CS300) every time JDC visited a friend's house in Kyamyorwa village in Muleba District, in the Kagera Region of Tanzania, he was asked to test at least two of the residents for malaria (mostly, of course, young children) and, most of the time, one or both were positive for *P. falciparum*. Following the intervention in 2015, he was never asked to test anyone, despite being a reasonably constant presence in the household.

Women's Groups

There is also 'many a slip' 'twixt cup and lip'. Sometimes, no matter how self-sufficient an intervention is, unforeseen obstacles can upset things, as happened to 'Tupendane', a women's group in Ifakara, Tanzania (Charlwood, 2008).

Adulticides, even if effective against *Cx. quinquefasciatus*, a major nuisance in many towns in the tropics, do not work. This is 'the little mosquito – delicate dancing hum in the air' – as Ray Bradbury would have it, that buzzes in your ear before attempting to feed, 'small, high, hateful bugle in my ear', perhaps to determine that you really are asleep and will not engage in defensive behaviour. A sleepy arm in the dark against a flying mosquito, at the ready to escape at the slightest change in air pressure generated by the descending arm, is only ever going to produce a self-inflicted thick ear. Adulticides have little effect because the adult survival rate is generally very low, even in the absence of any intervention. In many cases, the great majority (around 80%) of insects are newly emerged (Charlwood, 1979). Although it might kill all of the older insects, this would still only be a 20% reduction in numbers. The insect is generally 'tolerant' of many insecticides, meaning that mortality increases regularly as dosage increases. They will keep emerging for long enough that people consider the technique ineffective. Larval survival is high, and densities can be enormous. Larval control is obviously the way to go. One of the ways of control that had been tested in wet pit latrines to good effect (Reiter, 1978; Curtis and Minjas, 1985; Maxwell *et al.*, 1990), but not in septic tanks, was the use of expanded polystyrene beads for larval control. We obtained some barrels of the beads and conducted an experiment in the early 1990s that showed that they were, indeed, also effective in septic tanks (Charlwood, 1994). A blanket of beads a few centimetres thick prevents gravid female mosquitoes from laying their eggs and drowns larvae, while at the same time allowing solid matter to pass through. The beads refloat in latrines that have temporarily been dry and, in theory, do not need to be applied more than once. Since they are in an enclosed environment, they do not pollute anything. They also reduce the smell. At the time of the experiment, Tanzania was in transition from one-party rule to multiparty democracy.

The beads come unexpanded in drums. The polystyrene looks a little bit like white sand. It contains a gas which expands when heated to 100°C and in so doing expands the bead. Adding the material to a pot of boiling water is a little bit of magic. However, unless it is expanded, the gas gradually leaks out, and the product loses its value. There were a couple of drums left over from the experiment. The local pit latrines were an obvious target for the beads. We ourselves did not have the time, manpower or funds to do this. Just outside the institute there lived a bevy of young, vivacious, single mothers making a hand-to-mouth living. The idea of earning some money, a rare opportunity as far as these young women were concerned, was delightful (as the Everly Brothers used to sing, 'Money's important when you haven't any'). And so 'Tupendane', which means 'We like each other' in Kiswahili, was born. It was suggested that the money they might earn from the first barrel be spent in any way that they liked, but they save enough from the sale of the second drum to buy another one and continue.

They took off like wildfire. One or two of the women went around taking orders, another couple was responsible for boiling the water and expanding the beads (equals buying and carrying the wood, etc..) and one other pair was responsible for their application. They had orders from all over town. People flocked to pay for it to be put into septic tanks (with the occasional pit latrine). JDC acted as the group's ombudsman. As ombudsman, it was his job to follow up on any complaints and to make spot checks to determine that the layer of beads applied was sufficiently thick to kill larvae and to last long enough for the investment to have been worthwhile.

These women had never been so happy. The first barrel was used in literally a few days. This was in May, the rainy season, when mosquito densities were very high. Given that this was Tanzania in a time of transition from a one-party state to a multiparty democracy, who should come along but someone from the District Health Committee (DHC) to tell the women that 'Health provision in Tanzania is free' and that they should, therefore, deliver it for free or face proceedings. So, of course, they stopped. JDC learnt that this committee was to meet in a couple of weeks and managed to get 'Tupendane' on to the agenda. He attended and described the process and what diseases would be prevented if a large number of households used it. The man who had stopped them in the first place slept through it all. The upshot was that everyone agreed that the women were doing a good job and should carry on. The committee did suggest, however, that since money was involved, the issue should perhaps also be brought to the attention of the District Development Committee (DDC). JDC, of course, volunteered to go and tell them too, but was told, 'No, no. No need to bother, the committee has seven members, three of which are here now. Your case will be well represented, and this will really be just a formality'. It was only on leaving, after sitting through a number of other, less interesting, topics, that JDC happened to ask when the next DDC meeting would be. The informant rubbed his chin for a few seconds. 'Mmm, April, I think'. We were in the second week of May! The women did eventually get permission to continue, but by this time, they had all started doing other things, and anyway, their hearts were no longer in it.

So, *Cx. quinquefasciatus* continues to live, and 'Tupendane' died. Not long after they had done so, people were coming from Dar es Salaam to see this, somehow, famous women's group! On the other hand, where malaria vectors and *Cx. quinquefasciatus* occur together (as they do in São Tomé), the presence of the latter may encourage people to use their nets (even if they are not impregnated with insecticide), and so these mosquitoes may be unwitting accomplices in malaria control!

Internetworks

The day-biting *Aedes aegypti* and *Aedes albopictus* are less affected by many of the techniques used to control the major vectors of malaria, as indeed the invasive vector *Anopheles stephensi* may be less affected by conventional control than the major African vectors *Anopheles gambiae* and *Anopheles funestus*. This is because *An. stephensi* is an urban vector that is able to breed in artificial containers, such as water tanks, and that tends to bite outside in the early part of the night. Nevertheless, techniques that are effective against *Aedes* may well be useful against this vector. People, not just entomologists, living in cities (particularly those who do a considerable amount of trade with Ethiopia, Sudan and Djibouti) should be made aware of the possibility of *An. stephensi* invasion and know what to look for. Posters, or an app describing the things to look for and who to call should they be encountered, could be on display in high-risk areas. Before the entomologists actually find them, increased search intensity by the public might be useful. The establishment of a network in which people can send photographs taken on their phone of mosquitoes for identification (as it is available in Portugal: *www.mosquitoweb.pt*) may facilitate the ongoing search for *An. stephensi* (which really should start now). It should not be difficult for people to scan their legs every so often as they, anyway, sit outside in the early evening. If people are aware of the threat, they may be more likely to respond to it (by looking). Professionals can verify the presence of the mosquito and attempt control, again using local resources (community involvement), if possible. Presumably, locating and controlling potential and actual breeding sites would be more important than using adulticides.

Fred Soper (he of *An. gambiae* in Brazil fame) made his reputation by tracking down 'Typhoid Mary' using case histories over the previous 10 years from households from which she, Mary Mallone, had worked (Marineli *et al.*, 2013). Although she was healthy, she was a source of infection in others. The poor woman was incarcerated, in quarantine, for 20 years. Prison and isolation should have been the ultimate solution. Simple precautions would have prevented future outbreaks. Indeed, all control efforts should be designed to work in a situation where such carriers, be they a source of Covid-19 or malaria, exist in the population. Efforts should be made to reduce the number of sources by offering treatment, for free, to passive carriers. Such passive carriers, even those who don't help in control or elimination efforts by getting tested and treated (or at least adopting a lifestyle that reduces the possibility for transmission), are not criminals and, unlike poor Mary, should not be treated as such. This is where information should be available to the concerned citizen. Information on the potential effects of increased passive surveillance on transmission (resulting in a reduction in the number of reservoir hosts) should be available. It should be presented in such a way that the effect that this reduction in passive carriers has on the prospects for control and should over time be updated to include estimates of the prevalence of such carriers based on routine data collected over time. The closer to an estimate in real time to prevalence of such carriers, covering as small an area as possible, the better. This can be translated into a place on the graph produced by the models.

These same effects of reducing the infectious reservoir could also be presented for different vector control strategies – on, for example, the effect as more and more people sleep under mosquito nets; or the effect that identification of potential breeding sites and removal of actual ones has on transmission or, in the case of *Cx. quinquefasciatus*, mosquito density.

What the effect of such community-organized control might mean in terms of malaria transmission is worth modelling and might encourage larval control by community management. On Praia Burra on the island of Príncipe in the Gulf of Guinea, a water body with a very high density of *An. gambiae* larvae was a wheel rut in the middle of the village. A plastic cup, containing a number of larvae and pupae from this wheel rut, was set up in a cage on a table in the middle of the village. As soon as the villagers saw that larvae did, indeed, turn into malaria mosquitoes, they went and filled it in. They finished the job in an enthusiastic minute or two.

At the same time, studies in Praia Burra showed how adaptive the mosquito could be when appropriate water was available. The village was also the site of a completely different larval concentration.

A small stream entered the sea at one end of the village (stretched along the shore). Wave action periodically created a dam of sand where it entered the ocean. Over a considerable time, the stream turned into a lagoon. In the middle of the lagoon, reached by canoe, a large 'mat' of spirogyra occurred, and in among the plant's tendrils lived large numbers of *An. gambiae* larvae. Villagers, of their own accord, broke the dam (using the proverbial shovel), and the lagoon – which was really quite pretty – disappeared along with the mosquitoes.

As occurs with a diminishing infectious reservoir, the effects of source reduction will increase as the numbers of breeding sites are treated. In Massavasse, Mozambique, the only vector of malaria among many species collected was *An. funestus*. In a 3-year study, this was the only species of anopheline collected, in any number, resting inside houses and the only species that had members with a positive ELISA for sporozoites (Charlwood *et al.*, 2013). As is the case elsewhere, mosquitoes left houses at dusk – the males to swarm, the newly emerged females to mate and the gravid females to find an oviposition site and lay a batch of eggs. As described in the chapter on sampling, it is easy to catch these mosquitoes without the need for any sophisticated equipment. They only last for half an hour and pose no risk to the collector. Under such circumstances, exit collections at dusk provide a suitable sample of adult malaria vectors in places like Massavasse. All collections (light trap, tent trap, resting, exit and larval collections) indicated that *An. funestus* had a focal distribution. Larvae were found in a limited number and type of habitat. One might expect the number of newly emerged adults to decrease as a result of larval control. A greater proportion of the exit collection would be gravid females, and overall numbers would decline. These are all indicators that most people would be able to understand. One person could easily be trained in species identification, sex and gonotrophic condition and in the production of a paper chart, showing these data. Since the numbers involved are simple, data processing can be done both on-site and at a central database (in which several such sets of data are collected and analysed). Training villagers and having them undertake appropriate methods of larval control, the results being monitored by changes in exit collection, is the sort of community-organized control project that should be encouraged.

It is also as well to remember that correlation does not equal causation. Massavasse also provided a lesson in a different way. During the study, the interior walls of houses in the village were sprayed (with bendiocarb), and subsequently, numbers of *An. funestus* decreased, as did the proportion of the collection from indoor light traps. However, numbers of other mosquitoes caught in tent traps also decreased. Because they did not enter houses (and were not caught in light traps indoors), these mosquitoes would have been unaffected by the spray. One reason for their decline may have been because villagers had failed to pay the previous year's water bill, and so the supply of water to the canals was restricted, reducing the available sites for the development of many mosquito species, including probably *An. funestus*.

Even with relatively clear-cut data sets, more than one cause might be operating. For example, there was a good negative correlation between bed net use and malaria prevalence over a considerable time period in São Tomé (Fig. 13.1; Charlwood *et al.*, 2005).

The obvious interpretation might be that the nets were responsible for the decline in malaria. However, alternative interpretations are also possible. At the start of the study, it was common practice for people to sit in the open, under houses built on stilts, while watching Brazilian soap opera on TV. This exposed them to malaria transmitted by the exophagic *Anopheles coluzzii* that is found on the island. During the years that the study took place, the number

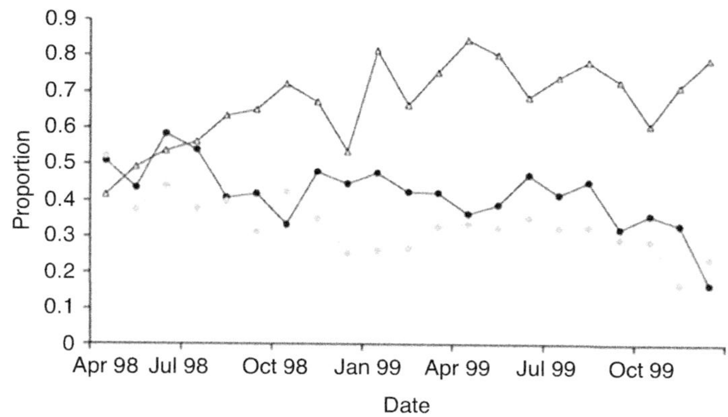

Fig. 13.1. Longitudinal variation of bed net use and malaria prevalence in bed net users and non-users. Key: (———) prevalence in non-net users; (···■···) prevalence in net users; (—△—) proportion of attendees using bed nets. (From Charlwood *et al.*, 2005.).

of people that owned their own TVs increased substantially, which meant that more people watched their own TVs indoors. This would have reduced their exposure to the vector. Unfortunately, data on TV ownership, or evening activities, were not collected.

Ignorance can also affect traditional anti-vector spraying programmes. Even in towns, many people have their gardens, and in rural areas, people live on what they produce. People are not unaware of the apparent advantages of insecticides against insect crop pests. One source of cheap insecticide, sold on the sly during house spraying campaigns, is what otherwise should be used to spray the interior walls of houses. In Massavasse, the village headman brought a packet to me and asked about dosage for his vegetables.

The diversion of insecticide has a number of possible effects. Of necessity, less insecticide reaches the walls than projected (or recorded). This reduces the dose that the mosquitoes have to withstand in order to survive. The use of the insecticide on crops means that it will get washed into water bodies, and consequently, mosquito larvae will also be challenged 24/7 with a very low dose of the insecticide. Together, these effects are likely to lead to resistance, in which case, even the properly applied dosage on house walls is of little use. People are not likely to know the appropriate dosages against agricultural pests. Because of its scarcity, they are also likely to underdose against agricultural pests, giving rise to resistance in these too! It is possible that if the person responsible for the spraying was local, rather than a stranger, people would be more reticent about such illicit purchases. Indeed, people might be more likely to insist that they do a proper job.

As Marcel Proust wrote, 'Pain we obey'. Once people put a value on something (either as a cost, as in a fine, or as a benefit, such as a child's education), then they act to preserve it. There was a habit among many residents of Praia Burra on the island of Príncipe to defecate on the beach, in the hopes that the incoming tide would remove any faeces. The threat of a US$200 fine for anyone who continued to do this stopped the practice (and improved the beach) immediately (Fig. 13.2).

Benefits Versus Costs

Better than a cost, however, surely is a perceived benefit. In Brazil in the 1980s, the Instituto Naçional de Pesquisas da Amazonia (INPA) supported a preschool, initially designed to cater for 20 or so children whose parents were employees. The school also catered for children living close by in the favela of Coroado, and the school eventually had around 350 pupils (all getting a school meal). The only 'payment' that parents had to make was that they had to attend a monthly parents' evening. A few parents failed to attend. Sending the children home (and then allowing them to return after abject parental apologies) eliminated that problem in all but the parents who worked at the institute and whose children could not be sent home since they were there by right. With so many pupils, the physical size of the school increased from a single small house to a sizeable area including three or four classrooms and a playground. This made the school a potential target of squatters. During

Fig. 13.2. 'Pain we obey'. View of Praia Burra from the author's house. A summary fine of US$200 immediately prevented defecation on the beach. (Photo: J.D. Charlwood.).

a meeting describing the problem, hands were raised, and various men said that they were carpenters, and each could supervise and help build a fence around the school grounds. The following weekend, it was done in a day with a party in the evening. True empowerment. Communal approaches like this are not uncommon in many parts of the world. In Papua New Guinea and Mozambique, putting a traditional roof on a house is one such activity undertaken by many people together. The time of the activity (putting the roof on) is always in the hands of the people doing it, not the house owner. It is a resource that remains untapped in many parts of the world (and, of course, is being lost as tin roofs replace traditional ones).

How, though, do we put a value on things? In the marketplace, the existence of high-priced/out-of-reach objects (e.g. Leica among cameras, Ferraris and their ilk for cars, Rolex among watches – the list could go on to include shoes and handbags) makes the price we pay for our own, still often expensive but almost equivalent, versions of these things seem to be so much more of a bargain. The closer we get to our own idea of a Leica (in a very diverse market), the more we ourselves value it. It follows that if one wanted to establish a market for bed nets, then making one that is an absurd price (with mod cons added in, speakers at the end, a fan, etc..) might make the purchase of a perfectly good net (still from a range with different qualities and prices) seem worth it. Nets might offer a variety of possible insecticide combinations or, if preferred, be untreated. Other options might include material or pockets sewn in to hold phones or glasses.

Other techniques that might be suited to such an approach are techniques for improving mosquito proofing in houses. Improvements to housing will be driven by the market, as described in Chapter 8 of the present volume.

References

Adongo, P.B., Kirkwood, B. and Kendall, C. (2005) How local community knowledge about malaria affects insecticide-treated net use in Northern Ghana. *Tropical Medicine and International Health* 10, 366–378. DOI: 10.1111/j.1365-3156.2005.01361.x.

Banerji, D. (2002) Report of the WHO commission on macroeconomics and health: A critique. *International Journal of Health Services* 32, 733–754.

Charlwood, J.D. (1979) Estudos sobre a biologia e habitos alimentares de *Culex quinquefasciatus* Say em Manaus Amazonas Brasil. *Acta Amazonica* 9, 271–278.

Charlwood, J.D. (1994) The control of *Culex quinquefasciatus* breeding in septic tanks using expanded polystyrene beads in southern Tanzania. *Transactions of the Royal Society of Tropical Medicine and Hygiene* 88, 380.

Charlwood, J.D. (2008) Women's groups and the marketing of health interventions a Tanzanian experience. *PNG Medical Journal* 51, 102–104.

Charlwood, J.D., Alcântara, J., Pinto, J., Sousa, C., Rompão, H. *et al.* (2005) Do bednets reduce malaria transmission by exophagic mosquitoes?. *Transactions of the Royal Society of Tropical Medicine and Hygiene* 99, 901–904.

Charlwood, J.D., Macia, G.A., Manhaca, M., Sousa, B., Cuamba, N. *et al.* (2013) Population dynamics and spatial structure of human-biting mosquitoes, inside and outside of houses, in the Chockwe irrigation scheme, southern Mozambique. *Geospatial Health* 7, 309–320.

Curtis, C.F. and Minjas, J. (1985) Expanded polystyrene for mosquito control. *Parasitology Today* 1, 36.

Darlington, C.D. (1969) *The Evolution of Man and Society*. Allen & Unwin, London, 751 pp.

Dewey, J. (1946) *Problems of Men*. Philosophical Library, New York, 320 pp.

Djukanovic, V. and Mach, E.P. (1975) Alternative approaches to meeting basic health needs in developing countries. In: Newall, K.W. (ed.) *Health by the People*. World Health Organization, Geneva, Switzerland.

Dubos, R.J. (1959) *The Mirage of Health: Utopias, Progress and Biological Change*. Rutgers University Press, New Brunswick, New Jersey.

Feyerabend, P.K. (1991) *Three Dialogues on Knowledge*. Blackwell, Oxford, UK.

Jameson, F. (1991) *Postmodernism, or, the Cultural Logic of Late Capitalism*. Duke University Press, Durham, North Carolina.

Jasanoff, S. (2003) Technologies of humility: Citizen participation in governing science. *Minerva* 41, 223–244. DOI: 10.1023/A:1025557512320.

Jasanoff, S. (2007) Technologies of humility. *Nature* 450, 33. DOI: 10.1038/450033a.

Latour, B. (1998) *Remaking Reality: Nature at the Millennium*. Sciences, Paris, France.

Lauwereyns, J. (2010) *The Anatomy of Bias: How Neural Circuits Weigh the Options*. The MIT Press, Cambridge, Massachusetts.

Marineli, F., Tsoucalas, G., Karamanou, M. and Androutsos, G. (2013) Mary Mallon (1869–1938) and the history of typhoid fever. *Annals of Gastroenterology* 26, 132–134.

Maxwell, C.A., Curtis, C.F., Haji, H., Kisumku, S., Thalib, A.I. *et al.* (1990) Control of Bancroftian filariasis by integrating therapy with vector control using polystyrene beads in wet pit latrines?. *Transactions of the Royal Society of Tropical Medicine and Hygiene* 84, 709–714.

Napier, A.D., Ancarno, C., Butler, B., Calabrese, J., Chatler, A. *et al.* (2014) Lancet commission on culture and health. *The Lancet* 384, 1607–1639.

Orwell, G. (1945) Notes on nationalism. *Polemic*. Available at: https://www.orwellfoundation.com/the-orwell-foundation/orwell/essays-and-other-works/notes-on-nationalism/ (accessed 13 September 2024).

Radin, P. (1933) *The Method and Theory of Ethnology: An Essay in Criticism*. McGraw-Hill, New York.

Reiter, P. (1978) Expanded polystyrene beads: An idea for mosquito control. *Annals of Tropical Medicine and Parasitology* 72, 590–595.

Schellenberg, J., Abdulla, S., Minja, H., Nathan, R., Mukasa, O. *et al.* (1999) KINET: A social marketing programme of treated nets and net treatment for malaria control in Tanzania, with evaluation of child health and long-term survival. *Transactions of the Royal Society of Tropical Medicine and Hygiene* 93, 225–231.

Vickers, G. (1968) *Value Systems and Social Process*. Tavistock Publications, London.

White, K.L. (2004) Some thoughts for academic revisionists to consider. *British Medical Journal* 328, 1377.

Wijkman, A. and Rockström, J. (2011) *Bankrupting Nature: A Report to the Club of Rome*. Routledge, Oxford, p. 206 .

The Role of Politics in Vector Control

14

J. D. Charlwood*

Global Health and Tropical Medicine, Instituto de Hygiene e Medicina Tropical, Lisbon, Portugal

Abstract

It is through political leadership that changes can take place. Politicians need success in order to justify their decisions and the resources allocated to health and vector control. However, if control has been achieved, it becomes more difficult to justify the continued investment. This influences various philosophical and practical aspects of healthcare, particularly in the context of malaria and vector-borne diseases. We advocate for a holistic approach that addresses not only material poverty but also the poverty of satisfaction and dignity. Critiquing the reliance on GDP as a measure of well-being, we call for a shift towards valuing education, public debate and integrity. The discussion includes proposals for decentralized healthcare systems, such as health vouchers, to promote diversity and individual autonomy. Political decisions significantly influence disease control outcomes, as evidenced by the case of DDT reintroduction in The Gambia. Additionally, the unintended consequences of interventions, such as distributing low-quality impregnated mosquito nets in Tanzania, is described. It emphasizes the importance of community engagement, informed decision making and responsible communication to combat misinformation and promote social justice in the digital age.

> The great mystery of modernity is that we think of certainty as an attainable state. (Jasanoff, 2007)

> In society, as in nature, we depend always on an order that we did not invent, and cannot re-create, our task can only be to remove the obstacles that we have ourselves put in the way of its natural healing, and, were this not enough, to provide prophylaxis against hazards generated by our own virtuosity. (Gray, 2009)

*Corresponding author: jdcharlwood@gmail.com

DOI: 10.1079/9781789248845.0014

Introduction

'If a job is worth doing, it is worth doing badly' has always been the argument I give for doing the job that I do. Perhaps to, at least, do no harm, perhaps to help reduce material poverty. But as Robert Kennedy said about the United States in 1968:

> Even if we act to erase material poverty, there is another greater task. It is to confront the poverty of satisfaction – purpose and dignity – that afflicts us all'. Too much and for too long, we seemed to have surrendered personal excellence and community values in the mere accumulation of material things. Our Gross National Product now is over $800 billion dollars a year, but that Gross National Product – if we judge the United States of America by that – counts air pollution and cigarette advertising and ambulances to clear our highways of carnage. It counts special locks for our doors and the jails for the people who break them. It counts the destruction of the redwood and the loss of our natural wonder in chaotic sprawl. It counts napalm and counts nuclear warheads and armoured cars for the police to fight the riots in our cities. It counts Whitman's rifle and Speck's knife, and the television programs which glorify violence in order to sell toys to our children. Yet, the gross national product does not allow for the health of our children, the quality of their education or the joy of their play. It does not include the beauty of our poetry or the strength of our marriages, the intelligence of our public debate or the integrity of our public officials. It measures neither our wit nor our courage, neither our wisdom nor our learning, neither our compassion nor our devotion to our country, it measures everything in short, except that which makes life worthwhile.

As pointed out by Lauwereyns (2010), 'We must relinquish individual power and authority and somehow come to terms with the idea that to be part of an open society implies the assumption of responsibility and the readiness to be judged by our peers'. Our thought, like our very physical senses, is itself a product of a history with which it engages. Unfortunately, 'mechanisms of bias are so ingrained in human thought that we encounter them in many different guises in a wide variety of contexts and situations' (Lauwereyns, 2010) or, as Paul Simon would have it in his song 'The Boxer', 'A man hears just what he wants to hear and disregards the rest'.

Orwell (1945) wrote, 'There is no law, there is only power. I am not saying that this is a true belief, merely that it is the belief that all modern men do actually hold. Those who pretend otherwise are either intellectual cowards, or power-worshipers under a thin disguise, or have simply not caught up with the age they are living in'. The situation has not changed. The Roll Back Malaria (RBM) group of the WHO, on the anniversary of the discovery of the role of the mosquito in malaria transmission, quoted Arundhati Roy, who wrote 'The Pandemic Is a Portal', in *The Financial Times* on 3 April 2020,

> Historically, pandemics have forced humans to break with the past and imagine their world anew. This one is no different. It is a portal, a gateway between one world and the next. We can choose to walk through it, dragging the carcasses of our

> prejudice... our data banks and dead ideas. Or we can walk through lightly, with little luggage, ready to imagine another world. And ready to fight for it. (Roy, 2020)

The RBM considered that malaria control, in particular, vector monitoring and control, had arrived at such a portal. The present book includes our responses to this challenge.

It means a change in the power relationships within, what Illich (1976) calls 'the new world-spanning church of medicine'.

It remains the opinion of the author that 'ultimately, the way to improve health standards in the world is to make people aware that they can look after themselves and to give them the wherewithal for doing so' (Charlwood, 1984). It also remains true that, 'if the amount of transmission of all species of malaria in a village can be effectively reduced by the judicious use of a pick and shovel, then the time to use them is now'. It remains important to monitor their effects so that their usefulness can be assessed. The conservative thinker John Gray wrote in 2009, 'Reform of contemporary medical institutions and practice cannot avoid taking the road of deprofessionalization, of curbing and limiting professional monopoly over health care'. Galen, the most influential doctor from Roman times to the Middle Ages, on his return from Aquileia, 'made a rare and startling admission: the regular people who had survived the disease (the Antonine plague), because of their experience and common sense, sometimes offered better advice than professional doctors' (Elliott, 2024). An extended health voucher or health credit scheme of the sort proposed below, though not without difficulties, could assist in promoting this objective. Gray also says, and I agree with him, that 'The perception of both health and the definition of illness are personal judgements rooted in cultural traditions, it would devolve healthcare decisions to the level of specific individuals in definite communities'.

A health credit that was usable in a diversity of medical traditions would meet Paul Feyerabend's desideratum that 'health and sickness are to be determined by the traditions to which the healthy or sick person belongs and within this tradition again by the particular ideal of a life a person has formed for himself'. If someone prefers to use their credit to go to a traditional healer, then that is up to them. Or if they prefer to put their credits towards a 2-month vacation in the Bahamas rather than another round of chemotherapy, which will only delay the inevitable, that is also up to them.

'Our ignorance as to what are the best remedies for our ailments justify the most liberal form of voucher, or health credit scheme'. But 'it would be a serious error to regard this as a panacea for all the dilemmas of health care' (Gray, 2009), just as, in the financial sphere, blockchain solutions should not be regarded as a panacea. For all such systems to be effective requires an educated public.

So how can these opinions and possible approaches be enacted in the sphere of malaria (and perhaps dengue) control and healthcare in general, not just in Africa but everywhere? As pointed out by DeLanda (2004), 'The role of the thinker is not so much to utter truth or establish facts but to distinguish among the large proportion of true facts those that are important and relevant from those that are not'. I don't claim to be much of a thinker ('bear of little brain'), and so, even as far as vector control is concerned, I continue to find it difficult to distinguish which facts are the really important ones, but having citizens participating in governing science (Jasanoff, 2007) would seem to be one of them.

Problems with Capitalism

Our economic and financial system – capitalism – shapes a range of behaviour patterns incompatible with sustainability. As the economic historian Fernand Braudel pointed out, capitalism is not the same as markets. He described long phases of economic history with 'material life' of basic subsistence as the bottom layer of development. Above this lies 'economic life', in which people encounter one another as equals in relations of exchange, or markets. Markets are characterized by transparency so that both prices and activities are visible to everyone and, because of competition, profits are small. Above this lies 'capitalism', a zone of anti-markets that are opaque and that involve monopoly, concentration of power and wealth and the kind of profit that is only achieved by escaping the norms of 'economic life'. Traders in the market engage with one another using a shared set of rules. Capitalists exploit their unrivalled control over time and space to impose their own rules on everyone. This directly affects the future of humanity on our planet. Inherent to it is the need for constant exponential growth, by definition unsustainable in a finite world. In addition to creating the conditions for boom-and-bust cycles, the system leads to short-term thinking, concentrates wealth and erodes equality, which promotes selfish and non-collaborative behaviours.

'Money talks' (or perhaps, as Bob Dylan would have it, 'Money swears'); it is not a passive commodity. Most of this money has no use. It is estimated that the total value of economic activity in the world is US$105 trillion. The value of the financial derivatives that result from this activity is US$667 trillion. This activity is a zero-sum game, in which winners are matched by losers. It adds no value. It does nothing and overall creates no benefit for society.

As Stendhal wrote in 1830 in his novel 'The Scarlet and Black', '*Bringing in money* – that is the magic phrase determining everything in Verrières; by itself alone it represents the usual subject of thought of more than three-quarters of its population. *Bringing in money* – is the decisive reason for everything in this little town you thought so pretty' (Stendhal, 1969). One reason for some of the present problems can be attributed to the present financial *monoculture* considered as a closed system by most economists. Monocultures are more prone to collapses and large fluctuations than diverse ecosystems. Closed systems, if left to their own devices, tend towards static equilibrium. The economy is, however, better viewed as an open system of complex flow networks in which money flows between and among different economic agents. What is needed is a *financial ecosystem* that balances the two complementary but equally important properties of efficiency and resilience. When, as happens at the moment, too much emphasis is placed on efficiency, diversity (a component that provides resilience) is lost, and the system is liable to sudden collapses. 'Confronting the threat of both climate and ecological breakdown will require us to remake the infrastructure and arrangements of the entire global economy. This is a Herculean task, and our time is short' (Buller, 2022). The development and use of electronic ADA (blockchain stakeholder) currencies that can act as a parallel currency (which eliminates banks and requires 100 times less energy than Bitcoin) increases diversity and hence stability in the economic sphere. How such currencies might generate a parallel economy is described by Lietaer *et al.* (2012). How such a system might benefit control of the invasive vector *Anopheles stephensi*, every malaria vector control person's nightmare, has been described by Charlwood (2023). Thus, a separate form of currency based on a system like ADA of the Cardano blockchain is a primary requirement. Despite the problems that affect the token-based system, including blockchain currency, how to establish such a system should be a healthcare priority. It would assist in diversifying, and so making more stable, fiscal exchange. A non-capitalist system could be tailored by the users (i.e. villagers with a mobile phone) throughout the world. How such a system might be developed for the reuse of old LLINs (as covering over the opening between roof and walls) is described in Chapter 8 on housing.

ADA from Cardano

The following is an annotated version of information available from the Cardano website (cardano.org).

Cardano is a decentralized third-generation proof-of-stake blockchain platform. Blockchain technology, otherwise known as distributed ledger technology (DLT), provides a decentralized and accessible data structure for various records. Such records might include financial payment and transaction details, as well as other types of information – from commerce to internet of things (IoT) records. A blockchain is a type of database or ledger that is duplicated and distributed to all participants within the blockchain network. It is made up of a set of interconnected nodes that store data or items of value in blocks. These blocks are validated, cryptographically secured and linked to each other in chronological order in a chain. The information stored in the validated blocks of the blockchain is permanently inscribed in the distributed ledger.

As a blockchain stores data in a decentralized manner, it is independent of centralized, controlling entities or middlemen. This provides enhanced transparency of data storage and its management. An important feature of blockchain is that it stores records immutably, which means that they cannot be changed, forged or deleted, as this will break the chain of records.

Blockchains not only provide an immutable and secure database but also act as a functional environment to transact funds, create digital currencies and process complex deals using digital agreements.

A cryptocurrency is a digital asset, which is stored on the ledger and is designed to serve as a medium of exchange for goods or services. It is otherwise called crypto. Cardano is home to the ADA cryptocurrency. It has a strong focus on sustainability, scalability and transparency. It is a fully open-source project that aims to deliver an inclusive, fair and resilient infrastructure for financial and social applications on a global scale. One of its primary goals is to bring reliable, secure financial services to those people who do not currently have access.

Cardano also supports the creation of native tokens – digital assets that are created for specific purposes – in the present projects case 'tokens' for a variety of uses. This means that users, developers and businesses can use the Cardano blockchain to create tokens that represent a footprint of value (whether defined by the community, market state or self-governed entity). A token can be fungible (interchangeable) or non-fungible (unique) and can act as a payment unit, reward, trading asset or information holder.

Cardano is a proof-of-stake blockchain. In contrast to proof-of-work blockchains, Cardano requires much less energy and computational power. The Bitcoin network is secured through computers doing ever-more-energy-intensive computations – proof of work – which is unsustainable in the long term. Cambridge University has an online tool that shows the computers powering Bitcoin already consume more electricity than some countries, like Switzerland, for example.

The text critiques capitalism, distinguishing it from markets and highlighting its unsustainable nature due to its reliance on constant exponential growth and concentration of wealth. It argues for a shift towards a financial ecosystem balancing efficiency and resilience, advocating for the development of electronic ADA currencies based on blockchain technology as a means to achieve this balance. Such currencies could promote diversity and stability in economic exchange, benefiting both financial systems and initiatives like malaria vector control. Cardano is a decentralized proof-of-stake blockchain platform with a focus on sustainability, scalability and inclusivity. The website explains the features of blockchain technology, including decentralized data storage, immutability of records and support for cryptocurrencies and native tokens. Cardano's proof-of-stake mechanism is highlighted for its energy efficiency compared to proof-of-work blockchains like Bitcoin. Overall, the text presents ADA currencies as a potential solution to address the shortcomings of capitalism and foster a more sustainable and equitable financial system.

As Bacon pointed out in 1620 (Bacon, 2000), 'Human knowledge and human power come to the same thing because ignorance of cause frustrates effect'. *The Yellow Emperor's Classic of Internal Medicine* (quoted by Dubos, 1959) stated that 'the ancients followed Tao and the law of the seasons under the guidance of their sages who were credited with the realization of the value of education in the prevention of disease'.

It is a truism that the early years of a person's life help define them more than any other. An expansion and improvement of the school system, particularly preschool (also, perhaps, using a fiscal system based on blockchain currencies), is required, as is approaching education in the way advocated by Illich (1970).

Presently, as a 'reward' for refereeing a paper with some major publishers, 30 days access is given to their database of published work that is normally behind a firewall of some sort. After 30 days this access runs out, presumably to encourage people to referee further papers. A suggestion would be to make the days non-consecutive so that the 30 days can be accessed according to need. And if a referee does not think that they will ever need the 30 days then they could sell them on the open market for ADA, that is eventually fungible for cash. It would diminish the present monopoly control of the price of access, which is good for the consumer. Institutional payment may not then be necessary since individuals could purchase access on the open market, presumably for a reasonable price.

Realpolitik and Vector Control

Two examples of the long-term effects of realpolitik and ideology on vector control and their sometimes, ruinous consequences are those of Venezuela and São Tomé and Príncipe (STP).

Venezuela

During the Second World War in the Pacific, as many troops were lost to vector-borne disease, particularly malaria, as enemy gunfire. A new, and secret, method of vector control was developed by the Americans. During the war, Venezuela had supplied the USA with oil and was considered a good ally, and so, when Ernesto Gabaldon, the head of the Department of Disease Control, asked for access to this wonder compound, it was granted, and a campaign against the coastal vectors *Anopheles darlingi* (now called *Nyssorhynchus darlingi*), *Anopheles aquasalis* and *Anopheles albimanus* was initiated. The compound dichlorodiphenyltrichloroethane (DDT), when applied at 2 g/m^2 to the interior walls of dwellings, was extremely effective against these vectors, and by 1954, Venezuela (closely followed by Surinam, where Giglioli had been responsible for the campaign) was declared free of malaria (Giglioli, 1951; Gabaldon and Berti, 1954). Despite a number of political upheavals, successive governments recognized the value of malaria control. However, it was considered that, since the problem appeared to be solved, the necessity for continuing operations was not apparent. The introduction of the Chavez government with a radically different agenda to previous governments meant that much of the previous system was dismantled, but little was put in its place to replace it. Today, Venezuela accounts for 47% of the region's malaria cases (Oletta *et al.*, 2018), with an unprecedented 365% increase in malaria cases between 2000 and 2015 (Conn *et al.*, 2018), once again displaying the ability of the disease to rebound once control efforts are relaxed.

As pointed out by Briceño-Leon *et al.* (2019),

> In the Venezuelan context, the politics of vector control, and the political narratives underpinning them, have relied both on proclamations of success as well as threats of the recurrence of disease. Vector control policies have to be proven successful, for that is their purpose, and political leadership needs success in order to justify the resources allocated and reap the fruits. But if success has been achieved, why should we continue to invest in vector control? This dilemma has arisen since the 1950s in the discussion between eradication and disease control in the World Health Organization (Litsios, 2005). The thesis of eradication expressed a valuable social ambition and a trust in technology, but it created an illusion of permanence that has proven to be epidemiologically incorrect and politically dangerous. Political management of the dilemma between success and permanent risk in vector-borne diseases has not been adequately communicated to the political leadership or to society.

When asked for reasons for the subsequent failure of the vector control campaign, Gabaldon was wont to reply, 'Air conditioning'. By this, he meant that overabundance of resources led to harmful consequences; namely, in the proliferation of unnecessary tasks, wastes of money and corruption and accompanying declines in motivation and purpose based on notions of social betterment.

'Overabundance, at least without civic oversight and good public management, promotes a culture of passivity' (Briceño-Leon *et al.*, 2019). It can also lead to a waste of valuable resources. This is not a problem limited to Venezuela. Following the tsunami of 2004 in Indonesia (when 150,000 people died), malaria was considered a risk (due to potential expansion of breeding sites of the vector, *Anopheles sundiacus*), and so, health centres where malaria was not a risk were stacked full of Artemisinin suppositories, an expensive treatment for acute malaria that could have been used to save many lives elsewhere (Sudomoa *et al.*, 2010). As pointed out in the introduction, in reviewing obstacles to the eradication of malaria, Feachem *et al.* (2019) said that it is operational obstacles, rather than technical ones, that largely limit the success of disease control programmes in many countries. These obstacles include ineffective management, poor use of data, a staff that is apathetic (as pointed out by the former director of the Amani Field Station in Tanzania, Bagster-Wilson, 'Sun loving larvae are not found by shade loving personnel') and local communities that are not involved either in the decisions that might affect them or in the outcome of any control efforts.

How to improve this aspect of vector control and how to ensure that staff are 'value-able' in their own eyes is mandatory.

São Tomé and Príncipe

In STP, malaria control in the 1980s was undertaken by means of IRS with DDT and weekly prophylaxis of children with chloroquine (Ceita, 1986). The campaign was so successful that for 3 years, there were almost no cases of the disease on the archipelago. It was close to being eliminated. The campaign, like many others, was a high-profile vertical one. It was run by the Centro Nacional de Endemias (CNE). STP is a small country with, at that time, less than 140,000 inhabitants. Allegedly, the Director of the CNE had political aspirations and, given the size of the country and the high profile of the campaign, hoped for elected office. He would have been a threat to the established order. For whatever reason, the shipment of DDT for the campaign in 1985 was delayed, and so, the spray round for that cycle was delayed, reducing protection indoors. After a number of years of IRS, people had also started to refuse entry to spraymen, the reason being given that the spray killed cockroaches, which were then eaten by the cats, which then died, with the result being that the rat population increased. At the same time, an immigrant or visitor from Angola harbouring a chloroquine-resistant strain of *Plasmodium falciparum* arrived on the island. The breakdown in control and the exposure of the, by now, relatively non-immune population to a difficult-to-treat parasite resulted in a major epidemic with such a high proportion of the population dying that, in the millennial edition of *Guinness Book of Records*, the island was considered to be the most dangerous place on earth. The epidemic was eventually brought under control, as much by climatic factors (a dry wet season for the year 1986) as by control measures. Despite the success of the early campaign, the population of *Anopheles coluzzii* had not been affected by the IRS. This appeared to be because the mosquito adapted to becoming exophagic and zoophilic by feeding on pigs and dogs outside the relatively well-built houses (Sousa *et al.*, 2001).

By the 1990s, malaria was again endemic on the island, but despite large populations of the vector *An. coluzzii*, transmission was less intense than it might otherwise have been, largely because of low human blood index in mosquitoes (Sousa *et al.*, 2001). Treatment remained complicated, but as predicted by Hagmann *et al.* (2003), the introduction of ACTs changed that.

In 1997, STP was one of the few countries to recognize Taiwan as an independent nation. As a result, the Embassy of the People's Republic of China was withdrawn from the archipelago, and a Taiwanese embassy took its place. One of the main activities undertaken by the Taiwanese was malaria control, in particular, IRS with fenitrothion. This produced a marked reduction in malaria incidence, and by 2015, the archipelago was again on the path to elimination.

In 2016, however, STP withdrew its recognition of Taiwan as a nation, and the roles were reversed, with the People's Republic of China once again establishing an embassy on the island as the Taiwanese departed. Rather than continue with the programme established by the Taiwanese, a new system of malaria control was introduced. In place of the vertically oriented, independent approach adopted by the Taiwanese, a more horizontal one was introduced. Both approaches have their advantages and disadvantages.

Vertical Versus Horizontal Control

A vertical system operates independently of any other health provision. The aims can be relatively easily articulated, and campaigns can be conducted in a 'military' style with clear chains of command. The system allows for little or no freedom for the recipients (the population affected) to express their opinions, nor does it empower recipients who, therefore, take a much more passive role in proceedings – there is little sense of ownership. Indeed, as pointed out by Banerji (2002), the emphatic recommendations for perpetuating vertical programmes against major communicable diseases (tuberculosis, AIDS and malaria) by the WHO's Commission on Macroeconomics and Health on the grounds that such programs have proved convenient to 'donors' reveal the real motivations for an almost openly ideology-driven agenda.

A horizontal system, when it is well established, has a number of potential advantages over a vertical one. It does, however, rely on the recipients to feel a sense of ownership and responsibility for the whole process. This can easily be undermined by a number of factors. If a system changes from a vertical one (as they all were initially), then it becomes difficult to eliminate a passive attitude. In addition, horizontal systems depend on the goodwill and good relations between the different professionals, such as those responsible for treatment of a disease and those responsible for its prevention.

The distinction between vertical and horizontal approaches to control reflects attitudes towards the elimination of poverty in general. There are those, like Fred Soper, who consider that malaria is a cause of poverty and that malaria elimination will also reduce poverty. They favour the vertical approach. Those who considered malaria to be an effect of poverty, independent of any other cause, favour the horizontal approach of using Primary Health Care (PHC) to reduce malaria.

Thus, the situation in STP is similar to that envisaged by the WHO during its first malaria elimination effort in which vertical programmes of malaria control became horizontal ones based on PHCs delivering treatment. In STP, a further, experimental, control measure was added to the mix by the introduction of larval control using an advanced geo-monitoring tool. Thus, a form of IVM was introduced in two districts. Unfortunately, they both suffered from malaria epidemics during the trial.

As pointed out by Briceño-Leon *et al.* (2019), 'The dilemma between authoritative military programs, which can be very efficient in their execution, and educational and participatory programs, which can guarantee the sustainability of the achievements, remains a relevant paradox'.

Wider public engagement would improve our capacity for action. Questions such as 'What is the purpose (of an intervention)?', 'Who will be hurt?', Who will it benefit?' and 'How can we know?' need to be asked of policy makers. Changes in political ideology and available resources will affect control decisions. For example, in Venezuela, revenue from oil exports meant that vector control was for a time, if anything, overfunded. This meant that control was undertaken in areas that were well known to be malaria free.

DDT in The Gambia

A further illumination of the role of politics in vector control was provided first hand to the author in relation to the question of the reintroduction of DDT for IRS in The Gambia in 2003. In the 1990s, The Gambia was a 'poster child' for the efficacy of impregnated mosquito nets (with chemoprophylaxis) against malaria (and indeed all-cause child mortality) (Alonso *et al.*, 1991). The successes observed provided a major impetus to the technology becoming accepted by the WHO and for its subsequent adoption throughout the tropics. By the early 2000s, increasing cases of malaria had apparently led to the Malaria Control Program to consider reintroducing DDT, and I was asked to go and evaluate their proposal. It was an assignment that I tried to avoid (although I had undertaken the fieldwork for my PhD there and had a real 'epiphany', it had not been an untrammelled pleasure). I pointed out (quite correctly) that there were others who knew the country better and who were in a better position to evaluate the proposal. I also pointed out that DDT was very much a political 'chestnut' and that decisions would be likely to be made on political grounds rather than scientific ones.

The reason why the decision might be a political rather than one based on scientific evidence is because DDT was surrounded by controversy. From its discovery as an insecticide, it had been widely used, particularly in agriculture. It is a chemical that stays in the environment for many years, is deposited in animal fat, and increases in concentration as the 'fat' goes up the food chain. Above all, in the USA, this affected larger raptors, such as the condor, which started laying eggs with fragile shells that broke when the bird sat on them. Following a national outcry in the USA, especially after the publication of Rachel Carson's *Silent Spring* (1962) – silent because the birds are all dead – DDT, with one exception, was banned for use. The one exception was its use for malaria control. It was argued that, given its low cost, and because it was used inside houses, which reduced the chances of environmental contamination, its use for IRS could be continued. DDT when used for IRS had had a large impact on malaria transmission. It certainly had an effect when used against a pyrethroid-resistant *Anopheles funestus* in South Africa (Sharp *et al.*, 1998). More recently, however, Corin and Weaver (2005), from an analysis of the situation in South Africa, concluded, 'Our assessment concluded that the net benefits of DDT indoor residual spraying for malaria control are at best marginal and recommends consideration of less toxic, cost-effective alternatives such as insecticidal nets'. Thus, spraymen would carry sacks of it and mix it on-site (Fig. 14.1). It's very longevity means that it needs to be applied less frequently than other insecticides, which makes it suitable for IRS. This also means that unused stocks have to be disposed of with care. Unlike other, more potent, insecticides, a considerable amount of DDT has to be applied for it to be effective (2 g/m^2), so stocks can be quite large.

Because such large quantities are used, the benefit of low cost was actually outweighed by the added transport and storage costs associated with DDT (Schofield, 1992), which led the eminent entomologist Curtis (1994) to question whether it should be used at all. DDT is also abrasive, and nozzle tips rapidly degrade. The supply of spare parts for equipment is a major problem for control programmes, and bad nozzles and leaking pumps were a common feature of many control programmes. More potent insecticides, which are applied at much lower doses, come in sealed gelatine bags that are dropped directly into the pump used for the IRS.

In Brazil in the 1970s, it was possible to watch *An.* (now *Nyssorhynchus*) *darlingi* enter an Amazonian house, sprayed that day, fly directly underneath an occupied hammock, feed and leave without touching the walls. When first

Fig. 14.1. The preparation of DDT into a wettable powder requires a considerable amount of water. Here, a sprayman in Brazil prepares the mixture beside a small stream in the Amazon. (Photo: J.D. Charlwood.).

reported, this was considered to be detrimental to the insecticide's efficacy (Hayes and Charlwood, 1977). Interestingly, it was later considered that it was DDT's true repellent effect (rather than it just being an irritant) that was responsible for the original reduction in malaria cases in Brazil (Roberts and Alecrim, 1991).

My reluctance to go to The Gambia only reinforced the WHO's determination to have me go. Although I had a wonderful time there, my misgivings regarding the introduction of DDT were largely borne out. The assignment involved visiting field sites throughout much of the country, examining malaria data and determining attitudes towards DDT and IRS both at such sites and in the headquarters in Banjul. The problem that the programme was facing was immediately obvious. The visit occurred before the development of Long-Lasting Insecticide-treated nets. At the time, nets required retreatment every 6 months. However, only 5% of nets were being retreated, and many nets were not repaired if they developed holes. Thus, the protection provided by the nets was compromised. All of the people working in the field were against the reintroduction of DDT ('Over my dead body' was one comment). The obvious solution to the problem of increasing malaria was to improve retreatment rates. Suggestions were prepared as to how this might be achieved (through the use of village-based entrepreneurs who would receive a small fee for repairing and retreating old nets). The Gambia at the time wanted to be recognized as a supplier of organically grown groundnuts. The road systems were bad, so accidents to lorries transporting DDT could not be ruled out. Contamination with DDT, even in very small amounts, would have destroyed this market. The insecticide has a long half-life in the environment, which was one reason for the prohibition for its general use. How to destroy old stocks was a headache for many countries.

When recommendations were given at a debriefing meeting in the capital it transpired that the decision to import the DDT had already been taken. It turned out that the insecticide had been offered as a donation from a Southeast Asian country that basically had no other way to get rid of its old stock (a win-win, as far as they were concerned). Whether other inducements to accept the insecticide were offered was not disclosed.

Paved with Good Intentions

Even when health planners may have 'fairness' in mind when considering their decisions, those decisions may do no one any good. In 1995, a cautionary tale regarding mistaken assumptions surrounding the newly introduced impregnated mosquito bed nets was published (Charlwood, 1995). This was towards the end of the 'Delight' phase of vector control when something new seemed to work. In early trials, it was considered that the insecticide was the main agent responsible for an effect. The nets needed to be retreated every 6 months. This meant that having a net with holes, as they generally were in Africa and elsewhere, was less important than it might otherwise be. Thus, when UNICEF bought nets and insecticide to treat them to Morogoro, Tanzania, the quality of the net took second place. Net strength is measured by 'denier'. A net of a minimum of 100 denier is now the norm. Those sold by UNICEF were 40 denier. I purchased two bales of 500 nets, and my house girls established their own business selling impregnated nets. If you took 10 nets (as technicians going to the field often did), you only paid for 9. We devised an automatic bioassay for reimpregnation schedule determination and a scheme for free reimpregnation (paying for reimpregnation, which originally was free, was the main problem in many schemes). Within 1–2 months of use, the vast majority of the nets were, indeed, full of holes. I tore mine the first time I used it. The insecticide was deemed responsible, since almost all of the (100 denier) mostly unimpregnated bed nets that we had previously sold (as part of the Kilombero Malaria Project; KMP) remained intact after 4 years of use; indeed, they became famous for their longevity. Even free offers could not induce any of our UNICEF customers to reimpregnate what remained of their nets. And who can blame them?

It was pointed out, and remains true, that 'the people of Ifakara would much rather have, and are prepared to pay for, good-quality non-impregnated bed nets than bad-quality once-impregnated ones' (Charlwood, 1995).

Forty kilometres from the town of Muleba on the shores of Lake Victoria, there are two villages, Kyamyorwa A and the smaller Kyamyorwa B, some 5 km apart, that are serviced by a single health post (Fig. 14.2).

There are frequent and easy *chapas* (local taxis) between the villages. Halfway between them, and 1.5 km off the main road, is an abandoned school. Someone must have looked at the map and thought that is was the 'fairest' place to put the Health Centre to serve both villages. The result is that it is much more difficult to get to than either village. Almost no one attends, despite malaria being highly endemic in the villages (Protopopoff *et al.*, 2018). Villagers go to shops selling treatment instead. It would obviously have been most sensible to locate the post in one or the other village. The smaller Kyamyorwa B had much more malaria than Kyamyorwa A, and so based on cases, this might have been the place to put it.

Social Media

Social media, when used in a positive way, can help to disseminate information about vectors and help to stimulate action by individuals and communities. This is a potential resource that remains largely untapped. More equal societies are healthier ones.

But 'the road to hell is paved with good intentions', and the internet has resulted in increasing inequalities between the rich and poor. Social media in particular is a Pandora's Box.

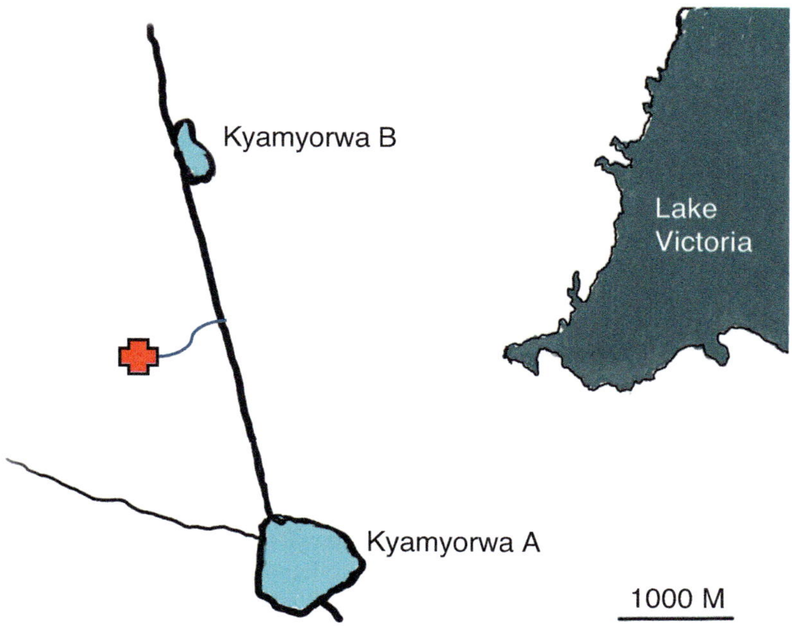

Fig. 14.2. Map of Kyamyorwa A and B, showing the location of the health post designed to serve them both.

Pandora's Box

Pandora was the Greek version of the Biblical Eve, and like Eve, fashioned by Almighty God, in this case Zeus. She was given a box by the gods on her marriage to Epimethus, the 'dull-witted' brother of Prometheus. Prometheus was the god who gave man fire and whom the others despised because of that. It contained the 'gifts of the gods' to humanity, but the box was not to be opened. Pandora, like Eve, could not, of course, contain her curiosity and opened it. Out came the gift of the gods – 'death, disease and many other evils, indeed all the evils of the world'. Some versions of the myth indicate that Hope was trapped inside when Pandora, realizing what she had unleashed, closed the lid, but in others, Hope actually manages to make it out.

Opening up the World Wide Web in the hopes that it will improve life has been a 'Pandora's Box'.

It has helped facilitate environmental destruction by, for example, making airlines more efficient, resulting in more aeroplanes flying and more CO_2 being emitted. It has strengthened global trade with a number of unforeseen consequences, and it has ushered in *techno-feudalism*, in which cloud capitalists obtain rent from advertisers (Amazon can charge 40% of the value of an item sold for advertising it) (Varoufakis, 2023).

Harm may also come from the propagation of false information, leading to baseless conspiracy theories. Barak Obama, the former President of the USA, has warned that, in order to reduce the perils of disinformation, technology companies (such as Facebook) must be reined in to address the 'weakening of democratic institutions around the world'. He said, 'One of the biggest reasons for the weakening of democracy is the profound change that's taken place in how we communicate and consume information'. The medium is the message. This change has occurred, and is occurring, throughout the world, even in poorer nations, not just in those countries where the internet is universally available. The number of conspiracy theories concerning Covid-19 and their effects on transmission rates are well known. Such theories can be applied to almost any enterprise. Vector control may not be exempt.

'From Rumour's tongues they bring smooth comforts false, worse than true wrongs'. Graham Greene in his book *Ways of Escape* (1980) describes how the film director Alex Korda, when on his boat, liked to play an old sailor in a T-shirt and a battered yachting cap with white stubble on his chin. One night in Naples, in a waterfront bar frequented by GIs, he persuaded the barman and the American soldiers standing around that Coca-Cola rendered men impotent. Greene says, 'My word would have carried no weight with them, nor a film producer's, but from this old seadog who had learnt his wisdom 'on the seven seas'...'.

Someone similar may have started a rumour in Muleba District. As Shakespeare said, 'Open your ears; for which of you will stop the vent of hearing when loud rumour speaks?' Sleeping under a mosquito net was thought to cause men to become impotent – and, of course, if someone believed this to be true, it probably did! Even the regional medical officer had heard of this rumour. But as Greene points out, it is as much to do with the messenger as the message itself.

A more insidious and widespread example is, or hopefully was, the rumour concerning AIDS sufferers and more recently people with TB. This required a huge publicity campaign (advising on condom use), with much of the money spent on these education efforts being designed to reduce prejudice among the general population against sufferers. This was expensive and used money that, without such prejudice, could have been spent on drugs, drug development and avoidance techniques. Prejudice makes 'informed consent' more difficult. 'Critical acumen is clearly essential to a democracy' (Solomon, 1999). People should also be in the position to 'speak truth to power'.

References

Alonso, P.L., Lindsay, S.W., Armstrong, J.R., Conteh, M., Hill, A.G. *et al.* (1991) The effect of insecticide-treated bed nets on mortality of Gambian children. *Lancet* 337(8756), 1499–1502.

Bacon, F., Jardine, L. and Silverthorne, M. (eds). (2000) . Translated by M. Silverthorne. *The New Organum*. Cambridge University Press, Cambridge, UK. (Originally published in 1620).

Banerji, D. (2002) Report of the WHO commission on macroeconomics and health: A critique. *International Journal of Health Services* 32, 733–754.

Briceño-Leon, R., Guevara, M. and Terán, I. (2019) Politics as disease in Venezuela. Vector control before and after the Bolivarian revolution. In: Bardosh, K. (ed.) *Locating Zika*. Routledge, Abingdon-on-Thames, UK, pp. 104–127.

Buller, A. (2022) *The Value of a Whale: On the Illusion of Green Capitalism*. Manchester University Press, Manchester, UK.

Carson, R. (1962) *Silent Spring*. Houghton Mifflin, New York.

Ceita, J.G.V. (1986) Malaria in São Tomé and Príncipe. In: Buck, A. (ed.) *Proceedings of the Conference on Malaria in Africa. Practical Considerations on Malaria Vaccines and Clinical Trials*, United States Agency for International Development, Washington, DC, pp. 142–156.

Charlwood, J.D. (1984) Which way now for malaria control?. *PNG Medical Journal* 27, 159–162.

Charlwood, J.D. (1995) Bednets, a cautionary tale. *Parasitology Today* 11, 184–185.

Charlwood, J.D. (2023) A self-sustaining system to reduce migration of *Anopheles stephensi* in East Africa. *Acta Tropica* 245, 106981.

Conn, J.E., Grillet, M.E., Correa, M. and Mureb Sallum, M.A. (2018) Malaria transmission in South America – Present status and prospects for elimination. In: Manguin, S. and Dev, V. (eds) *Towards Malaria Elimination – A Leap Forward*. IntechOpen, London. DOI: 10.5772/intechopen.76964.

Corin, S. and Weaver, S. (2005) A risk analysis model with an ecological perspective on DDT and malaria control in South Africa. *Journal of Rural and Tropical Public Health* 4, 21–32.

Curtis, C.F. (1994) Should DDT be recommended for malaria vector control?. *Medical and Veterinary Entomology* 8, 107–112.

DeLanda, M. (2004) *Intensive Science and Virtual Philosophy*. Continuim, New York.

Dubos, R.J. (1959) *The Mirage of Health: Utopias, Progress and Biological Change*. Rutgers University Press, New Brunswick, New Jersey, 282 pp.

Elliott, C. (2024) *Pox Romana: The Plague That Shook the Roman World*. Princeton University Press, Princeton, New Jersey.

Feachem, R.G.A., Chen, I., Akbari, O., Bertozzi-Villa, A., Bhatt, S. *et al.* (2019) Malaria eradication within a generation: Ambitious, achievable, and necessary. *The Lancet Commissions* 394(1023), 1056–1112.

Gabaldon, A. and Berti, A.L. (1954) The first large area in the tropical zone to report malaria eradication: North-Central Venezuela. *American Journal of Tropical Medicine and Hygiene* 3, 793–807.

Giglioli, G. (1951) Eradication of *Anopheles darlingi* from the inhabited areas of British Guinea by DDT residual spraying. *Journal of the National Malaria Society* 10, 142–161.

Gray, J. (2009) *Gray's Anatomy*. Allen Lane, Penguin Books, Strand, London.

Greene, G. (1980) *Ways of Escape*. The Bodley Head, London.

Hagmann, R., Charlwood, J.D., Gil, V., Rosario, V.E. and Smith, T. (2003) Malaria and its possible control on the island of Príncipe. *Malaria Journal* 2, 9.

Hayes, J. and Charlwood, J.D. (1977) *Anopheles darlingi* evita o DDT numa area de malaria resistente à drogas. *Acta Amazonica* 7, 289.

Illich, I. (1970) *Deschooling Society*. Harper and Row, London.

Illich, I. (1976) *Limits to Medicine: Medical Nemisis: The Expropriation of Health*. Penguin, London.

Jasanoff, S. (2007) Technologies of humility. *Nature* 450, 33.

Lauwereyns, J. (2010) *The Anatomy of Bias. How Neural Circuits Weigh the Options*. MIT Press, Cambridge, Massachusetts.

Lietaer, B., Arnsperger, C., Goerner, S. and Brunnhuber, S. (2012) *Money and Sustainability: The Missing Link*. Triarchy Press, Charmouth, UK.

Litsios, S. (2005) The health, poverty, and development Merry-Go-Round. In: Gunn, S.W.A., Mansourian, P.B., Davies, A.M., Piel, A. and Sayers, B.M. (eds) *Understanding the Global Dimensions of Health*. Springer, Boston, Massachusets. DOI: 10.1007/0-387-24103-5_2.

Oletta, J.F., Walter, C., Orihuela, Á.R., Pulido, M., Carvajal, A.C. *et al.* (2018) Análisis preliminar del Informe Mundial de malaria, caso de Venezuela. Available at: https://www.ovsalud.org/publicaciones/publicaciones-salud/epidemia-malaria-venezuela/ (accessed 23 September 2024).

Orwell, G. (1945) Notes on Nationalism. *Polemic*. Available at: https://www.orwellfoundation.com/the-orwell-foundation/orwell/essays-and-other-works/notes-on-nationalism

Protopopoff, N., Mosha, J.F., Lukole, E., Charlwood, J.D., Wright, A. *et al.* (2018) Effectiveness of a long-lasting PBO treated insecticidal net and indoor residual spray interventions, separately and together, against malaria transmitted by pyrethroid resistant mosquitoes: A community randomised factorial design trial. *The Lancet* 391, 1577–1588.

Roberts, D.R. and Alecrim, W.D. (1991) Behavioral response of *Anopheles darlingi* to DDT-sprayed house walls in Amazonia. *Bulletin of the Pan-American Health Organisation* 25(3), 210–217.

Roy, A. (2020) The pandemic is a portal. *Financial Times*.

Schofield, C.J. (1992) DDT and malaria vector control. *International Pest Control* 4, 2–4.

Sharp, B.L., Ngxongo, S.M., Botha, M.J., Ridl, F.C. and Sueur, D. (1998) An analysis of 10 years of retrospective malaria data from the Kwa Zulu areas of Natal. *South African Journal of Science* 84, 102–106.

Solomon, R.C. (1999) *The Joy of Philosophy: Thinking Thin Versus the Passionate Life*. Oxford University Press.

Sousa, C.A., Pinto, J., Almeida, P.G., Ferreira, C., Rosario, V.E. *et al.* (2001) Dogs as a favoured host of *Anopheles gambiae* sensu stricto (Diptera,Culicidae) of São Tomé, West Africa. *Journal of Medical Entomology* 38, 122–125.

Stendhal (1969) Translated by M.R.B. Shaw. *Scarlet and Black*. Penguin, London..

Sudomoa, M., Ariantia, Y., Wahidb, I., Safruddinb, D., Pedersen, E.M. *et al.* (2010) Towards eradication: Three years after the tsunami has malaria transmission been eliminated from the island of Simeulue. *Transactions of the Royal Society of Tropical Medicine and Hygiene* 104, 777–781.

Varoufakis, Y. (2023) *Technofeudalism; What Killed Capitalism*. Vintage Publishing, New York.

Index

Printed and bound by CPI Group (UK) Ltd, Croydon, CR0 4YY

13/09/2026

14961107-0003